Springer-AAS Acoustics Series

Series Editor

This series publishes peer reviewed high quality monographs and contributed volumes on all topics in acoustics. Books in this series range from those focused on a particular aspect of acoustics and vibration to practical handbooks covering a range of topics. The advantage for authors is that the inclusion of a book as part of this series will demonstrate high quality content of the book. While this series encourages authors and topics that are relevant to the Australasian region, proposals for contributions to the series are not be restricted to this region.

Valentyn Skalskyi · Zinoviy Nazarchuk

Magnetoelastic Acoustic Emission

Theory and Applications in Ferromagnetic Materials

Australian Acoustical Society

Springer

Valentyn Skalskyi
Karpenko Physico-Mechanical Institute
National Academy of Sciences of Ukraine
Lviv, Ukraine

Zinoviy Nazarchuk
Karpenko Physico-Mechanical Institute
National Academy of Sciences of Ukraine
Lviv, Ukraine

ISSN 2948-2062 ISSN 2948-2070 (electronic)
Springer-AAS Acoustics Series
ISBN 978-981-99-4031-8 ISBN 978-981-99-4032-5 (eBook)
https://doi.org/10.1007/978-981-99-4032-5

This Springer imprint is published by the registered company Springer Nature Singapore Pte Ltd.
The registered company address is: 152 Beach Road, #21-01/04 Gateway East, Singapore 189721, Singapore

Preface

In most industries, in aerospace, chemical, energy, oil refining, pipeline transport, and mechanical engineering, it is important to diagnose the state of products, structural elements, as well as equipment that operates in hydrogen-containing media. It is especially important to identify, in addition to areas of accumulation of small defects or local plastic deformations, the places of material hydrogen charging since their propagation can lead to accelerated initiation of macrocracks and hence to accidents and catastrophic consequences for both the production and the environment.

Modern non-destructive test methods used for the technical diagnostics of such objects, including ultrasound and X-ray flaw detection, require appropriate surface treatment of the object of control, are quite time-consuming, and are difficult to apply under the operation of the equipment. The method of acoustic emission is a promising one for solving this problem. However, its traditional implementation requires the application of additional external mechanical stress or deformation to products or structural elements. Because this load necessary to provoke the propagation of small defects can sometimes be much higher than the optimal stress regimes in the material, the practical application of acoustic emission diagnostics is somewhat limited. To expand these boundaries, as well as to identify the places of hydrogen charging of ferromagnetic elements of structures and products more effectively, it is proposed to use the phenomenon of generating magnetoelastic acoustic emission signals under the influence of an external magnetic field. The latter is known to cause the movement of the walls of magnetic domains (Barkhausen effect). These processes occur most intensively in the vicinity of individual defects or their clusters—where there are significant gradients of mechanical stresses. In addition, the parameters of magnetoelastic acoustic emission signals are significantly affected by changes in the structure of the material and the level of hydrogen degradation during operation.

The research results presented here make it possible to propose a new approach to the development of effective methods and means of non-destructive magnetoacoustic testing of products and structural elements of long-term operation, which work under the influence of hydrogen-mechanical factor. The authors developed such methods and conducted experimental studies of the peculiarities of generating magnetoelastic acoustic emission signals during the action of the above physical factors on ferromagnets and quantified their parameters during the characteristic for each factor influence on the change of Barkhausen jumps. The new analytical dependences that relate these parameters with the magnetic field, mechanical stresses, changes in material structure, and the presence of hydrogen are important here. They are tested experimentally, considering the structure of the material, its magnetic and mechanical characteristics, and the degree of hydrogen degradation.

The above presented allows us to construct methodological bases and technical means for assessing the hydrogen degradation of structural ferromagnetic alloys on existing industrial equipment by the method of magnetoelastic acoustic emission. The monograph is addressed not only to scientists and engineers—specialists in the field of acoustics and non-destructive testing, but also to graduate students and senior students of higher education institutions interested in the problems of technical diagnostics of the state of objects, which operate in contact with hydrogen environment.

Lviv, Ukraine

Valentyn Skalskyi
Zinoviy Nazarchuk

Contents

Abbreviations

AE	Acoustic emission
AES	Acoustic emission signal
BE	Barkhausen effect
BJ	Barkhausen jumps
BN	Barkhausen noise
CGR	Crack growth rate
HC	Hydrogen corrosion
HD	Hydrogen destruction
HDT	Hydrogen degradation of transformation
HE	Hydrogen embrittlement
LMA	Light magnetization axis
MAE	Magnetoelastic acoustic emission
MO	Molecular orbital
NDT	Non-destructive testing
OK	Object of control
SCC	Stress corrosion cracking
SIF	Stress intensity factor
SSS	Stress-strain state
α	Morrow coefficient; critical indicator
α_i	Angle between the local (random) axes of easy magnetization
β_{ij}	Angle between the magnetic moments $\vec{i}$ and $\vec{j}$
χ	Magnetic susceptibility of the substance
δ_w	Domain wall width
δ_p	Crack opening displacement under the action of a load of intensity p
$\delta(...)$	Dirac delta function
$\varepsilon_0(\rho)$	Maximum amount of deformation
ε_{ijk}	Obliquely symmetrical Kronecker symbol
ε_{zz}	Component of the strain tensor
$\Delta\varepsilon$	Value of the plastic deformation range
$\Delta\varepsilon_{th}$	Threshold deformation range in the hydrogenated material

$\Delta\varepsilon_{rs}$	Transformational deformation
Φ'	Induced signal
$\Phi_{1,2}^{(d)}(\theta)$	Angular distribution of radiation for longitudinal and transverse wave
φ	Angle; scalar magnetic potential
Γ	Effective viscosity
γ	Ratio of charge to electron mass
γ_w	Surface energy of the domain wall
$\gamma_s^{(H)}$	Density of the static component of the dissipating energy of plastic deformations in the material
$\gamma_c^{(H)}$	Energy fracture density of the material
λ	Lamé constant
$\lambda_{\alpha\beta\gamma\delta}$	Magnetoelastic tensor
λ_s	Magnetostriction constant
μ_0	Magnetic permeability of vacuum
μ	Relative magnetic permeability of the material; shear modulus; Lamé constant
υ	Domain movement velocity; signal amplitude; volume fraction of non-magnetic inclusions; signal voltage
υ_q	External normal to the surface Σ at point q
$\nu(l)$	Rate of microcrack growth to macroscopic size
θ	Angle between the direction of magnetization and the vector normal to the surface
ρ	Dislocation density; radius, environment density
$\rho(\vec{r})$	Density of bulk magnetic charges
σ	Surface charge density, uniaxial stress
$\sigma_{\alpha\beta}$; σ_{ij}	Mechanical stress tensor
σ_i	Internal elastic stress; main stresses
σ_{Tc}	Plastic yield stress
σ_{int}	Mean value of fluctuations of internal mechanical stresses
σ_{ic}	True ultimate strength
σ_a	Amplitude of stresses in the prefracture zone
τ	Critical indicator
∇^2	Laplace operator for spatial coordinates
∇	Gradient operator
Δ	Distance between the center of the disk-like crack and the edge of the semi-infinite crack
A_p	Cross-sectional area of the coil
A	Cross-sectional area of the sample; exchange interaction constant
a	Ratio of exchange energy constants and anisotropy energy; crack radius
a_i	Parameters of material; environment; load
$\vec{B}$	Magnetic induction vector
B_0	Absolute value of the magnetic induction vector
$\vec{b}$	Burgers vector

$C_{\alpha\beta}$	Function that determines the angular distribution depending on the type of dislocation
C	Amount of hydrogen in the fracture area
C_*	Hydrogen concentration
C_{cr}	Critical hydrogen concentration
C_{H}^{S}	Hydrogen concentration on the specimen surface
c	Coefficient of proportionality
c_{pqrs}	Elastic constants
c_1	Longitudinal wave rate
c_2	Transversal wave rate
c_{R}	Rayleigh wave rate
D	Distribution of dislocation positions; dispersion
$D_{\mathrm{H}}(T)$	Hydrogen diffusion coefficient
d	Specimen thickness; distance
E	Energy of ferromagnetic material
E_{ex}	Exchange energy
E_{m}	Magnetostatic energy
E_{an}	Energy of magnetic anisotropy
E_{dis}	Disordering energy
$E(\{h(\vec{r},t)\})$	Total energy functional
$\vec{e}$	Random place function depending on the grain
$F(\omega)$	Spectral function
$F(m)$	Magnetic field potential
$f(m)$	Uncorrelated random field
$f_p(x)$	Strength of the domain wall fixation
f_{sh}	Scale function
f_v; f_g; f_{energy}	Some universal functions
G	Shear modulus
$g(\vec{r})$	Mean value of the interaction energy
g_0	Interaction in the absence of inclusions
$\vec{H}$	Magnetic field intensity vector
$\vec{H}_{\mathrm{dem}}$	Demagnetizing field
$H_{\mathrm{dem}}^{(\alpha)}$	α-component of the demagnetizing field
$H(\ldots)$	Heaviside function
J_s	Saturation magnetization
$\vec{J}$	Hydrogen flow, magnetization
$J_0(\ldots)$, $J_1(\ldots)$	Bessel functions of zero and first order
K_{il}	Symmetric tensor, which describes anisotropy of the material
K_0	Magnetic anisotropy constant in the uniaxial case
$K_0(\ldots)$	Modified Hankel function
K_{I}	Stress intensity factor
$K_{\mathrm{I}\max}^{\rho}$	Stress intensity factor at the notch tip
$K_{\mathrm{I}cr}$ (or C_{cr})	Critical value of SIF (or concentration)
$K_{\mathrm{I}th}$	Lower value of the threshold stress intensity factor
K_{H}	Coefficient of hydrogen solubility on the plate surface

k	Demagnetizing factor; Boltzmann constant
L	Specimen width; jump length
l	Crack length
$\vec{\mathrm{M}}$	Magnetizing vector; surface force vector
$\vec{M}(\vec{r})$	Continuum field
M_α	α-component of the magnetizing vector
M_s	Saturation magnetization
$\vec{\mathrm{M}}_0$	Magnetic dipole moment
m	Magnetoelastic energy
N	Number of turns; number of magnetic moments
N_1	Number of load cycles
$\vec{n}$	Normal vector to the surface
P	Hydrogen pressure
p, q	Fourier transform components along the z and y axes, respectively
p	Plasticity restriction coefficient
p_{H_2}	Partial pressure of hydrogen in the environment
R	Universal gas constant
r_y	Size of the plasticity zone
$(r_i,\ h_i)$	Coordinates of the center of attachment
S_{dw}	Area of the undeformed domain wall
S	Value of Barkhausen's jump
$\vec{s}(\vec{r}_i)$	Set of spins
T	Jump duration; absolute temperature
$T_0,\ S_0$	Critical cut parameters
t	Point of time
t_n	Hydrogen diffusion time in the region of the hydrogen maximum
t_p	Time of plastic deformation
u_r	Component of the displacement vector
V	Volume; crack growth rate
V_{H}	Partial molar volume of hydrogen in metal
$W_c^{(\mathrm{H})}$	Static fracture energy of hydrogenated material
$W_{f1}(x)$	Distribution of energy of plastic deformations
$w_{th}^{(\mathrm{H})}$	Energy of plastic deformations dissipation
x	Wall position
Δx	Interval of domain boundary jump

Chapter 1
Some Concepts on Remagnetization of Ferromagnets

In this chapter, the basic concepts of remagnetization of ferromagnets are presented. In particular, the nature of ferromagnetism, the domain structure of ferromagnets, and their magnetization and demagnetization are analyzed. Theoretical aspects of the magnetic interaction of ferromagnetic atoms, magnetic anisotropy, and magnetostriction are described. The physics of magnetic domains and features of the structure of domain walls are highlighted.

1.1 Nature of Ferromagnetism

Electric current that passes through the conductor creates a magnetic field in the surrounding space. The intensity of this field is determined by the law of Biot–Savart–Laplace:

$$\vec{H} = \int_{l} \mathrm{d}\vec{H}, \tag{1.1}$$

$$\mathrm{d}\vec{H} = \frac{\mu_0}{4\pi} \frac{\vec{I}}{r^3} \left[\vec{r} \mathrm{d}\vec{l} \right], \tag{1.2}$$

where $\vec{I}$ is the current strength; $\mathrm{d}\vec{l}$ is the vector of the conductor element, numerically equal and drawn in the direction of the current; $\vec{r}$ is the radius vector drawn from this conductor element to the considered point of the field; $\mu_0 = 4\pi \times 10^{-7}$ in V s/(A m) $\equiv 4\pi \times 10^{-7}$ H/m; $r = |\vec{r}|$; symbol [] means vector product.

In SI units, the magnetic field strength is measured in amperes per meter (A/m). This field creates an infinitely long thin conductor, along which the current $I = 1$ A flows.

V. Skalskyi and Z. Nazarchuk, *Magnetoelastic Acoustic Emission*, Springer-AAS Acoustics Series, https://doi.org/10.1007/978-981-99-4032-5_1

For a constant magnetic field, Eqs. (1.1) and (1.2) are a consequence of Maxwell's differential equations:

$$\mathrm{rot}\vec{E} = -\frac{\partial \vec{B}}{\partial t}; \tag{1.3}$$

$$\mathrm{rot}\vec{H} = \frac{\partial \vec{D}}{\partial t} + \vec{j}; \tag{1.4}$$

$$\mathrm{div}\vec{D} = \rho; \tag{1.5}$$

$$\mathrm{div}\vec{B} = 0, \tag{1.6}$$

where ρ is the density of the exterior electric charge; $\vec{j}$ is the electric current density; $\vec{E}$ is the electric field strength; $\vec{H}$ is the magnetic field intensity; $\vec{D}$ is the electric induction; $\vec{B}$ is the magnetic induction.

Regarding the needs of further presentation, the essence of these equations is in the curl nature of the magnetic field and in the absence of free magnetic charges in nature. In view of this, the magnetic dipole, a system of two disconnected magnetic poles, is considered to be an elementary particle in the modern theory of magnetism (Vonsovsky 1952). The main characteristic of a magnetic dipole is the dipole magnetic moment $\vec{M}$.

In the absence of an external magnetic field, the ferromagnet is a set of macroscopic regions, domains, magnetized to saturation, in which the magnetization $\vec{J}_s$ is oriented so that the total magnetic moment of the whole sample is zero (Vonsovsky 1971; Vonsovsky and Shur 1948).

To explain the magnetization of individual domains to saturation, Weiss in 1907 hypothesized the existence of an internal "molecular" field in ferromagnets (Weiss 1907) (the idea of the internal field was put forward earlier by Rozing (1892) but finalized by Weiss). The formal theory of ferromagnetism developed based on these ideas makes it possible to explain a few experimental facts. From this theory follows the existence of a certain temperature (Curie point) in each ferromagnet, above which the spontaneous magnetization $\vec{J}_s$ tends to zero. This theory also gives a generally correct dependence $\vec{J}_s$ on temperature. However, for quantitative agreement with the experimental data, it must be assumed that the molecular field inside the ferromagnet has a huge value of about 10^5 A/m.

The presence of such a giant field inside the ferromagnet is hardly probable. If such a field existed, it is completely unclear how the external fields, a million times smaller than the internal ones, can remagnetize the ferromagnet. Dorfman's study (Dorfman 1927) on the deviation of fast α-particles passed through the ferromagnetic foil showed that the deflecting field was less than 10^3 A/m, i.e., not less than 1% of the molecular field of Weiss.

In addition to Dorfman's experiments, the following facts speak of the non-magnetic nature of forces that cause spontaneous magnetization:

1. The energy of magnetic interaction per 1 atom has a value of about 10^{-16} erg. This energy of thermal motion kT corresponds to a temperature of 1 K. This means that ferromagnetism should disappear at temperatures close to 1 K. Real ferromagnets have a Curie point $T_s \sim 10^3$ K.
2. The magnetic energy of a real ferromagnetic crystal depends on its orientation (Akulov 1939) when the energy of the "molecular" field is completely isotropic.

All these facts forced to abandon the magnetic concept of the internal field of ferromagnets, and Frenkel (1928) expressed the opinion that the forces that lead to the existence of ferromagnetism should be not magnetic but electric in origin.

The creation of a consistent theory of ferromagnetism based on the "electrostatic" concept proved to be impossible based on classical physics. The elementary magnetic moments that form the magnetization of ferromagnets, as shown by Einstein–de Haas experiments (1915), are the moments of electron spins. Consideration of the role of spin magnetic moments in the occurrence of the ferromagnetic state is possible only from the standpoint of quantum mechanics. It also follows from general theoretical considerations that classical theory for explaining the nature of ferromagnetism is unsuitable. As shown by Van Leeuwen (1921) and Terletsky (1939), the magnetic moment of any magnet in an external magnetic field, which is considered a statistical ensemble of elementary charges moving according to the laws of classical physics, is zero in the steady state.

The quantum theory of ferromagnetism was developed by Frenkel (1928) and Heisenberg (1928) and further clarified by Blokh (1936). As is known, in quantum mechanics, when considering interatomic interaction, in addition to Coulomb energy, additional, so-called exchange interaction energy should be considered (Blokhintsev 1976). Contrary to Coulomb energy, the exchange energy significantly depends on the mutual orientation of the magnetic moments of electron spins. Just in this dependence, Frenkel and Heisenberg suggested seeking the cause of ferromagnetism. The presence of a strong electrical connection between the electrons of the crystal, with an account of the exchange part of the interaction energy under certain conditions, leads to a more energetically advantageous state with a parallel orientation of the magnetic moments of the electron spins. If these conditions are satisfied, the crystal has a spontaneous magnetization $\vec{J}_s$, i.e., it is a ferromagnet.

For the part of the energy W that depends on the exchange forces, Heisenberg found the following expression:

$$W = -\left(m^2/n\right) \cdot A, \tag{1.7}$$

where n is the total number of electrons in the crystal; m is the number of electrons with parallel spin orientation; A is the exchange integral, the value of which depends on the structure of the electronic shells of crystal atoms.

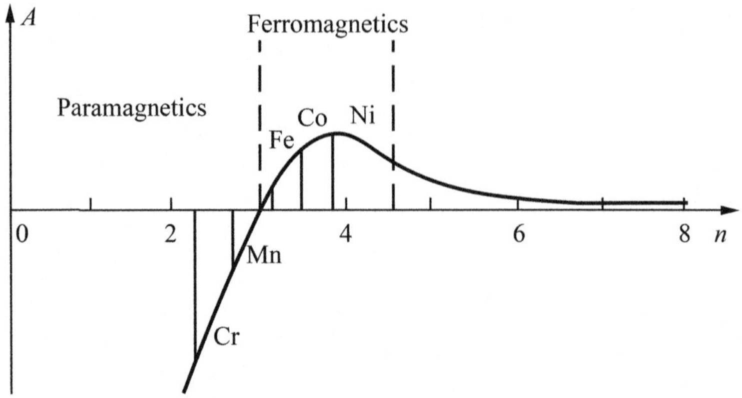

Fig. 1.1 Dependence of the exchange integral A on the distance n between adjacent atoms in the lattice to the radius of the unfilled electron shell

When $A > 0$ the minimum energy corresponds to the magnetized state. This means that a necessary condition for ferromagnetism is condition $A > 0$. Estimation of the order of W shows that this energy can ensure the existence of ferromagnetism.

A necessary condition for ferromagnetism is a positive value of the exchange integral, and the following two conditions are its criterion: 1—there must be an internal unfilled electron shell; 2—the radius of this shell should be several times smaller than the distance between adjacent atoms in the lattice (Krinchik 1976). These conditions are met for the group of iron and rare earth elements (Fig. 1.1).

Thus, the magnetic moments of electron spins are the elementary carriers of ferromagnetism. The nature of this phenomenon is purely quantum and caused by electric exchange interaction. The presence of a strong exchange interaction leads to the fact that in some materials in a certain temperature range the parallel arrangement of the electron spin moments becomes more energetically advantageous. This leads to spontaneous magnetization and, consequently, to the existence of ferromagnetism.

1.2 Domain Structure of Ferromagnets

If the exchange interaction was the only type of interaction in the ferromagnetic crystal, then in the absence of an external magnetic field, the ferromagnet would be magnetized to saturation as a whole, i.e., there would be no domain structure. Other types of interactions in ferromagnets need to be considered to determine the cause of domains appearance.

In addition to the exchange energy, in the external magnetic field absence, the most important role is played by: (1) the energy of the demagnetizing field—the magnetostatic energy of the form; (2) the energy of the magnetic interaction of spin moments and orbital electrons—the energy of magnetic anisotropy. Even though both magnetic energies are about 10^3 times smaller than the exchange energy, it is the magnetic interaction that causes the domain structure and all other phenomena associated with the existence of a hysteresis loop in ferromagnets.

Consideration of all the types of interactions demonstrates that the more the energy-efficient is not the state in which the whole crystal is uniformly magnetized to saturation, but the state in which the ferromagnetic sample is divided into separate regions with such a distribution of the magnetizations $\vec{J}_s$ in them that the resulting magnetization of the whole sample is equal to zero. The first quantitative theoretical substantiation of the hypothesis of areas of spontaneous magnetization (domains) was given by Frenkel and Dorfman (1930). Even though (in addition to the exchange energy) they took into account only the energy of the demagnetizing field

$$W_{\text{dem}} = NI^2/2, \tag{1.8}$$

where N is the demagnetizing factor of the sample (Arkadiev 1913), they managed not only to justify the division of the ferromagnet into domains (Fig. 1.2), but also to obtain an expression for the width of the domains:

$$l = (l_0 d)^{1/2}, \tag{1.9}$$

where d is the linear size of the sample, $l_0 \sim 10^{-4}$ cm.

This means that for sample sizes $L \sim 1$ cm, the width of the domains should be $l \sim 10^{-2}$ cm. These estimates correlate well with the experimental data.

A more rigorous theory of the domain structure of ferromagnets, which considers the energy of magnetic anisotropy (according to Akulov, the latter is defined by the expression (Akulov 1939))

$$W_{\text{anis}} = K_0 + K_1\left(\alpha_1^2\alpha_2^2 + \alpha_2^2\alpha_3^2 + \alpha_1^2\alpha_3^2\right) + K_2\alpha_1^2\alpha_2^2\alpha_3^2, \tag{1.10}$$

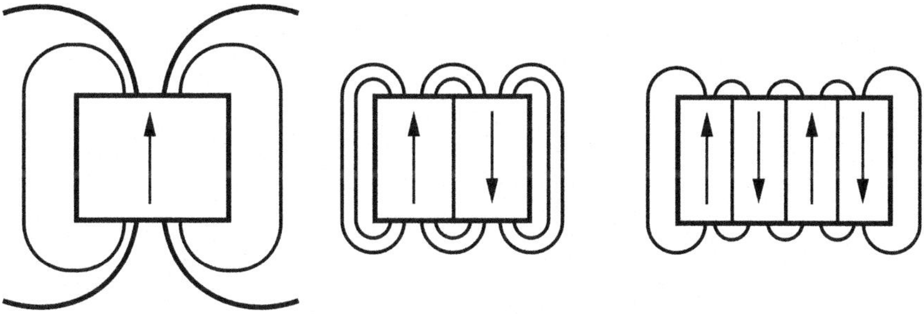

Fig. 1.2 Division of a ferromagnetic crystal into domains that help to reduce the magnetostatic energy of the demagnetizing field

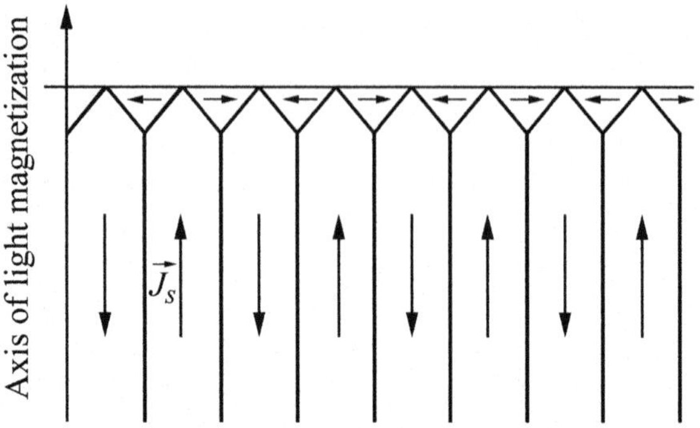

Fig. 1.3 Domain structure of a ferromagnet with one axis of light magnetization

where K_0, K_1, K_2 are constants of crystallographic magnetic anisotropy; α_1, α_2, α_3 are guiding cosines of magnetization $\vec{J}_s$ were built by Landau and Lifshitz (1935; Lifshitz 1945). They showed that the magnetization in individual domains is directed along the axes of light magnetization (ALM). In the case of a perfect uniaxial ferromagnetic crystal, the domains should have the shape shown in Fig. 1.3.

Near the surface, the shape of the domains becomes such as to reduce the energy of the demagnetization field W_{dem} even due to some increase in the energy of the magnetic anisotropy W_{anis}. This leads to the formation of boundary domains in the form of triangular prisms, in which the direction $\vec{J}_s$ is perpendicular to the ALM, and to the closure of the magnetic flux, which corresponds to the minimum free energy of the crystal. As the surface of triangular regions grows, the shape of the domains presented in Fig. 1.4 becomes more advantageous.

Landau and Lifshitz have also found an expression for the thickness of the boundary layer between domains and the law of magnetization change in this layer.

Note that the presence of a transition boundary layer of finite thickness δ is unfavorable for the energy of magnetic anisotropy W_{anis}, which increases with δ. Much more advantageous for the minimum W_{anis} would be the absence of a long domain boundary, i.e., the jump of the vector $\vec{J}_s$ from the certain direction to the opposite one.

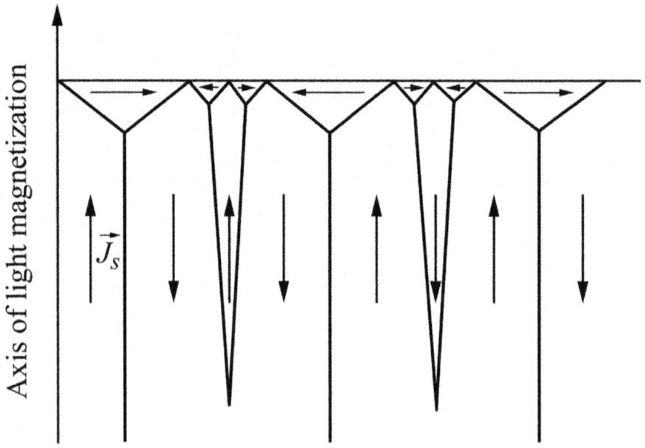

Fig. 1.4 Fine domain structure of a ferromagnet

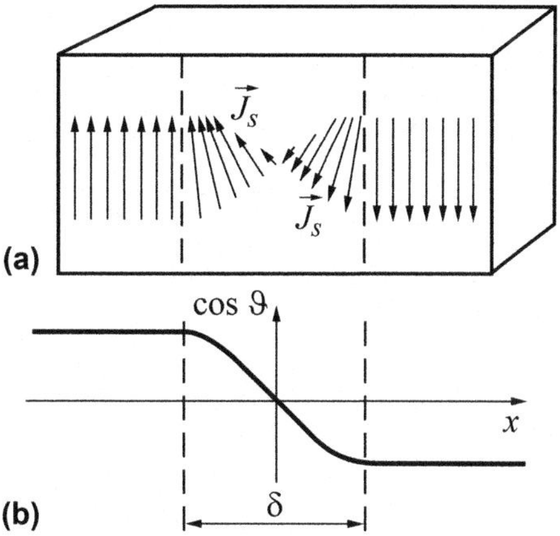

Fig. 1.5 Structure of the boundary layer between the domains in the ferromagnetic crystal (**a**) and dependence $\cos\vartheta$ on x (**b**)

However, such a jump is disadvantageous in terms of exchange forces, as it causes a significant increase in the exchange energy. Therefore, the exchange forces make the transition smoother, that is, such in which the magnetization would change smoothly. The competition between these two interactions determines both the thickness δ of the transition layer and the nature of the magnetization $\vec{J}_s$ change in it (Fig. 1.5). In the transition layer the vector $\vec{J}_s$ is gradually rotated, and its projection on the surface of the sample changes according to the law

$$\tau_t = \tau_s \cos\vartheta, \quad \cos\vartheta = -\,\mathrm{th}(2x/\delta), \tag{1.11}$$

where x is the distance from the middle of the boundary layer.

Landau and Lifshitz obtained a general formula for the thickness of the boundary layer

$$\delta \approx \left(A/K_{\mathrm{eff}}a^3\right)^{1/2}, \tag{1.12}$$

where A is the exchange integral; K_{eff} is the effective constant of magnetic anisotropy; parameter a has the dimension of length and the order of the constant of the crystal lattice. At room temperature, formula (1.12) gives for δ a value of about 10^{-6} cm.

For the surface energy density of the boundary layer, the same authors obtained the expression

$$\gamma \approx (K_{\mathrm{eff}}A/a)^{1/2}. \tag{1.13}$$

For iron at room temperature $\gamma = 1$ erg/cm^2. Note that the value of γ and its dependence on the coordinates play a crucial role in the magnetization processes, especially in the processes of shifting the domain boundaries. In the case of a multi-axial ferromagnet, in addition to the above-mentioned so-called 180° boundaries between domains, the regions with mutually perpendicular directions of spontaneous magnetization, the so-called 90° boundaries (or 90° neighborhood), are also possible.

Subsequently, the theory of Landau and Lifshitz was developed in the works by Kittel (1946) and Néel (1944), who, without changing its basic provisions, made several clarifications.

In real crystals, the picture is significantly complicated (Vonsovsky and Shur 1948; Krinchik 1976; Grechishkin 1975) due to structural defects and internal stresses. But the boundary between the domains is always located so that the increase in free energy of the crystal, which it contributes, is minimal. This means that in the case of 180° boundaries, domains are usually located in places with minimum internal stresses, where the value K_{eff} is minimal and then, γ is the minimum value. In the case of 90° boundaries, the domains are located mainly where the stress changes sign, because the change in the sign of internal stresses corresponds to the change in the directions of the axes of light magnetization in neighboring domains, which determine the directions of magnetization.

1.3 Magnetization and Demagnetization of Ferromagnets

Obtaining a magnetic field. Various methods are used to excite the magnetic field. The choice of the method that is best suited in each case is determined by the required intensity and uniformity of the magnetic field, as well as the volume of the working space.

The magnetic field inside the coil with the air core, which occurs due to the passage of electric current $\vec{i}$ through it, for a given number of turns and fixed shape of the coil will always be proportional to the current:

$$\vec{H} = C\vec{i}, \tag{1.14}$$

where C is the coil constant.

In the case of an infinitely long solenoid, with account that

$$\vec{H} = n\vec{i}, \tag{1.15}$$

where n is the number of turns per unit length, we have:

$$C = n. \tag{1.16}$$

The most common source of a uniform magnetic field are Helmholtz coils, which are made in the form of a pair of identical round coil rings of radius R, which have N turns located at a distance r from each other. The most homogeneous field is

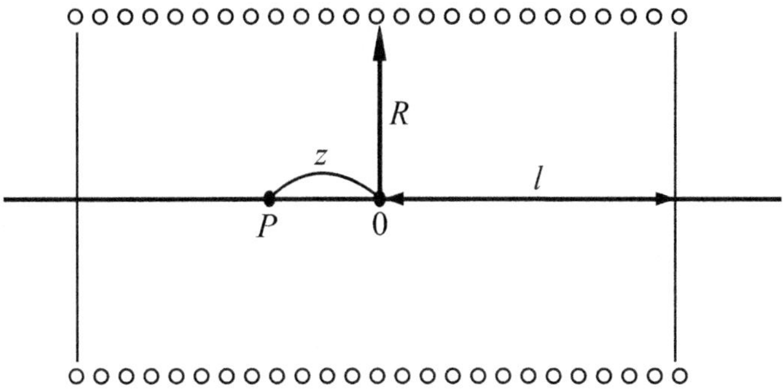

Fig. 1.6 Solenoid in the form of a single-layer coil

provided in the middle between the coils, and the values of the constant C are found by the ratio:

$$C = \frac{8N}{5\sqrt{5}R} = \frac{0.716}{R}N. \tag{1.17}$$

Helmholtz coils are usually used to produce relatively weak fields in large volumes, for example, to compensate the Earth's magnetic field.

If to consider a solenoid of finite length, then its constant C at some point of observation P, located on its central axis (Fig. 1.6) will be equal to (Tikadzumi 1987):

$$C = \frac{n}{2}\left(\frac{l+z}{\sqrt{R^2+(l+z)^2}} + \frac{l-z}{\sqrt{R^2+(l-z)^2}}\right), \tag{1.18}$$

where $2l$ is the length of the solenoid; z is the distance from the center of the solenoid to the point P; R is the radius; n is the number of turns per unit length.

For a thick multilayer solenoid, which has a winding with outer and inner radii R_2 and R_1 (Fig. 1.7), the constant C is calculated by the formula:

$$C = \frac{n}{R_2 - R_1}\left[(l+z)\ln\left(\frac{R_2+\sqrt{R_2^2+(l+z)^2}}{R_1+\sqrt{R_1^2+(l+z)^2}}\right) + (l-z)\ln\left(\frac{R_2+\sqrt{R_2^2+(l-z)^2}}{R_1+\sqrt{R_1^2+(l-z)^2}}\right)\right]. \tag{1.19}$$

The maximum magnetic field strength in a solenoid is determined by the maximum current that can be passed through it. The nature of the distribution of the magnetic field strength H inside the solenoid is shown in Fig. 1.8.

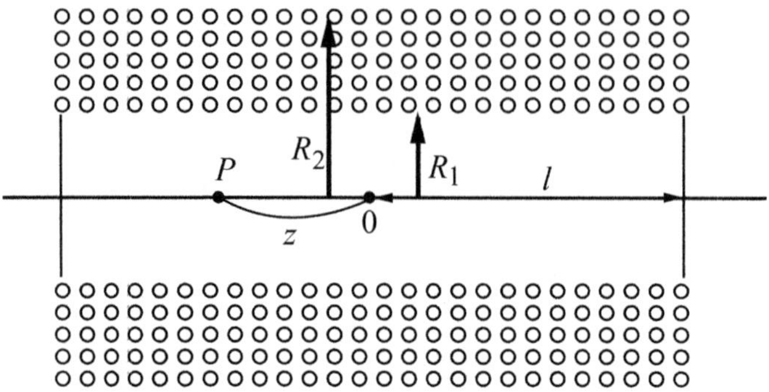

Fig. 1.7 Solenoid as a multilayer coil

Fig. 1.8 Distribution of magnetic field lines inside the solenoid (**a**) and distribution of the magnetic field strength in the solenoid (**b**)

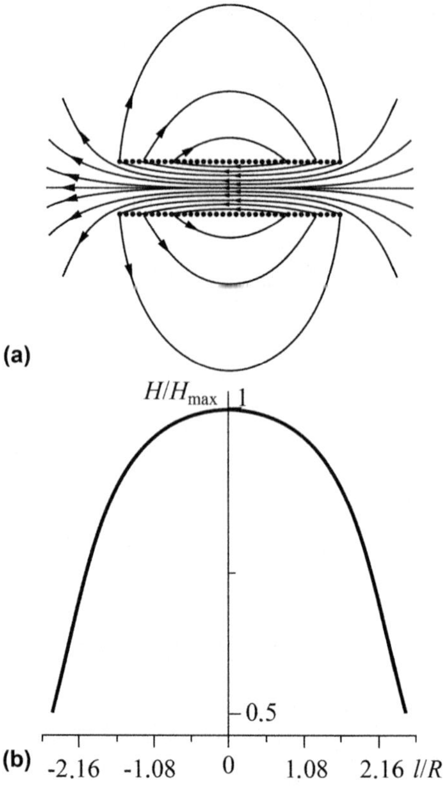

As we see (Fig. 1.8b) the highest magnetic field strength is reached in the center of the solenoid. Figure 1.8a shows the magnetic field lines inside and outside of such a solenoid (Parcell 1983). One can see that some power lines do pass through the windings. The cylindrical current layer serves as a discontinuity surface for the magnetic field. If to look in detail at the field in the immediate vicinity of the conductors, one would see not infinitely steep bendings, but very complex wave paths of power lines around individual turns and through them.

Thus, the current flows in a spiral in the solenoid, and because there are many turns, and they are stacked close to each other, we consider the model of the solenoid as a set of rings with current.

Excitation of rapid periodic changes in magnetic field strength. With rapid periodic changes in the magnetic field strength, the type of loop that expresses the dependence $B = f(H)$ differs from the static hysteresis loop, which is obtained by slow changes in the field strength. In this case, magnetic induction is a function not only of the field strength, but also of its time derivatives. This is due to the eddy currents generated in the ferromagnetic material and the magnetic ductility (aftereffect). The area of the dynamic loop, which reflects the real dependence $B = f(H)$, is determined by the total loss per unit volume of ferromagnetic material for remagnetization and eddy currents for one period of field strength.

The value of the demagnetization coefficient N varies greatly depending on the size and configuration of the sample, the nature of the external magnetic field, the characteristics of the ferromagnet.

Contrary to the static case, the dynamic mode is characterized by a limited time of the magnetizing magnetomotive force action. If in the static regime the transient processes in the material have time to be completed before the cessation of the action of the magnetomotive force (MMF), then in the dynamic mode the transient processes can often not be completed during the action of the MMF. The main indicators that determine the mode of remagnetization in alternating magnetic fields are the magnitude of the amplitude of the magnetizing MMF and the time of its action. In addition, the character of changes in the MMF over time, which can be both periodic and aperiodic, is of great importance. Figure 1.9 shows various possible cases of changes in the magnetizing MMF over time (Troitsky 2002).

As we see in Fig. 1.9a–e the MMF change is periodic, in which there are no pauses between the action of the MMF of opposite polarities. In Fig. 1.9f–h the MMF change is of an aperiodic, or as it is often called, pulse character.

There is also symmetrical and asymmetrical character of the action of MMF, which may differ in both amplitude and duration of action. For the symmetrical periodic character of the MMF action we have: $F_{\mathrm{m}}^{+} = F_{\mathrm{m}}^{-}$; $t^{+} = t^{-} = T/2$ $t^{+} = t^{-} = T$. For the asymmetric periodic character of the MMF action—$F_{\mathrm{m}}^{+} \neq F_{\mathrm{m}}^{-}$ or $t^{+} \neq t^{-} \neq T/2$, but $t^{+} = t^{-} = T/2$. Here F_{m}^{+} and F_{m}^{-} are the maximum values of the magnetizing MMF of positive and negative polarities; t^{+} and t^{-} are the value of the action time of the MMF of these polarities. Symmetrical or asymmetrical character of the action can also occur during the aperiodic changes in the MMF, with the only difference that $t^{+} + t^{-} < T$, since $T = t^{+} + t^{-} + t_{\mathrm{n}}$ (Fig. 1.9f, g), where t_{n} is the pause time.

The MMF shape change can be of rectangular, sinusoidal, trapezoidal, triangular character, and so on. However, the shape of magnetizing MMF begins to appear only at small amplitudes of MMF, when the generated magnetic field is commensurable

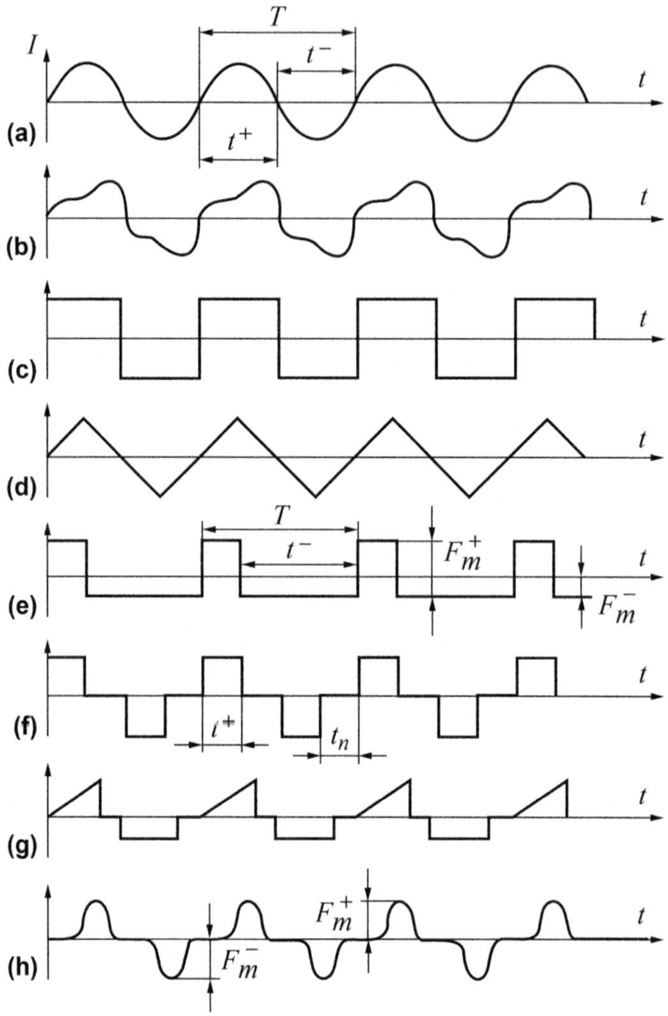

Fig. 1.9 Regularities of time change of current $I(t)$ in the magnetizing

with the magnitude of the coercive force of the material ($H_m \leq H_c$). For sufficiently large amplitudes of the MMF, when the generated magnetic field significantly exceeds the coercive force ($H_m >> H_c$), the shape of the MMF has little effect on the remagnetization mode, especially if the rise time of the MMF is significantly less than the MMF duration.

During remagnetization in the rapidly changing magnetic fields, the configuration of the hysteresis loops changes compared to the static loops (Fig. 1.10) (Troitsky 2002). Such hysteresis loops are called dynamic.

The change in the shape of the hysteresis loops is due to time effects (eddy currents, magnetic aftereffect—ductility), which largely depend on the electric parameters of

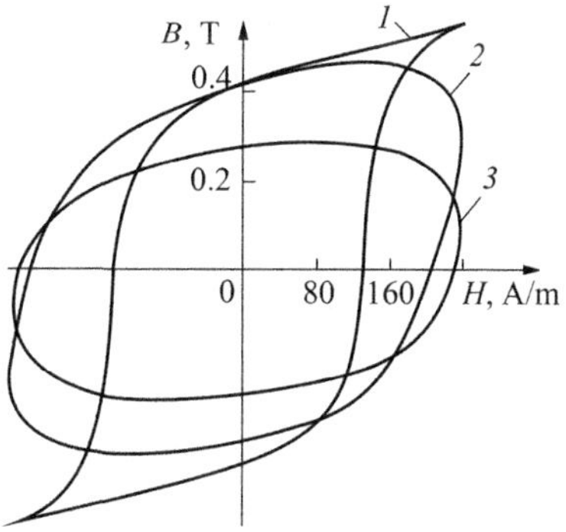

Fig. 1.10 Hysteresis loops: *1* quasi-static magnetization; *2* dynamic magnetization with frequency f_1; *3* dynamic magnetization with frequency $f_2 > f_1$

the magnetic material. In its turn, the degree of influence of eddy currents depends on the frequency and character (periodic or pulsed) of magnetization, its mode (symmetric or asymmetric), and so on. The dynamic parameters of magnetic materials are significantly different from static ones. The value of the coercive force H_c, for example, increases as the hysteresis loop expands and the absolute magnetic permeability of the material μ_a decreases.

Any change in the magnetic flux inside a closed circuit induces MMF in it, which causes an electric current. Currents in a conductive medium induced by alternating magnetic flux are eddy currents (they are also called Foucault currents). Because of the small value of the electrical resistance of metal magnetic conductors, eddy currents can reach significant values. In its turn, eddy currents create their own magnetic field, which is counter-directed to the external magnetic flux that causes them. The interaction of counter magnetic fluxes leads to a surface effect (skin effect), i.e., to a sharp change in the depth of penetration of an alternating electric field into the depth of the ferromagnet. Thus, with a strong surface effect, the alternating magnetic field is mainly concentrated in a thin surface layer.

Ferromagnetic material is heated by eddy currents, and its effective cross-section is reduced due to the surface magnetic effect. The value of the commutated magnetic flux also decreases and thus contributes to the detection of small surface defects.

Alternating voltage, which has different amplitudes and phases, causes variations in magnetic permeability. With increasing magnetic field strength (current in the coil winding), the magnetic permeability first increases, and then, reaching a maximum, decreases.

The total magnetic flux Φ coupled to the coil turns is not proportional to the current I. Therefore, the inductance of the coil

$$L = \Phi / I \tag{1.20}$$

and its inductive resistance

$$x = \omega L \tag{1.21}$$

are variables. The curve $L = f(I)$ repeats the curve $\mu = f(i)$ of magnetic permeability.

At the sinusoidal voltage applied to the coil, in case of neglect of its active resistance, the instantaneous value of the current will be

$$I = \frac{\Phi}{L} = \frac{\Phi \sin \omega t}{L} = \frac{Um \sin \omega t}{\omega L} \tag{1.22}$$

and will vary according to the non-sinusoidal law since the inductance L is a nonlinear quantity. In practice, this nonlinearity is often neglected, and the non-sinusoidal current of the calculation coil is replaced by an equivalent sine wave. However, when monitoring, using primary converters that respond to the instantaneous current value, it is important to know the character of the changes and the momentary values of the circuit parameters. The curve of momentary current values $I = f(t)$ can be constructed if the curves of dependences $\Phi = f(I)$ and $\Phi = f(t)$ are known.

In the presence of hysteresis in each half-period, the process of magnetization and demagnetization takes place on the mismatched branches of the loop. Therefore, the curves of current and magnetic flux pass through zero values not simultaneously (Fig. 1.11).

Thus, the character of the magnetic field change in the magnetizing device must always be chosen considering the conditions of the experiment and the technical characteristics of the means involved.

Magnetic hysteresis. A typical feature of materials with spontaneous magnetization $\vec{J}_s$, which primarily include ferromagnets, is the presence of the magnetization curve and the hysteresis loop (Vonsovsky 1971), which are shown schematically in Fig. 1.12.

In the absence of field ($H = 0$) the sample is demagnetized, its magnetization $\vec{J}$, which is defined as the magnetic moment of a unit volume, is zero, then for simplicity we will consider samples of unit volume. As the field strength $\vec{H}$ increases, the magnetization $\vec{J}$ increases and reaches in field $\vec{H}_s$ the value of the saturation magnetization $\vec{J}_s$. If the value does not change in the interval, it can be accepted

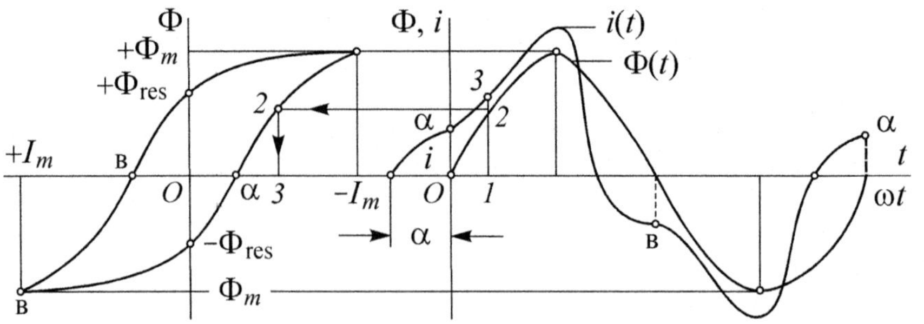

Fig. 1.11 Character of the change in magnetic flux $\Phi(t)$ and current in the coil $i(t)$ during its sinusoidal remagnetization

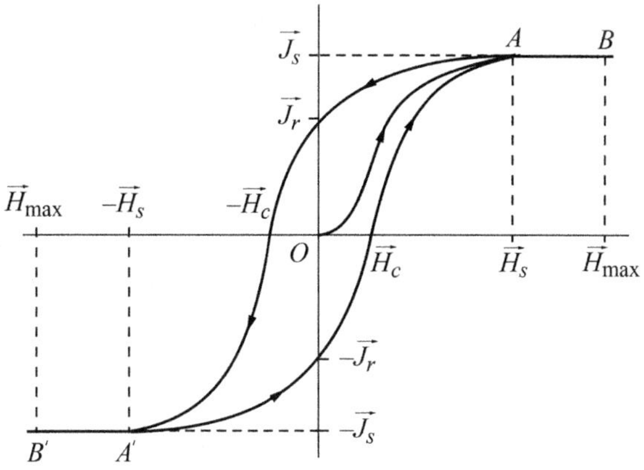

Fig. 1.12 A typical magnetic hysteresis loop and its main parameters

as a spontaneous (involuntary) magnetization of the material. The curve OAB in Fig. 1.12 is the initial magnetization curve. When reducing the field from $\vec{H}_s$ to zero, the sample does not come to a state with $J = 0$; on the contrary, at $H = 0$ the sample has a residual magnetization $\vec{J}_r$. This means that the change of $\vec{J}$ lags the change in magnetic field strength $\vec{H}$. Only in some field with opposite direction (negative) $-\vec{H}_c$ magnetization $\vec{J} = 0$. This field $\vec{H}_c$ is called the coercive field or coercive force (Vonsovsky 1971). With a cyclic change of the field $\vec{H}_{max} \to 0 \to (-\vec{H}_{max}) \to 0 \to \vec{H}_{max}$ the magnetization vector describes along the closed curve $AJ_rA'(-J_r)A$, usually symmetric with respect to the origin. This curve is called the magnetic hysteresis loop with magnetization. Its main characteristics (parameters) are: $\vec{J}_s$, $\vec{J}_r$, $\vec{H}_c$ and the area S of the loop. The latter is proportional to the work done by the external field to remagnetize the sample.

Spontaneous magnetization is determined by quantum exchange interactions between the electronic shells of atoms in the crystal lattice (Vonsovsky 1971; Nazarchuk et al. 2013). This is a fundamental characteristic of the material. At the known temperature we consider $\vec{J}_s = \text{const}$. Values $\vec{J}_r$, $\vec{H}_c$, S are structurally sensitive parameters. They can be changed in a wide range (hundreds or thousands of times) by various treatments (thermal, thermomagnetic, mechanical, etc.) of the material. The values of hysteresis parameters determine the field of technology where one or another magnetic material is used. Thus, magnetically soft materials (generator magnetic conductors, transformer cores, etc.) require as little $\vec{H}_c$ value as possible, and magnetically hard materials (permanent magnets) require as greater $\vec{H}_c$ value as possible. As a result, in modern magnetic materials the values of the coercive force $\vec{H}_c$ can differ by 5–6 orders of magnitude (Mishin 1991).

In a ferromagnetic crystal, under the action of magnetic anisotropy forces (Vonsovsky 1971; Nazarchuk et al. 2013), the vectors $\vec{J}_s$ are oriented along certain axes. There may be several such axes. Each of them is called the light magnetization axis (LMA). For example, a cobalt crystal has one LMA, iron—three, nickel—four. For simplicity, we will consider only ferromagnetic crystals with one LMA.

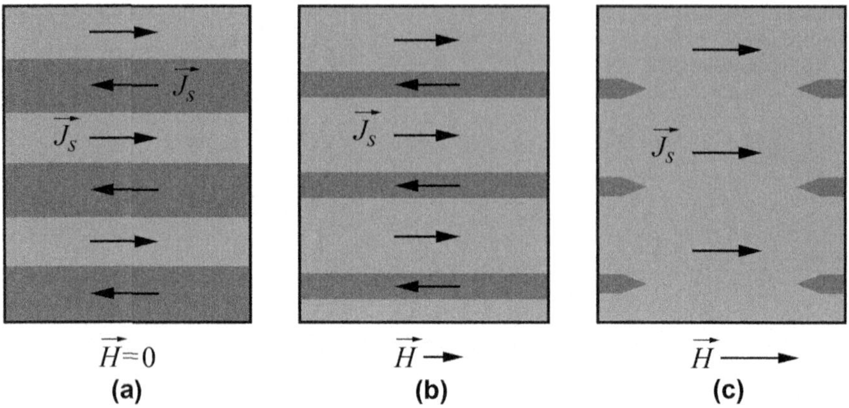

Fig. 1.13 Scheme of the domain structure of a magneto uniaxial crystal in the demagnetized state (**a**) and change of its structure during the magnetization process (**b**, **c**)

In the absence of the magnetic field, the crystal splits into magnetic regions—domains with opposite orientations of spontaneous magnetization vectors, as shown in Fig. 1.13a.

When the vector sum $\sum \vec{J}_s = 0$, the sample is demagnetized, i.e., $\vec{J} = 0$. For switching on the field directed along LMA (Fig. 1.13b), there is a shift of domain borders, the volume (on the flat scheme—width) of domains with $\vec{J}_s$, oriented to the right, increases, the volume of domains with the opposite direction $\vec{J}_s$ decreases. The total magnetization arises. With an even larger magnetic field, the boundaries of the decreasing domains close in the middle, the band domains become wedge-shaped, and those, in turn, decrease with growth, contracting to the edges of the sample, and disappear (Fig. 1.13c). The sample is magnetized to saturation $\vec{J} = \vec{J}_s$. If there are no defects in the crystal and nothing prevents the domain boundaries (walls) from moving, then their displacement is opposite. For example, decreasing the field $\vec{H}$ from the value corresponding to Fig. 1.13b to zero, the walls of the domains will return to their original state (Fig. 1.13a), passing the same positions that they consistently occupy during the increase of the magnetic field. There is no hysteresis. The displacement of the walls becomes significantly irreversible when there are defects in the sample, or when there is a qualitative change in the type of domain structure, for example, during the transition from Fig. 1.13b to c. Let's focus on the first case as simpler one.

Figure 1.14a presents the photo of two domains: light, with the vector directed "to us", and dark with $\vec{J}$ spontaneous magnetization directed "from us". Domains in Fig. 1.14 are detected by the magneto-optical effect, which consists in the fact that during the reflection of plane-polarized light from different domains, the plane of polarization turns at different angles, for example, $+\alpha$ and $-\alpha$. This makes it possible to adjust the microscope analyzer and obtain a contrasting picture of dark and light domains.

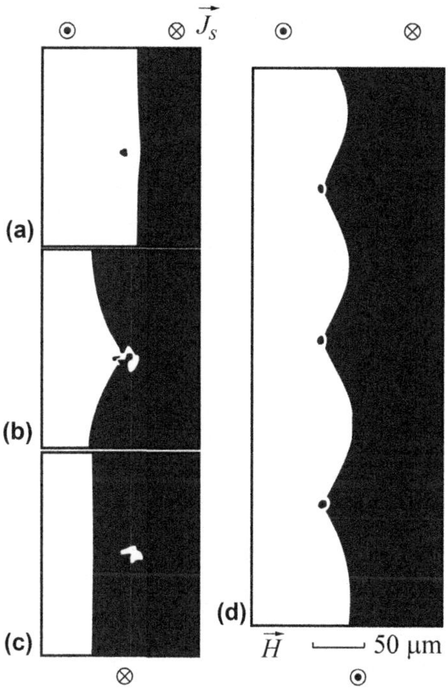

Fig. 1.14 Shift of the domain wall through the region of the crystal containing one (**a**–**c**) or several (**d**) defects. Magnetization in the light domain is directed "to us", in the dark—"from us"

In Fig. 1.14a, a defect is visible to the left of the wall. The field $\vec{H}$ oriented "from us" is included, and the wall begins to move to the left. As long as it does not meet defect, the shift is opposite. When the wall reaches the defect, it is seen that with increasing field $\vec{H}$ strength, its upper and lower edges are shifted, and the middle section stops at the defect (Fig. 1.14b). With further $\vec{H}$ growth, the domain wall bends more and more and eventually there is a breakdown: the wall jumps to the left of the defect (Fig. 1.14c). At the same time, the area of the black domain increased by a jump, which means that the jump magnetization $\vec{J}$ occurred.

Then, with increasing intensity of the magnetic field $\vec{H}$, the shift of the wall to the left is again opposite. If to reduce $\vec{H}$, thereby forcing the wall to move to the right, one can see that its bend, again caused by the presence of a defect, will be directed in the other direction compared to Fig. 1.14b. This means that the jump-like transition from the state shown in Fig. 1.14b to the state shown in Fig. 1.14c is substantially irreversible.

Determining the change in the area of the "black" and "light" type of domain, one can build dependence $I(H)$ during the boundary movement to the left and right. A small hysteresis loop will appear. The boundary hysteresis loop consists just of such local loops caused by the irreversible shift of domain boundaries (Fig. 1.12).

Consider why the domain boundary stops at the defects. Figure 1.14 shows that the domain wall is like an elastically tension film and has a reserve of energy. The energy of a wall per unit area is called the surface density of the ultimate energy

γ. If the wall is outside the defect, its total energy is $\varepsilon_1 + \gamma S$, where S is the wall area. Assume that the defect is a non-magnetic spherical inclusion of diameter D. Then, being on the defect, the wall has the energy $\varepsilon_2 + \gamma(S - (\pi/4)D^2)$, $\varepsilon_2 < \varepsilon_1$. In other words, the defective wall is in the energy potential hole. The harder the wall is fixed, the deeper the hole, the greater field one needs to apply to "pull" the wall out. Obviously, several defects will prevent the wall from shifting more than any of them. This is clearly shown in Fig. 1.14d, where a few artificially created defects hold the wall like clothespins.

In the theoretical model of non-magnetic spherical inclusions distributed in a sample with a certain bulk density ρ, the following formula for the coercive force is obtained

$$\vec{H}_c \approx \frac{\gamma \rho^{2/3}}{\vec{J}_s D}. \tag{1.23}$$

Material characteristics γ and $\vec{J}_s$ may be different for different materials. For example, for iron $\gamma = 0.002\ \mathrm{J/m^2}$, and for $SmCo_5$ alloy $\gamma = 0.1\ \mathrm{J/m^2}$. It is clear that in order to obtain a magnetically soft material it is necessary to take an alloy with a small value of γ, high $\vec{J}_s$ and as few defects as possible that prevent the shift of domain boundaries. On the contrary, in the case of magnetically hard material, a small $\vec{J}_s$, high value of γ and a sufficient number of defects are required to provide high coercive force (Mishin 1991; Kandaurova and Onoprienko 1986).

Note that in Figs. 1.13 and 1.14 the external magnetic field is directed along the LMA. If this is not the case, and the field $\vec{H}$ forms an angle with the LMA, the pressure from the field to the wall decreases. It will be determined by the component of the field $H \cos\varphi$ (at $\varphi = 90°$ the wall will not shift at all). Hence the law of anisotropy of coercive force in the form

$$H_c(\varphi) = \frac{H_c(0)}{\cos\varphi}, \tag{1.24}$$

is obtained, where $H_c(0)$ is the value of the coercive force in the orientation of the field along the LMA. Dependence $H_c(\varphi)$ is shown in Fig. 1.15a.

Peculiarities of magnetization of ferromagnets. Magnetization depends on the magnetic field intensity and the shape of the body. If a ferromagnetic body is placed in a homogeneous magnetic field $\vec{H}_e$, an additional magnetic field $\vec{H}_0$ appears on its surface, which inside the body has a direction opposite to the external field and magnetization (Fig. 1.16) (Troitsky 2002; Kiefer and Pantyushin 1995). Thus, the external magnetic field will be superimposed on the external demagnetization field.

In this case, the true field $\vec{H}$ inside the body will be equal to the difference:

$$\vec{H} = \vec{H}_e - \vec{H}_0. \tag{1.25}$$

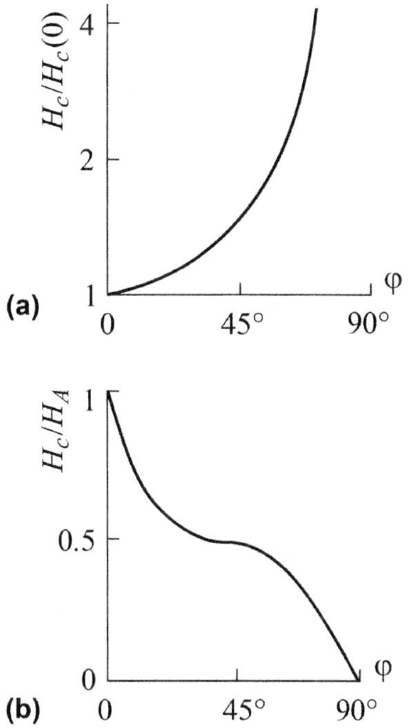

Fig. 1.15 Anisotropy of the induced coercive force $H_c/H_c(0)$ and H_c/H_A in the case of two mechanisms: shift of domain boundaries (**a**) and uniform rotation of spontaneous vectors of magnetization $\vec{J}_s$ (**b**)

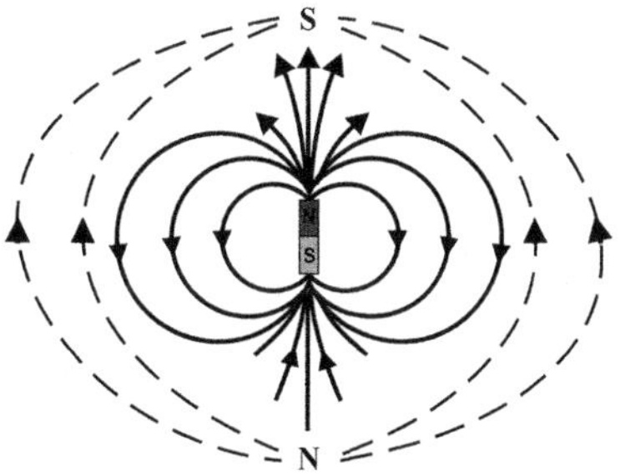

Fig. 1.16 Poles formation during the introduction of an open-shape ferromagnetic body into an external magnetic field (dashed lines)

In most cases, many magnetizing parts can be represented as a sphere or an ellipsoid. Thus, a sphere of material with magnetic permeability μ, which is placed in an external homogeneous magnetic field, is polarized magnetically homogeneously. If the sphere is placed in a vacuum and $\mu > \mu_0$, where $\mu_0 = 4\pi \times 10^{-7}$ Hn/m is the magnetic permeability of the vacuum, the field vector $\vec{H}_0$ determined by the magnetization of the sphere inside it will be directed against vector $\vec{H}_e$ of the external magnetic field. The field $\vec{H}_0$ is called the demagnetization field. Its intensity inside the body is less than the external field, and (Troitsky 2002):

$$\vec{H}_0 = \frac{\mu - \mu_0}{\mu + 2\mu_0} \vec{H}_e. \tag{1.26}$$

Outside the sphere, the field due to its magnetization has the same value as the current field in a very small, closed circuit, located in the center of the sphere and has a magnetic moment $\vec{P}$, equal to the geometric sum of the magnetic moments of all elementary currents in the sphere volume:

$$\mu_0 \vec{P} = 4\pi \mu_0 R^3 \vec{H}_0 = 4\pi R^3 \mu_0 \frac{\mu - \mu_0}{\mu + 2\mu_0} \vec{H}_e, \tag{1.27}$$

where R is the sphere radius.

The final intensity $\vec{H}$ and the final magnetic induction inside the sphere are equal to:

$$\vec{H} = \vec{H}_e - \vec{H}_0 = \frac{3\mu_0}{\mu + 2\mu_0} \vec{H}_e; \tag{1.28}$$

$$\vec{B} = \frac{3\mu}{\mu + 2\mu_0} \mu_0 \vec{H}_e = \frac{3\mu}{\mu + 2\mu_0} \vec{B}_e. \tag{1.29}$$

The larger μ, the stronger the demagnetization field $\vec{H}_e$ and the weaker the field $\vec{H}$, but the stronger the field $\vec{B}$. In the limiting case for $\mu \to \infty$ we have (Troitsky 2002):

$$\vec{H}_0 = \vec{H}_e; \;\; \vec{H} = 0; \;\; \vec{B} = 3\vec{B}_e; \;\; \vec{B}_e = \vec{B}/3. \tag{1.30}$$

Thus, the magnetic flux inside the body is significantly greater than the external induction.

Figure 1.17 (Rudyak 1986) shows the external homogeneous field around the ellipsoid, the field of the vector $\vec{H}_0$, which is determined by the magnetization of the ellipsoid and is related to the conditional idea of induced magnetic masses, the final fields of the vector $\vec{H}$ and vector $\vec{B}$.

For components of ferromagnetic materials both $\vec{H}_0$, and the magnetization $\vec{J}$ are proportional to the intensity $\vec{H}_e$ of the external magnetic field. That is, one can write:

$$\vec{H}_0 = N\vec{J}, \tag{1.31}$$

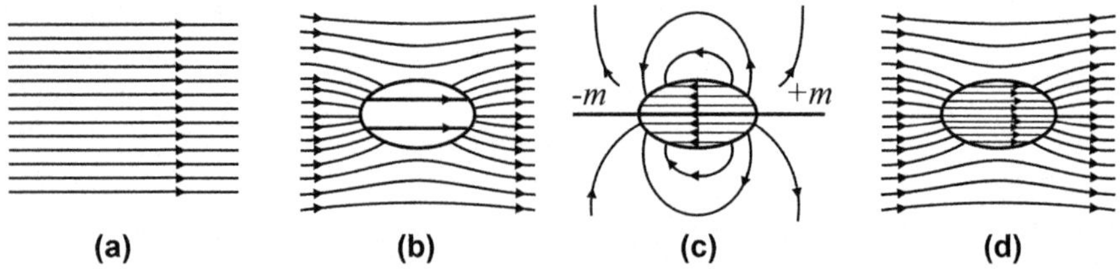

Fig. 1.17 Scheme of distribution of the ellipsoid magnetic fields: **a** external homogeneous field; **b** resulting field; **c** field of the induced magnetic masses; **d** field of the magnetic induction vector

where

$$\vec{J} = \vec{B}/\mu_0 - \vec{H}, \tag{1.32}$$

and N is the coefficient of proportionality, which is called the demagnetization coefficient. It determines, for a given magnetization, the value of the demagnetization field intensity. As can be seen from Fig. 1.17, the demagnetization factor depends on the shape of the body that is magnetized.

Elliptical bodies possess an important property—to be magnetized uniformly in a uniform external magnetic field—which is widely used in magnetometry. The demagnetization coefficient in formula (1.31) can be calculated accurately only for ellipsoids and their partial cases: spheres, plates, infinite cylinders with elliptical or circular cross-section. Bodies that differ in shape from ellipsoids are magnetized inhomogeneously, even in a uniform external magnetic field. Moreover, an ellipsoid body unevenly introduced into the magnetic field can give a partial pole magnetization, i.e., the spotted magnetization can be called the second demagnetizing factor.

Thus, the true intensity $\vec{H}$ of magnetic field acting on the ferromagnet is less than the external field intensity $\vec{H}_e$. The demagnetizing factor N depends very much on the relative length of the sample, i.e., on the ratio of its length to the transverse dimensions. The value of N decreases with increasing length of the sample and for practical calculations is taken from the known literature sources (Tikadzumi 1987; Troitsky 2002; Kiefer and Pantyushin 1995; Klyuev et al. 2006).

Considering the above, it is possible to obtain the following relationship between the internal and the external fields $\vec{H}$ and $\vec{H}_e$:

$$\vec{H} = \vec{H}_e \frac{1}{1 + \chi N}, \tag{1.33}$$

where χ is the magnetic susceptibility of matter, which is determined only by its physical nature. The χ value is related to the magnetic susceptibility of the body χ_e by the ratio:

$$\chi_e = \frac{\chi}{1 + \chi N}. \tag{1.34}$$

Based on this, we obtain the expression for the relative magnetic permeability of the body:

$$\mu_e = \mu \frac{4\pi}{4\pi + N(\mu - 1)}, \tag{1.35}$$

where μ is the relative magnetic permeability of the material.

Thus, the magnetic permeability of the body μ_e and its susceptibility depend not only on the physical nature of the material, but also on the shape of the body. Both values for a body of a certain shape are smaller than for the material of this body.

Non-ellipsoidal samples of finite dimensions in a homogeneous external magnetic field are magnetized unevenly. In this case, the demagnetization factor is a variable quantity. To describe the magnetic state of the material, two types of $\vec{J}$ and $\vec{B}$, are distinguished and, hence, there are two types of demagnetizing factors: ballistic N_{B} and magnetometric N_{M}.

Ballistic demagnetizing factor N_{B} is used when $\vec{J}$ is measured in the middle part of the sample. The magnetometric demagnetizing factor N_{M} is used in the case of averaging the values $\vec{J}$ and $\vec{B}$ over the entire body volume, which is typical for magnetometric measurements.

The value of N_{B} is always less than N_{M}, because during remagnetization, for example, in a homogeneous field of a cylindrical sample, the magnetization in its middle part is always greater than $\vec{J}_{\text{M}}$.

The following formulas can be used to calculate the values of N_{B} and N_{M} (Troitsky 2002):

$$N_{\text{M}} = \frac{4\pi}{\lambda^2}(2.72\lg\lambda - 0.69); \tag{1.36}$$

$$N_{\text{B}} = \frac{4\pi}{\lambda^2}(2.01\lg\lambda - 0.46), \tag{1.37}$$

where λ is the ratio of the length of the cylinder to its diameter.

Dependences (1.36) and (1.37) are valid if $\chi \to \infty$ for $9 \le \lambda \le 28$.

Based on the above mentioned, during precision research, the shape of the samples should be chosen so that the magnetization is uniform or close to it. This can only be achieved for samples of ellipsoidal and toroidal shapes with a small ratio of the width of the section to the radius of the toroid. In other cases, if the demagnetizing factor is known, the field $\vec{H}$ value must be calculated or measured.

The role of the demagnetizing factor increases with decreasing sample size. This factor is significant in calculating the modes of the samples.

In the case of a spheroid, when $\lambda < 1$, we have:

$$\frac{4\pi}{N} = \left(1 - \lambda^2\right)/\left[1 - \lambda/\sqrt{1-\lambda^2}\arccos\lambda\right]. \tag{1.38}$$

If $\lambda > 1$, we obtain:

$$\frac{4\pi}{N} = \left(\lambda^2 - 1\right)/\left[\frac{\lambda}{\sqrt{\lambda^2-1}}\ln\left(\lambda + \sqrt{\lambda^2-1}\right) - 1\right]. \tag{1.39}$$

For cases when $\lambda > 50$, expression (1.39) is simplified and takes the form:

$$\frac{4\pi}{N} = \frac{\lambda^2}{\ln 2\lambda - 1}. \tag{1.40}$$

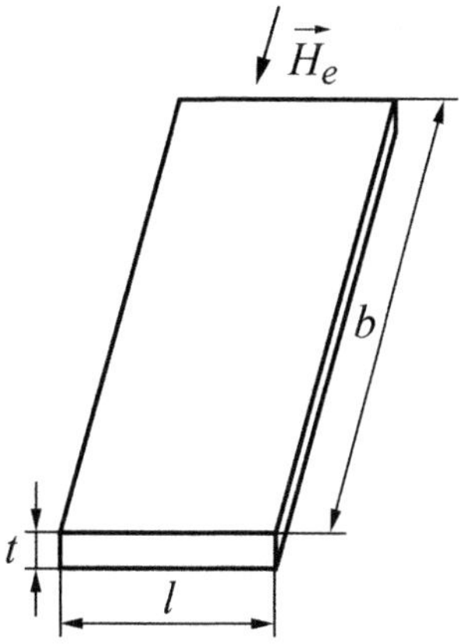

Fig. 1.18 A plate in an external magnetic field

In the dependences (1.38)–(1.40) the value of λ presents the ratio of the diameter to the length of the ellipsoid.

The demagnetizing factors of components, which in shape are approximately a partial case of the rotation ellipsoid, are different. Thus, for a thin plate (or disk) it is $N \approx 1.0$ ($\lambda \approx 0$). In the case of an infinitely long cylinder, the demagnetization factor along its axis is approximately zero.

Calculate the demagnetization coefficient for the samples used in the subsequent experiments. Consider the case of magnetization of a plate with a size of 35 × 240 mm and a thickness of 2 and 6 mm, which is magnetized by a field $\vec{H}_e$ parallel to its largest size (Fig. 1.18).

If the charge per unit length of the face parallel to the direction $\vec{H}_e$ is q, then the magnitude of the field strength generated by this face at a distance r ($r >> t$, where t is the thickness of the plate) is determined by the formula

$$H_0 = \frac{q}{2\pi \mu_0 r}. \tag{1.41}$$

Because the density of surface charges

$$\sigma = \mu_0 J_n, \tag{1.42}$$

where J_n is the magnetization along the normal to the longest face, then

$$q = \sigma t = \mu_0 J_n t.$$

For the central cross-section of the plate with a width equal to the distance from the edge, $r = b/2$. If a complete field is created by two faces, then

$$H_0 = -2\frac{\mu_0 J_n t}{2\pi \mu_0 (b/2)} = \frac{2t}{\pi b} J_n. \tag{1.43}$$

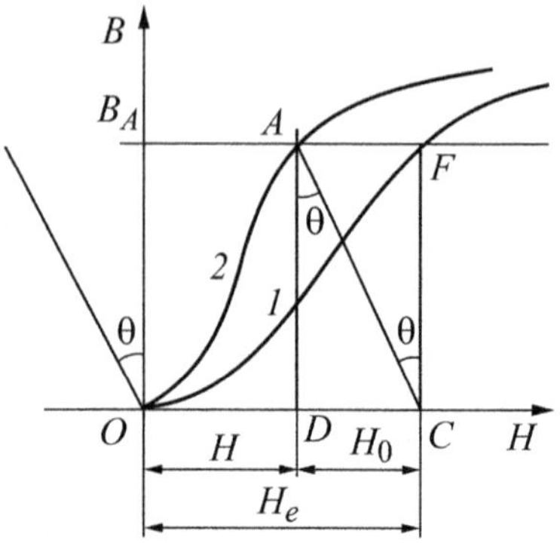

Fig. 1.19 Construction of the magnetization curve of a body (*1*) according to the known magnetization curve for material (*2*)

Considering formula (1.31) and putting

$$r = b/2, \tag{1.44}$$

we obtain

$$N = 2t/(\pi b). \tag{1.45}$$

Let's reconstruct the magnetization curve for material into the magnetization curve for a body (Fig. 1.19).

In the figure, such a rearrangement is shown for point A, which corresponds to the field $H = OD$. Draw a section AC so that $\mathrm{tg}\theta = N$. Then $OC = OD + DC = H + B_A N = H + H_0 = H_e$. Point F on the curve corresponds to the magnetization for the body of induction B_A. We see that the magnetization curve for the body will be lower, with the same dependence for the material, especially for large values of N.

The demagnetization coefficients for cylindrical and ellipsoidal samples are sought by formula (1.40). The dimensions of the ellipsoid and the cylinder correspond to the experimental ones. The diameter and length of the ellipsoid are 10 mm and 100 mm, and the cylindrical sample are 8 mm and 100 mm, respectively. The results of calculations of the demagnetization coefficient are given in Table 1.1.

Table 1.1 Demagnetization coefficients for bodies of different shapes

Shape of the body	N	θ (°)
Plate 2 mm	0.0053	0.3
Plate 6 mm	0.0159	0.9
Ellipsoid	0.001–0.173	0.1–10
Cylinder (field across the axis)	0.5	27
Cylinder (field along the axis)	0	0

As one can see, the demagnetization coefficient for the plate, ellipsoid and cylinder (along the axis of light magnetization) is very small, so it can be neglected. The demagnetization factor must be considered on extended objects. In this case the direction of the field action in relation to the axis of the object should be considered. As can be seen from Table 1.1, the angle θ for the cylinder under magnetic field action across its axis is significantly different from zero.

1.4 Theoretical Aspects of Magnetic Interaction of Ferromagnet Atoms

The elementary magnetic moments that make up the ferromagnet magnetization are mainly the spin moments of electrons. The study of their role in the occurrence of the ferromagnetic state is considered only from the standpoint of quantum mechanics. As is well known, it interprets interatomic interaction, considering, in addition to Coulomb energy, also additional, so-called exchange energy of interaction. Exchange energy is a purely quantum effect, and we will not consider it here to simplify the perception of the presented material.

There are two main causes of the appearance of electron magnetic moment: the orbital motion of the electron and its spin (Tikazumi 1983). The first appears due to the fact that the electron on the atom's orbit can be considered as a small, closed current circulating around the nucleus. Due to the mechanical moment or its intrinsic moment (i.e., spin), the spin magnetic moment occurs.

The magnetic moment of a multielectron atom is the sum of the magnetic moments of all the electrons, including both orbital and spin moments. Each electron makes an independent vector contribution to the total magnetic moment of the atom.

Since all filled orbits (shells) have zero total momentum, they also have zero magnetic moment. Atoms that have the only the filled orbits do not have constant magnetic moments and, accordingly, they cannot be paramagnetic.

Thus, the magnetization of most materials and ferromagnets is due to the magnetic moments of the atoms of which they are composed. The atomic nucleus has also a negligible magnetic moment, which makes almost no contribution to magnetization. In addition, the spin μ-meson rotation effect has recently been used to study the phenomenon of magnetism.

Consider one of the moments of the atom caused by the orbital motion of electrons. Since this motion of the electron corresponds to the electric current, the value of the magnetic moment is defined as (Tikazumi 1983):

$$M = -\frac{\mu_0 e \omega r^2}{2}, \tag{1.46}$$

where e is the electron charge; ω is circular frequency; r is the radius of the circle along which the electron moves around the nucleus; μ_0 is a magnetic constant.

The modulus of the mechanical momentum vector in this case will be

$$P = m\omega r^2, \tag{1.47}$$

where m is electron mass.

Then (1.47) can be written as

$$M = -\frac{\mu_0 e}{2m}\vec{P}. \tag{1.48}$$

Thus, the magnetic moment is closely related to the momentum of the electron: their values are proportional to each other, while the directions are opposite.

The motion of the electrons that make up the outer shell of the atom is quantized, so they must occupy only several discrete orbits. The momentum of one electron changes, assuming values that are multiples of the Planck constant divided by 2π. Therefore, the magnetic moment also changes, remaining a multiple of the Bohr magneton.

In addition to the orbital momentum already mentioned, the electron has a spin moment that corresponds to its own rotation and is not related to motion in space. The spin of the electron corresponds to its magnetic moment. The relationship between the spin magnetic moment and the spin momentum can be written as

$$\vec{M} = -\frac{\mu_0 e}{m}\vec{P}. \tag{1.49}$$

The spin magnetic moment is equal to the Bohr magneton. Combining the formulas for the relationship between momentum and magnetic moments, i.e., considering dependences (1.47)–(1.49), we write:

$$\vec{M} = -g\frac{\mu_0 e}{2m}\vec{P}, \tag{1.50}$$

where for spin $g = 2$, and for orbital motion $g = 1$.

If we write the coefficient at mechanical momentum as $\upsilon = -g\frac{\mu_0 e}{2m} = 1.105 \times 10^5 g$ (m/A c), then the dependence (1.50) takes the form of

$$\vec{M} = -\upsilon\vec{P}. \tag{1.51}$$

Note that the coefficient υ in formula (1.51) is called as magnetomechanical relation, and the coefficient g in Eq. (1.50) is called the g-factor.

The total momentum $\vec{j}$ of an electron is the sum of the orbital mechanical moment $\vec{i}$ and the spin $\vec{s}$ in one orbit. It corresponds to the quantum number of the total moment.

If many electrons move around the same nucleus, they interact with each other. If their distribution remains spherically symmetric, the sequence of energy orbits is not broken. However, if the shape of the orbit differs from the circular one, the situation changes: for example, a 3d-orbit is circular and a 4s-orbit is elliptical, and one part of it is near the nucleus. Thus, the 4s-electron, penetrating the inner electron cloud of the atom, feels the action of the intense electronic field of the unshielded nucleus (attraction), and therefore its energy decreases.

Elements that have unfilled electronic shells exhibit the specific magnetic, chemical and other properties, and are therefore called transitional (3d-, 4d- and 5d-transient elements). All the elements that exhibit strong magnetic properties belong to one of these groups.

Denote by s_i and l_i, s_j and l_j respectively, the spin and orbital moments of the ith and jth electrons belonging to the same unfilled shell. Stronger or weaker interactions are possible between any of these vectors. But still the interaction between the same moments of the electrons, i.e., between s_i and s_j, or between l_i and l_j will be predominant. The interaction between the spins of individual electrons leads to the resulting spin of the entire electron shell (Tikazumi 1983):

$$\vec{S} = \overset{\rightarrow}{\sum_i} \vec{s}_i. \tag{1.52}$$

In turn, the interaction of orbital moments gives the resulting orbital moment of the given shell:

$$\vec{L} = \overset{\rightarrow}{\sum_i} \vec{l}_i. \tag{1.53}$$

The energy of the spin–orbit interaction is equal to

$$\omega = \lambda \vec{L} \vec{S}, \tag{1.54}$$

moreover, the vectors $\vec{L}$ and $\vec{S}$ depending on the sign of the constant λ are oriented parallel or antiparallel to each other and in the sum give the full momentum (Fig. 1.20)

$$\vec{J} = \vec{L} + \vec{S}. \tag{1.55}$$

The parameter λ in expression (1.10) is positive when the number of electrons in the shell is less than half, and negative—when more than half.

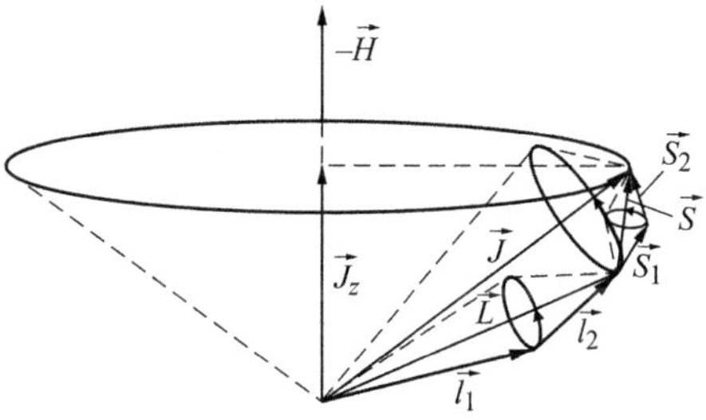

Fig. 1.20 Interaction between spin and orbital moments

If the mechanical moments $\vec{L}$ and $\vec{S}$, are determined for the atom, the magnetic moments associated with them can be found. The orbital magnetic moment will be (Tikazumi 1983)

$$\vec{M}_{\mathrm{L}} = -M_{\mathrm{B}}\vec{L}, \tag{1.56}$$

where M_{B} is the Bohr magneton.

For the spin magneton we have

$$\vec{M}_{\mathrm{S}} = -2M_{\mathrm{B}}\vec{S}. \tag{1.57}$$

Then the resulting magnetic moment is defined as

$$\vec{M}_{\mathrm{R}} = \vec{M}_{\mathrm{L}} + \vec{M}_{\mathrm{S}} = -\vec{M}_{\mathrm{B}}\left(\vec{L} + 2\vec{S}\right). \tag{1.58}$$

Usually $\vec{L}$ and $\vec{S}$ are not collinear, and in this case, as shown in Fig. 1.21, vector $\vec{L} + 2\vec{S}$ does not coincide in direction with $\vec{J}$. But since both $\vec{L}$ and $\vec{S}$ process around $\vec{J}$, then, as a result, the vector $\vec{L} + 2\vec{S}$ also performs precession around $\vec{J}$. Therefore, the average magnetic moment is directed along $\vec{J}$, and its magnitude can be given in the form

$$\vec{M}_{\mathrm{S}} = -gM_{\mathrm{B}}\vec{J}. \tag{1.59}$$

The moment expressed by formula (1.59) is called the magnetic moment of saturation, and the value of effective magnetic moment is found by the expression

$$\vec{M}_{\mathrm{eff}} = g\vec{M}_{\mathrm{B}}\sqrt{\vec{J}\left(\vec{J}+1\right)}. \tag{1.60}$$

During the action of the magnetic field on the atom, it turns out that the angular momentum can be oriented relative to the field only in some discrete directions. This phenomenon is known as spatial quantization, and the number *m*, which changes the angular momentum, is the magnetic quantum number.

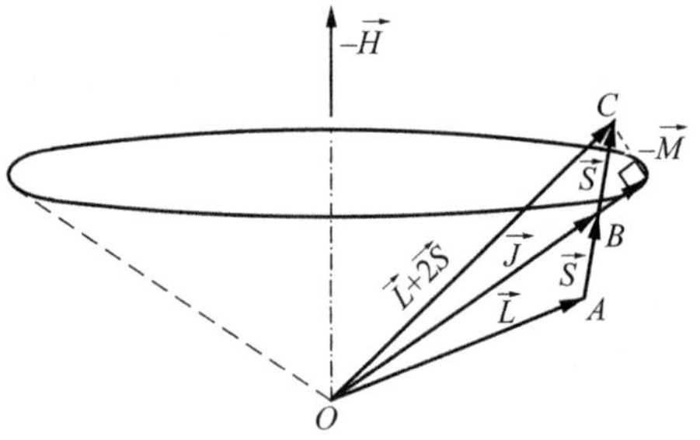

Fig. 1.21 Relationship between the vectors of the angular momentum and the magnetic moment

Below the Curie point, the molecular field, orienting the spins in parallel, creates a spontaneous magnetization, which is commonly referred to J_S. The J_s value is equal to the saturation magnetization, which is obtained because of magnetization to the saturation of ferromagnets with an external field.

Summarizing the above presented, we can say that the magnetic properties of magnets are associated with the spin and orbital motion of the electron, and because of the electron and the external magnetic field interaction there appears a change in the total magnetic moment of the multielectron atom.

1.5 Magnetic Anisotropy and Magnetostriction

The theoretical considerations described above make it possible to understand the nature of one of the most fundamental properties of a ferromagnet—magnetic anisotropy and magnetostriction, which arise due to spontaneous magnetization.

Magnetic anisotropy is a consequence of the predominant orientation of the spontaneous magnetization of the ferromagnet along special crystallographic axes which are typical to this material. In fact, this is a phenomenon of change in the internal energy of the ferromagnet depending on the orientation of the spontaneous magnetization in the crystal. There are a few external causes of magnetic anisotropy: deformation, heat treatment of the material, and so on.

In its pure form, when there is no manifestation of any of the special factors, the internal energy of the magnet reflects the symmetry of the crystal. Such magnetic anisotropy is called as magnetocrystalline. Of course, the simplest case of magnetic anisotropy is uniaxial.

The internal energy, which depends on the direction of spontaneous magnetization, is called the energy of magnetic anisotropy, and in those cases when it reflects the symmetry of the crystal, it is called the energy of magnetocrystalline anisotropy. Each of the magnetization axes of the ferromagnet has its own magnetization constant K_i, and in the case of cubic crystals (e.g., iron and nickel)—cubic anisotropy constants, the values of which are known for different materials (Tikadzumi 1987).

In the uniaxial orientation of the magnetization, the anisotropy energy is minimal, and in the absence of an external magnetic field in the steady state, the vector of spontaneous magnetization is directed along this axis. As already mentioned, such directions of stable spontaneous magnetization are called axes of light magnetization or simply light axes. Unstable directions of magnetization are called axes of difficult magnetization or difficult axes.

Since the phenomenon of magnetic anisotropy consists in the change in internal energy during the change inside the crystal of the orientation of the group of parallel spins that create spontaneous magnetization, to describe the energy of magnetic anisotropy it is necessary to consider the directions of crystallographic axes. If we denote by φ the angle that forms the axis connecting the pair of spins with their direction, the energy ω of the mentioned pair is written by the Legendre polynomial of the argument of $\cos \varphi$

$$\omega(\cos\varphi) = g + l\left(\cos^2\varphi - \frac{1}{3}\right) + q\left(\cos^2\varphi - \frac{6}{7}\cos^2\varphi + \frac{3}{35}\right) + \cdots \quad (1.61)$$

The first term of this series is a constant, which includes the energy of exchange interaction:

$$\omega_{ij} = -2J\vec{S}_i \cdot \vec{S}_j = -2JS^2\cos\varphi. \quad (1.62)$$

Here S is the spin value, φ is the angle between the $\vec{S}_i$ and $\vec{S}_j$ vectors. The second term of the series (1.61) describes the dipole–dipole interaction. If we denote the magnetic moment of the spins as μ, and write the coefficient l in the form

$$l = -\frac{3\mu^2}{4\pi\mu_0 r^3}, \quad (1.63)$$

then the expression for the magnetic interaction will correspond to the expression for the interaction of magnetic dipoles.

To substantiate the value of the magnetic anisotropy of a real ferromagnet, the coefficient l must be 100–1000 times higher than the energy of the magnetic interaction, which is determined by formula (1.63). The reason is that the orbital moments turn together with the spins. Then, due to the change in the overlap of the orbital wave functions, the electrostatic, and hence the exchange (pseudodipole) interaction, changes.

The third term of the series (1.61) has an even higher order of smallness and corresponds to the quadrupole interaction.

This model of magnetic anisotropy of a crystal magnet, built based on spin pair energy, well describes the effect of crystal symmetry on magnetic anisotropy or the mechanism of induced magnetic anisotropy (Tikadzumi 1987). It is also called the spin pair model. Thus, the spin pair model describes ferromagnets well in the case when the atoms with spins are located close to each other. However, in practice they are often surrounded by different ions. This circumstance has its effect on the magnetic anisotropy of ferromagnets.

For example, cobalt (one of the striking representatives of ferromagnets belonging to the metals of the 3d transition group) has a face-centered cubic lattice and is characterized by rather high constants of uniaxial magnetic anisotropy. The introduction to cobalt iron, manganese, and other additives causes a significant change in anisotropy. The addition of rare earth elements causes even greater magnetic anisotropy of cobalt (Tikadzumi 1987).

Cubic anisotropy is anisotropy of a higher order than uniaxial, so its value is smaller than of the uniaxial one. Alloys of iron and nickel are the most interesting here. If we take the Fe–Al alloy, the magnetic anisotropy constant K_1 gradually decreases as non-magnetic Al is added as an impurity.

In the Fe–Co alloy a similar tendency is observed with increase of cobalt as for the Fe–Si alloys. In the latter alloy, non-magnetic silicon, without changing the magnetic moment of iron, simply reduces the magnitude of the saturation magnetization. However, the coefficient of magnetic anisotropy decreases even faster. The constant K_1 increases slightly during addition of titanium to iron.

In the Ni–Fe alloy, depending on its cooling rate from 600 °C, the constant K_1 acquires a negative value and increases with the cooling rate and content of iron. As a result of the addition of nickel to cobalt, the absolute value of K_1 decreases and for the cobalt content of 35% it turns into zero. If the cobalt content is higher than 35%, then $K_1 > 0$; in the presence of 18% of this element K_1 is again equal to zero; with further cobalt content increase K_1 becomes negative. The Ni–Cr and Ni–V alloys behave in almost the same way.

It should be noted that the magnetoelastic energy makes a significant contribution to magnetic anisotropy, and at large values of the constants of magnetic anisotropy a large value of magnetostriction is observed. The latter is a phenomenon of distortion of the magnet external shape during its magnetization. The relative deformation $\Delta l/l$ associated with this shape distortion is usually very small and is of the order of 10^{-5}–10^{-6}, so it can be determined only by precision experimental methods. However, despite such a slight change in size due to magnetostriction, this phenomenon is essential when studying the domain structure and mechanism of magnetization. It has received many practical applications.

Magnetostriction is caused by the same factors as magnetic anisotropy, i.e., it is determined by the energy of the spin interaction. In the absence of magnetostrictive deformation of the crystal, i.e., when the distance between the spins is fixed, the energy of the spin interaction is determined by formula (1.61). In the case when the distance between the spins changes, the energy of their interaction can be written in the form (Tikadzumi 1987):

$$\omega(r, \cos\varphi) = g(r) + l(r)\left(\cos^2\varphi - \frac{1}{3}\right) + q(r)\left(\cos^4\varphi - \frac{6}{7}\cos^2\varphi + \frac{3}{35}\right) + \cdots \tag{1.64}$$

Since the interaction depends on the distance r between the spins (Fig. 1.22), then during the manifestation of ferromagnetic properties due to it the distortion of crystal lattice takes place.

The first term in dependence (1.64) shows that the exchange energy depends on r. Since this term does not depend on the spin direction determined by the angle φ, the crystal distortion corresponding to $g(r)$ is not related to the direction of the spontaneous magnetization vector. Thus, this term does not contribute to magnetostriction, which we understand in the usual sense. But it is essential for negative magnetostriction.

The second term is related to the dipole interaction, which also depends on r. Since this interaction also depends on the angle φ, the crystal distortion corresponding to such an interaction change with the change of the direction of the spontaneous magnetization vector. This will be ordinary magnetostriction. The same can be said

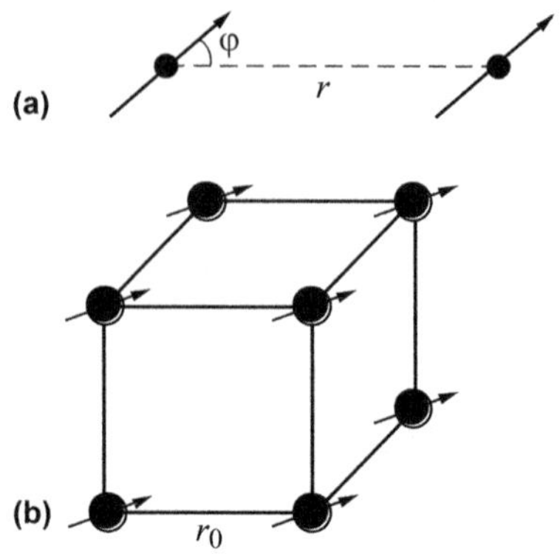

Fig. 1.22 Orientation of parallel spins and distance between them in the spin pair (**a**) and location of dipoles in a simple cubic lattice (**b**)

about the third and other terms. Considering that the following terms are usually small compared to the second, they can be neglected and the expression for energy can be written as

$$\omega(r,\varphi) = l(r)\left(\cos^2\varphi - \frac{1}{3}\right). \tag{1.65}$$

So, we see that the distance between the spins and the direction angle between them make a significant contribution to the energy of spin interaction. The change of these factors occurs under the influence of impurity atoms or ions, as well as under the action of an external magnetic field.

1.6 Physics of Magnetic Domains

The ferromagnet consists of domains surrounded by their walls. Each of domain has a certain direction of spontaneous magnetization. This state of the material is due to the optimal distribution of magnetostatic energy (Vonsovsky and Shur 1948), which is proportional to the size of the domain: the smaller the size, the lower the magnetostatic energy. At the same time, during the domains splitting, the total number of domain walls increases, which leads to an increase in their total energy. Therefore, the actual size of the domains is determined by the condition of the minimum sum of these two energies.

Such ferromagnetic materials as iron and nickel, which have a cubic crystal structure, except for domains (Fig. 1.23), which are magnetized perpendicular to the surface (180° domain), are characterized by the presence of closing domains (90° domain).

The latter exclude the appearance of magnetic poles. It is the closing domains that play a decisive role in the material magnetostriction. Then the size of the domains is determined by the balance of magnetoelastic energy and energy of the domain walls.

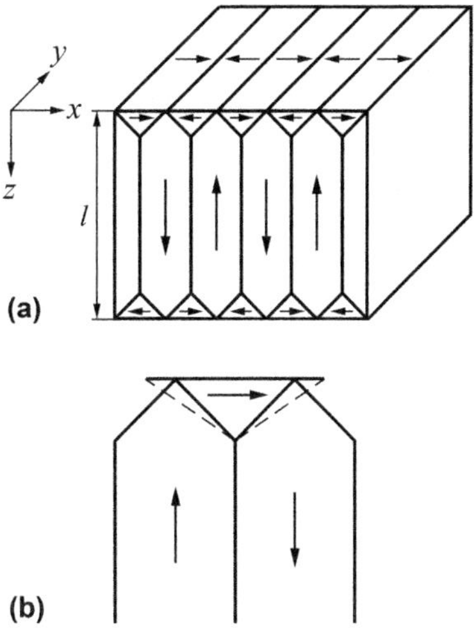

Fig. 1.23 Schematic division of a cubic monocrystal into 180° (indicated by vertical arrows) and 90° domains (**a**) and magnetostriction-induced hypothetical deformation of the closing domain (**b**)

For example, for an iron crystal with a domain width $d = 5.3 \times 10^{-4}$ m, the total energy is 6.1×10^{-2} (J/m^2).

During technical magnetization, the vectors of spontaneous magnetization line up in the same direction with increase of the external magnetic field. Then the spins inside the domain wall gradually turn continuously from the direction that had a spontaneous magnetization vector of one domain to the direction of the spontaneous magnetization vector of another. Spins cannot change direction abruptly because as the angle φ between adjacent spins increases, the energy of the exchange interaction increases very rapidly (proportionally to φ^2).

If the magnetic field intensity $\vec{H}$ is applied to a ferromagnetic material with a multidomain structure parallel to the magnetization vector of one of the domains, the spins inside the domain wall will experience the effect of the moment of force caused by this field. Then they will turn (albeit clockwise) as shown in Fig. 1.24 (Tikadzumi 1987). In this case, the domain wall will shift to the side, the volume of the domain in which the direction of magnetization is parallel to $\vec{H}$, will increase, and the volume of the domain with the opposite direction of magnetization will decrease. As a result, the average magnetization of the sample will increase in the direction $\vec{H}$.

This phenomenon can also be considered by replacing the action of the field $\vec{H}$ with the pressure p, which is applied to the domain wall (Tikadzumi 1987). If under the action of the field $\vec{H}$ the domain wall of area S is shifted by the distance s, it passes a volume equal to Ss. Considering that the spontaneous magnetization of the domains is equal to $\vec{J}_s$, we obtain the expression for the increase of the magnetic moment in the field direction $\vec{H}$ (Fig. 1.25), which is caused by the wall shift:

$$\vec{M} = 2\vec{J}_s Ss. \tag{1.66}$$

Therefore, the work performed by the magnetic field $\vec{H}$ is equal to

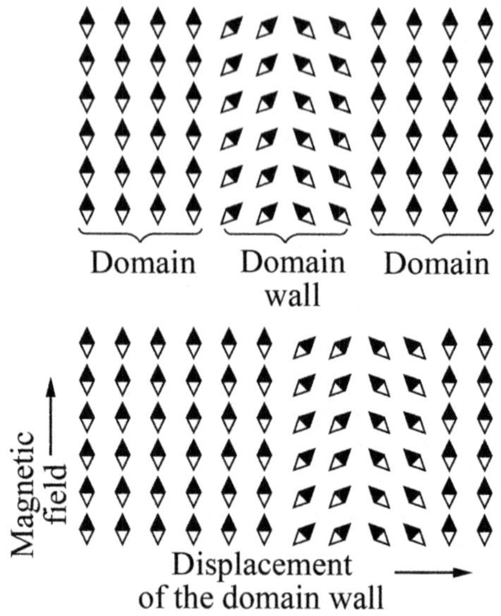

Fig. 1.24 Pressure that acts on the domain wall in ferromagnetic field

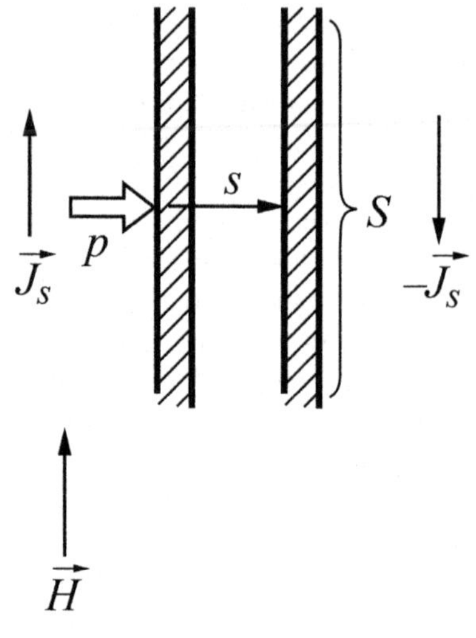

Fig. 1.25 Scheme of displacement of the domain wall under the action of an external magnetic field

$$W = \vec{M}\vec{H} = 2Ss\left(\vec{J}_s\vec{H}_s\right). \tag{1.67}$$

On the other hand, the same value can be expressed in terms of the pressure p. Since the domain wall area S is subjected to action of the force pS, the work will be

$$W = pSs. \tag{1.68}$$

Comparing (1.67) and (1.68), we obtain

$$p = 2\vec{J}_s\vec{H}. \tag{1.69}$$

Thus, we can assume that in the presence of the magnetic field $\vec{H}$ the pressure acts on the domain wall perpendicular to its surface (1.69). If the angle between the magnetic field and the vector $\vec{J}_s$ is equal to θ, the effective field that shifts the domain wall is a component that is parallel to $\vec{J}_s$, so expression (1.69) is written as

$$p = 2J_sH\cos\theta. \tag{1.70}$$

For a 90° domain wall, in the case when the magnetization vectors on both sides of it form angles θ_1 and θ_2 with the field $\vec{H}$, we obtain

$$p = J_sH(\cos\theta_1 - \cos\theta_2). \tag{1.71}$$

If to consider the domain structure of a cubic crystal with a cubic anisotropy constant $K_1 > 0$, then when placed in an external magnetic field $\vec{H}$, acting in the direction [100], 180° and 90° walls will shift under pressure. The latter will cause an increase of the volume of domains whose magnetization coincides with the direction $\vec{H}$ (in Fig. 1.26—shaded). As a result, all domains will merge into one, i.e., there will be a state of magnetic saturation. If nothing prevents the shift of the magnetic walls, it ends in a weak magnetic field. In this case, the magnetization curve increases rapidly from the coordinates origin (Fig. 1.27) and reaches saturation.

When the magnetic field is applied in the direction [110] (Fig. 1.26b), there is a shift of the domain walls, which leads to an increase in the volume of domains of

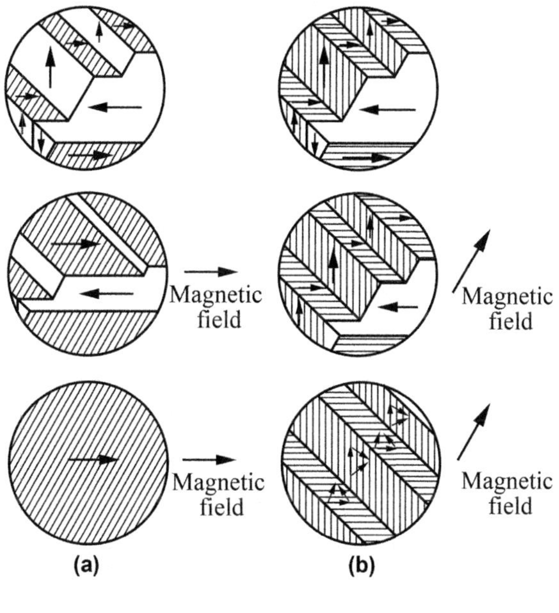

Fig. 1.26 Movement of domain walls and rotation of magnetization in cases when the vector is parallel to the crystallographic axis [100] (**a**) and [110] (**b**)

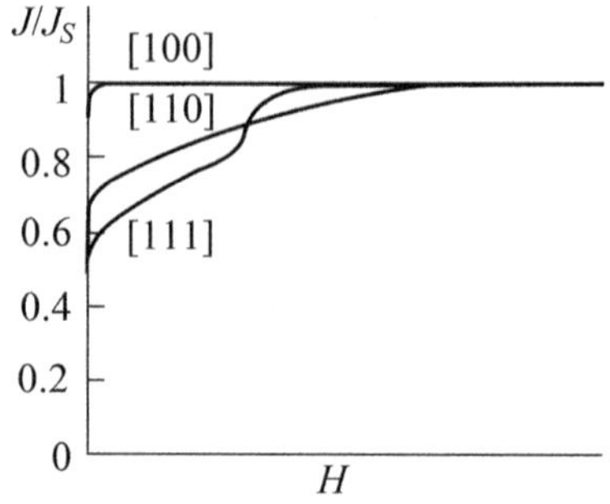

Fig. 1.27 Curves of monocrystal magnetization with anisotropy constant $K_1 > 0$ (the directions of magnetization are indicated near the curves)

two types, namely those whose magnetization has the nearest to field $\vec{H}$ direction [100] or [010]. As a result, domains with a different direction of magnetization are converted into domains of these two types. Since in all domains the vector of spontaneous magnetization forms an angle of 45° with the field $\vec{H}$, the component of magnetization parallel to the field is equal to

$$J_r = \frac{J_s}{\sqrt{2}} = 0.71 J_s. \tag{1.72}$$

As shown in Fig. 1.27, the curve [110] deviates from the y-axis at a point where J/J_s is approximately 0.7. If after that the external magnetic field continues to increase, the vector of each domain will rotate, moving toward the saturation magnetization. This process is called the rotation of magnetization.

To start the rotation of magnetization in the presence of strong magnetic anisotropy, a magnetic field of sufficient magnitude is required. For magnetization in the direction [111], at the moment when the displacement of the domain walls ends, the magnetization should be

$$J_r = \frac{J_s}{\sqrt{3}} = 0.58 J_s, \tag{1.73}$$

which is confirmed by the corresponding curve in Fig. 1.27. At the time when the rotation of the magnetization begins, the curve deviates from linearity.

For a crystalline ferromagnet in the demagnetized state, the magnetization is distributed in all directions equally (point O in Fig. 1.28). If to apply the field $\vec{H}$ in the positive direction, then due to the displacement of the domain walls, the vectors of local magnetization, oriented in the negative direction, will begin to rotate, and move into a semicircle (point B in Fig. 1.28). The vectors of local magnetization, the orientation of which is opposite to the field $\vec{H}$, begin to rotate with the force of the beginning of magnetization, because the domain walls, due to which such rotation occurs, are subjected to the greatest pressure. If the displacement of the domain walls is complicated due to obstacles of certain (finite) sizes, then first begin to move those walls to which the greatest pressure is applied. As a result, under a sufficiently strong

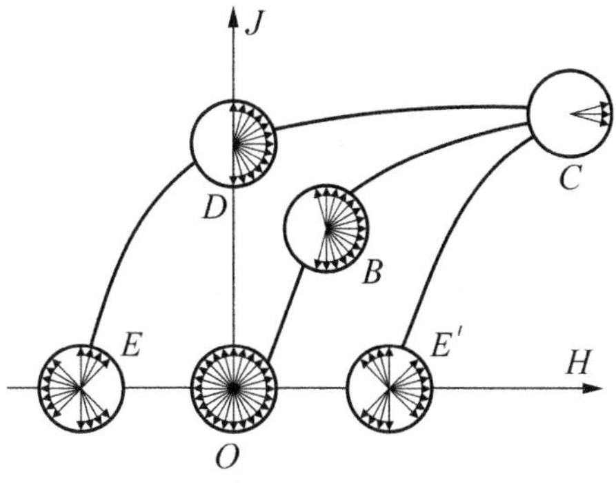

Fig. 1.28 Distribution of magnetization at different points of the magnetization curve

magnetic field, all local magnetization vectors are oriented in the positive direction. Then the distribution shown at point C is formed, where saturation occurs.

Now, if we reduce the magnetic field from saturation, the magnetization of different areas will begin to rotate to the directions of the axes of light magnetization, and at $H \to O$ the magnetization vectors will be within a semicircle, as shown in Fig. 1.28 at the point D corresponding to the residual magnetization. If to move from the residual magnetization state to the area of negative external fields, the local magnetization vectors, which are directed in the positive direction and have the lowest magnetization, will begin to rotate. When the field reaches a value that coincides with the value of the coercive force, a distribution corresponding to the point E will be established. If one now starts to increase the field, the magnetic saturation will be reached again, the transition to which will be carried out along the curve $E' \to C$.

According to the simplified scheme described above, the magnetization is distributed inside the magnet when the magnetic anisotropy is uniaxial and the direction of the light axis of magnetization changes gradually from section to section.

For materials with cubic magnetic anisotropy, the pattern of magnetization distribution within the volume is somewhat different. For example, when $K_1 > 0$, i.e., when the axes [100] are the axes of light magnetization, the magnetization vector in the state of residual magnetization is set along the axis [100], which is closest to the direction $\vec{H}$ (Fig. 1.29). In this case, the largest deviation of the magnetization is observed in crystal grains, in which the axes [111] are oriented along $\vec{H}$. In this case the angle between the vectors of magnetization of the grains and the direction of the field is 55° (Fig. 1.30).

When $K_1 < 0$, directions [111] are the axes of light magnetization. The magnetization distribution is like the case of $K_1 > 0$. In this case, if $\vec{H}$ is in the first quadrant, the magnetization turns, when approaching the residual magnetization state, to the direction [111]. The value of I_r in this case is greater than $K_1 > 0$. The reason is that although the maximum deviation angle of the local magnetization vectors is still 55°, there are now four axes of light magnetization. Then the probability of orientation in the direction of the field $\vec{H}$ is higher. In any case, for a magnet of the cubic syngony

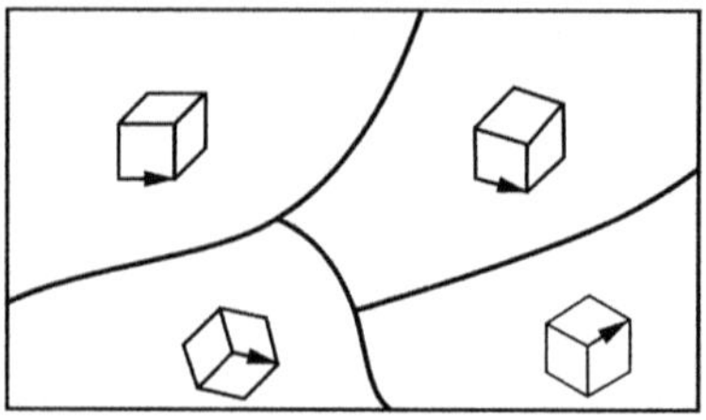

Fig. 1.29 State of the residual magnetization of the polycrystal of cubic syngony ($K_1 > 0$)

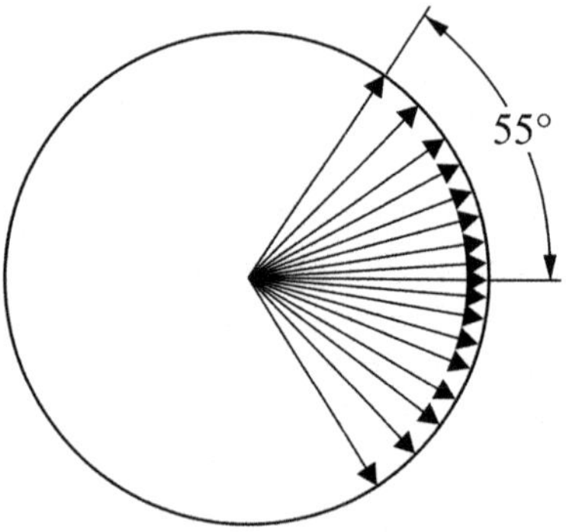

Fig. 1.30 Distribution of residual magnetization at $H = 0$ in polycrystal of cubic syngony ($K_1 > 0$)

in the absence of deformation, the residual magnetization is much greater than 50% and reaches 70–90%. Accordingly, the area of the hysteresis loop increases.

The cases discussed above refer to the absence of the influence of magnetic poles that occur at the boundaries of crystal grains. However, in reality, they are formed in the state of residual magnetization at the grain boundaries, and under the action of the magnetic demagnetizing field they create, some negative component of magnetization appears. This is due to the fact that in this state the magnetostatic energy is lower. As shown in Tikadzumi (1987), in the polycrystal the magnetostatic energy of such poles is 20–50% of the energy of the isolated crystal grain. Therefore, it is clear that the formation of the domain structure leads to a decrease in magnetostatic energy (the exception is the case when the crystal grains are so small that a single-domain structure occurs).

Thus, in the absence of an external magnetic field, a ferromagnet is a collection of microscopic areas—domains. Each of them is magnetized to saturation, and their magnetizations $\vec{J}_s$ are oriented so that the total magnetic moment of the whole sample is zero (Vonsovsky and Shur 1948; Rudyak 1986). The energy of the magnetic interaction per atom is of the order of 10^{-23} J. The magnetic energy of a real ferromagnetic crystal is anisotropic (Rudyak 1986), and the energy of the "molecular" field is completely isotropic. The domain structure is sensitive to the formation and development of magnet damage: rearrangement of dislocations, formation, and development of pores, microcracks, etc.

1.7 Peculiarities of the Structure of Domain Walls

At the boundary of two domains, the spins gradually change their orientation from a direction parallel to the magnetization vector of the first domain to a direction parallel to the magnetization vector of the second one. This transition layer is called the domain wall. The reason for the smooth rotation of the spins in the domain wall is that, as shows the formula:

$$\omega_{ij} = -2J\vec{S}_i\vec{S}_j, \tag{1.74}$$

the estimates of the exchange energy ω for the two spins $\vec{S}_i$ and $\vec{S}_j$, located at an angle φ to each other (J is the exchange integral), the spin energy is proportional to φ^2. It follows from the above that a sharp change in the direction of the spins in the wall causes a rapid change in exchange energy. Figure 1.31 presents a chart of the atomic layers of the domain wall. It is considered that the rotation angle in each of the N atomic layers is approximately the same and equal to $\varphi = \pi/N$. Then the exchange energy γ_* concentrated in the unit of surface area of such a domain wall is expressed by the formula

$$\gamma_* = \frac{N}{a^2}\omega_{ij} = \frac{JS^2\pi^2}{a^2N}, \tag{1.75}$$

where a is the atomic lattice constant; S is the quantum number of the spin moment of the angular momentum.

As we see, with increasing thickness N of the transition layer γ_* decreases. If to take into account the exchange energy only, the most advantageous is the spins rotation as slow as possible. From that point of view, an energy wall of infinite thickness would be the most energetically advantageous. If the dependence of ω_{ij} on φ were linear rather than quadratic, then γ_* from formula (1.75) would not depend on N and the spins could instantly rotate by 180° without the transition layer.

However, when the spins deviate from the axis of light magnetization, the energy of magnetocrystalline anisotropy increases. At the maximum deviation of the spins from this axis, the energy of the anisotropy increases by approximately K per unit volume, where K is the constant of magnetostrictive anisotropy. In the model shown in Fig. 1.31a there are N/a^2 atoms per unit area of the domain wall, so the corresponding volume will be $(N/a^2) \cdot a^3 = Na$. Therefore, the anisotropy energy per unit area of the wall is equal to

$$\gamma_\mathrm{a} = KNa. \tag{1.76}$$

As follows from formula (1.76), with decreasing N, i.e., with thinning of the domain wall, the anisotropy energy decreases. Therefore, the thickness of the domain wall in the real case is determined by the balance condition of two energies: exchange energy, a separate consideration of which shows that the wall should be thicker, and

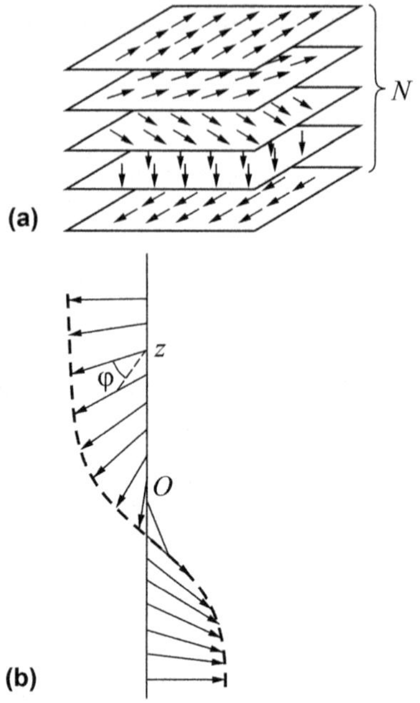

Fig. 1.31 Scheme of rotation of spins according to the thickness of the domain wall (**a**) and azimuthal rotation (**b**)

anisotropy energy, analysis of which shows that the wall should be as thin as possible. The balance condition is determined by the minimum of total energy

$$\gamma = \gamma_* + \gamma_a = \frac{JS^2\pi^2}{a^2N} + KNa. \tag{1.77}$$

Solution of the equation

$$\frac{\partial \gamma}{\partial N} = -\frac{JS^2\pi^2}{a^2N^2} + Ka = 0 \tag{1.78}$$

gives an expression for an approximate estimation of the domain wall thickness:

$$\delta = Na = \pi\sqrt{\frac{JS^2}{Ka}}. \tag{1.79}$$

Domain walls are divided into two types: those of 180°, in which the direction of magnetization changes during the transition from one side of the wall to another one by 180° and 90°, in which the direction of magnetization changes only by 90° (see Fig. 1.23). In materials whose axes of light magnetization form, as in iron, the family [100], there are six stable directions of magnetization (Tikazumi 1983): [100], [100], [010], [010], [001], [001]. In this case, the boundaries between the domains [100],

[100] are 180° walls, and, for example, between the domains [100] and [010]—90° walls.

The domain wall is strictly rectilinear. In the case of its bending (Fig. 1.32), the magnetization vectors on both sides of the wall will be directed to the same point, i.e., there will be magnetic poles. The demagnetizing field created by the magnetic poles will try to rotate the wall in the bending area, as a result of which it will straighten, as shown in Fig. 1.32a.

The above applies to the case when we considered the cross-section of the domain wall in the plane in which the magnetization vector lies. In the sample depth, the appearance of bends of the domain wall, as shown in Fig. 1.33, is possible. At such bends magnetic poles do not arise. However, during the domain wall bending, its total area increases, which means that something must cause such a bending. The wall can be caught by impurity inclusions, pores, etc. Its bended shape may be more stable due to internal stresses or due to inhomogeneities in the concentration of the alloy. The dependence of the surface energy of the domain wall on the direction in the crystal can also play a certain role.

A 90° domain wall is one for which the angle formed by the directions of magnetization of the domains on both sides of it is equal to 90° or close to this value. They, like 180° walls, are oriented so that the normal component of the magnetization during the transition from the left domain to the right one does not feel the jump, and

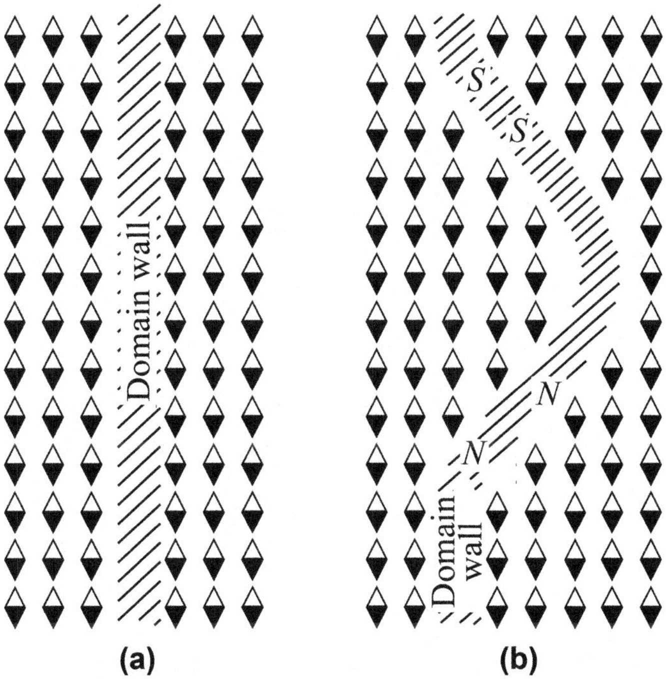

Fig. 1.32 Appearance of magnetic poles during bending of the domain wall

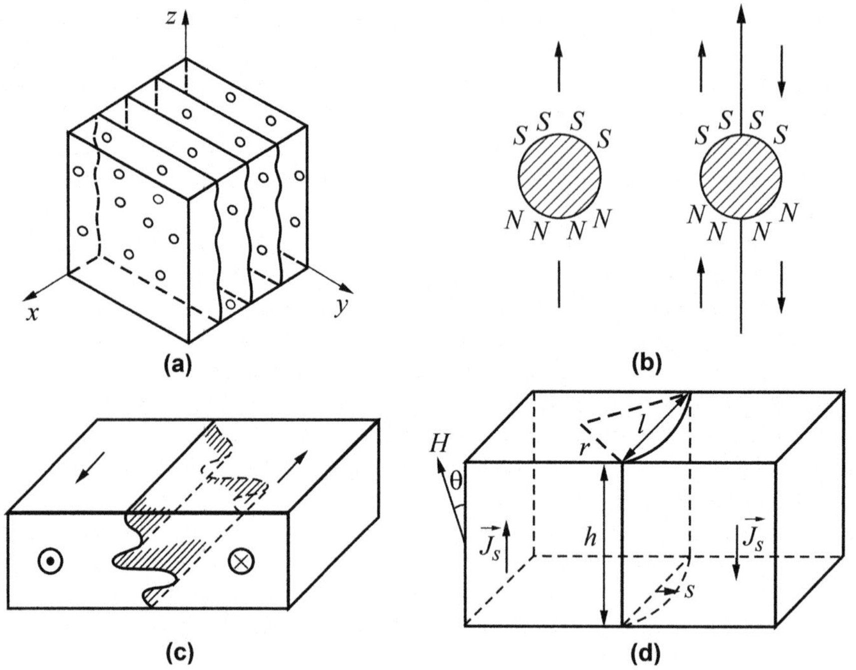

Fig. 1.33 Domain walls inside the crystal (**a**); magnetic bands appearing on the surface of the pore or non-magnetic inclusion inside the ferromagnet (**b**); the shape of the distorted 180° domain wall (**c**) and the inflation of the 180° domain wall under the action of a magnetic field (**d**)

the surface of the domain wall would not have poles. In this case, as for 180° domain walls, there is more than one direction for which the rule of continuity of the normal component is fulfilled. The wall can be bent anyway to pass through the bisector of the angle formed by the two directions of magnetization (Fig. 1.34).

As calculations (Tikadzumi 1987) have shown, the dependence of the energy γ per unit area of the domain wall on the angle φ, which determines its direction, is nonlinear. For $\varphi = 0$, the energy value is minimal, and the 90° wall tends to merge with the plane in which there are both magnetization vectors. The energy is minimal because a certain azimuthal angle of rotation of the spins is 90°. If $\varphi = 90°$, the total azimuthal angle of rotation is 180°, and the wall then has the highest energy γ.

Thus, the movement of domain walls is significantly affected by defects in the ferromagnet structure, spin interaction, and external magnetic field.

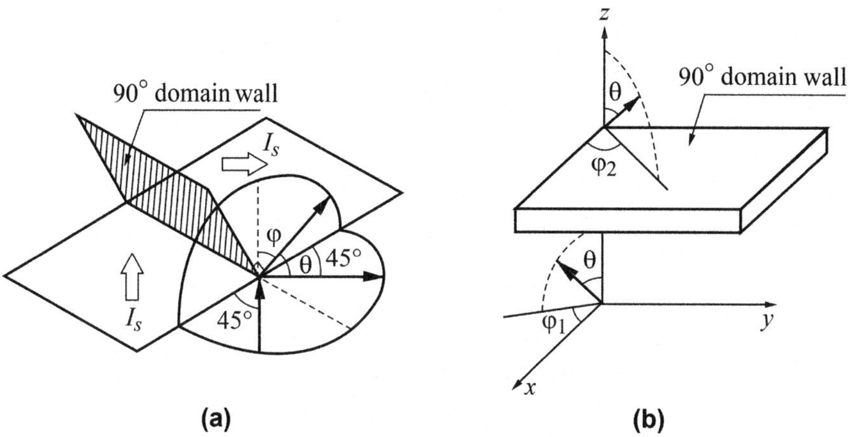

Fig. 1.34 Orientation of a 90° domain wall with no magnetic poles on its surface (**a**) and direction of the spontaneous magnetization vector on both sides of the 90° domain wall (**b**)

References

Akulov IS (1939) Ferromagnetizm (Ferromagnetism). State Publishing House of Technical and Theoretical Literature

Arkadiev VK (1913) Teoria elektromagnitnogo polia v ferromagnitnom metalle (Theory of electromagnetic field in a ferromagnetic metal). Zh Russ Fiz-tekh obshchestva (J Russ Phys-Chem Soc (Phys Part)) 45:312

Blokh F (1936) Molekuliarnaia teoriia magnetizma (Molecular theory of magnetism). United Scientific and Technical Publishing House

Blokhintsev DI (1976) Osnovy kvantovoi mekhaniki (Fundamentals of quantum mechanics). Nauka Publishing House

Dorfman J (1927) The intrinsic fields in ferromagnetic substances. Nature 119(2992):353

Einstein A, de Haas WJ (1915) Experimenteller Nachweis der Ampereschen Molekularströme (Experimental proof of Ampère's molecular currents). Deut Phys Ges Verh 17:152–170

Frenkel J (1928) Elementare Theorie magnetischer und elektrischer Eigenschaften der Metalle beim absoluten Nullpunkt der Temperatur. Z Phys 49(1/2):31–45

Frenkel J, Dorfman J (1930) Spontaneous and induced magnetisation in ferromagnetic bodies. Nature 126(3173):274–275

Grechishkin RM (1975) Domennaya structura magnetikov. Chast I (Domain structure of magnets. Part I). Kalinin State University Publishing House

Heisenberg W (1928) Zur Theorie des Ferromagnetismus. Z Phys 49:619–636

Kandaurova NI, Onoprienko LG (1986) Domennaia struktura magnitov. Osnovnyie voprosy micromagnetikov (Domain structure of magnets. Basic questions of micromagnetics). Ural State University Publishing House

Kiefer II, Pantyushin VS (1995) Ispytaniia ferromegnitnykh materialov (Testing of ferromagnetic materials). State Publishing House of Technical and Theoretical Literature

Kittel C (1946) Theory of the structure of ferromagnetic domains in films and small particles. Phys Rev J Arch 70:965–971

Klyuev VV, Muzhitsky VF, Gorkunov ES, Shcherbinin VE (2006) Nerazrushayushchii kontrol. T. 6, kn. 1: magnitnyie metody kontrolia (Non-destructive testing. Vol. 6, book 1: magnetic control methods). Machinebuilding Publishing House

Krinchik TS (1976) Fizika magnitnykh yavlenii (Physics of magnetic phenomena). Publishing House of Moscow State University

Landau LD, Lifshitz EM (1935) On the theory of the dispersion of magnetic permeability in ferromagnetic bodies. Phys Z Sowjetunion 8:153

Lifshitz EM (1945) O magnitnom stroyenii zheleza (About the magnetic structure of iron). Z eksp teoret fiz (J Exp Theoret Phys) 15(3):97–107

Mishin DD (1991) Magnitnyie materialy (Magnetic materials). Higher School Publishing House

Nazarchuk ZT, Andreikiv OY, Skalskyi VR (2013) Otsinyuvannia vodnevoi dehradatsii feromahnetykiv u mahnetnomu poli (Estimation of hydrogen degradation of ferromagnets in a magnetic field). Naukova Dumka Publishing House

Néel L (1944) Les lois de l'aimantation et de la subdivision en domaines élémentaires d'un monocristal de fer. J Phys Radium 5:241–279

Parcell E (1983) In: Shalnikova AI, Weisberg AO (eds) Elektrichestvo i magnetism (Electricity and magnetism). Nauka Publishing House

Rozing BL (1892) O magnitnom dvizhenii veshchestva (On the magnetic motion of matter). Zh Russ Fiz-tekh Obshchestva (J Russ Phys-Chem Soc (Phys Part)) 24:105

Rudyak VM (1986) Protsessy perekluicheniia v nelineinykh crystalakh (Switching processes in nonlinear crystals). Nauka Publishing House

Terletsky YP (1939) Obobshchennyye teoremy o nevozmozhnosti klassicheskogo obiasneniya magnetizma (Generalization of the theorem on the impossibility of classical explanation of magnetism). J Exp Theoret Phys 9(7):796–797

Tikadzumi S (1987) Fizika ferromagnetizma. Magnitnyie charakteristiki i prakticheskiie primeneniia (Physics of ferromagnetism. Magnetic characteristics and practical applications). Mir Publishing House

Tikazumi S (1983) Fizika ferrromagnetizma. Magnitnyie svoistva veshchestva (Physics of ferromagnetism. Magnetic properties of matter). Mir Publishing House

Troitsky VA (2002) Magnitoporoshkovyi control svarnykh soyedinenii i detalei mashyn (Magnetic powder inspection of welded joints and machine parts). Fenix Publishing House

Van Leeuwen H-J (1921) Problèmes de la theorie électronique du magnétisme. J Phys Radium Ser VI 2:361–377

Vonsovsky SV (1952) Sovremennoie ucheniie o magnetizme (Modern doctrine about magnetism). State Publishing House of Technical and Theoretical Literature

Vonsovsky SV (1971) Magnetizm. Mamagnitnyie svoistva dia-, para-, ferro-antiferro and ferromagnets (Magnetism. Magnetic properties of dia-, para-, ferro-, antiferro- and ferrimagnets). Nauka Publishing House

Vonsovsky SV, Shur YS (1948) Ferromagnetizm (Ferromagnetism). State Publishing House of Technical and Theoretical Literature

Weiss P (1907) L'hypothese du champ moleculaire et la propriete ferromagnétique. J Phys Théor Appl 6(1):661–690

Chapter 2
Barkhausen Effect and Emission of Elastic Waves Under Remagnetization of Ferromagnets

The traditional implementation of the AE method requires the application of an external load to the controlled object (Nazarchuk et al. 2017). Considering that the load required for the small defects' propagation can be much higher than the optimal loading modes of the object of control, the application of this approach in the practice of AE diagnostics is limited. To avoid this, in ferromagnetic materials, AE can be excited by an external magnetic field caused by the spontaneous movement of the walls of magnetic domains—the Barkhausen effect (BE).

Thus, by combining the signals of magnetoelastic AE with known physical phenomena, it is possible to create techniques for effective detection and quantification of the degree of degradation of structural materials, for example, under the influence of hydrogen-mechanical factors.

2.1 NDT Methods Used for Diagnostics of the State of Structural Materials

To ensure the reliable operation of responsible structures, operating under heavy loads, it is necessary to constantly study their technical state. The quality of the latter is determined mainly by the presence of defects such as cracks, i.e., defects that can propagate, reaching critical sizes. Modern non-destructive testing (ND) plays an important role in this case.

They are divided into types, united by common physical phenomena of the interaction of the physical field or substance with the object of control (OC) (Derzhavnyi Standart Ukrainy 1995). There are 13 types of NDT: magnetic, electric, electromagnetic, radio wave, thermal, optical, radiation, acoustic, penetrating substances, organoleptic, visual, visualization and photography in high-voltage fields, and electrodynamic.

V. Skalskyi and Z. Nazarchuk, *Magnetoelastic Acoustic Emission*, Springer-AAS Acoustics Series, https://doi.org/10.1007/978-981-99-4032-5_2

Each of the NDT types is divided into methods by the following features: the character of the interaction of physical fields or penetrating substances with OC; primary informative parameter; way to obtaining the primary information.

The character of the interaction of physical fields with OC is determined by the nature and parameters of physical field, material, structure, physical and geometric parameters of OC, shape, size and physical characteristics of OC defects, type of controlled feature, interaction conditions, and so on.

The primary informative parameter is a specific characteristic of the information carrier—physical field or penetrating substance—the changes of which are used to obtain information about the controlled feature of OC. The method of obtaining primary information is determined by the type of primary transducer or penetrating substance used to record and measure the primary information parameter.

The initiation and propagation of defects cause a few accompanying physical phenomena. The most interesting are those related to the transfer of energy or matter. They allow us to effectively detect defects as well as determine their parameters at certain, sometimes quite significant, distances from the location. These include radiation of heat, electrons, electromagnetic and elastic waves. The phenomenon of emission of elastic waves during deformation or fracture of materials, phase transformations and separation of secondary phase particles, magnetic or surface changes, etc. is called acoustic emission (AE) (Nazarchuk et al. 2017; Greshnikov and Drobot 1976).

The experience of the last ten years has demonstrated the great potential of the AE method. It is especially useful in conditions when it is impossible to visually control the appearance and movement of cracks due to their tunneling inside the material or when access to OC is extremely difficult. Remote control, high sensitivity, the ability to detect defects at distances that are orders of magnitude greater than the size of the damage, regardless of the shape and size of the OC, registration of the development of fracture in real time etc.—these are the advantages that put the AE method in one of the leading places among the promising methods of NDT.

Applying the AE phenomenon, it is possible to obtain additional information about dynamic magnetostriction phenomena during domain rearrangement in ferromagnets (Fowler et al. 1960). After all, it is then that the so-called magnetoelastic AE (MAE) occurs (Volkov et al. 1987).

The study of the Barkhausen effect makes a significant contribution to the understanding of physical processes that take place not only on the surface of ferromagnetic materials, but also in their internal volume. It is also used as one of the nondestructive methods of studying the ferromagnetic materials themselves (Lomayev 1977; Lomayev et al. 1984; Gorkunov et al. 1999; Durin and Zapperi 2006). The BE is especially widely used today in such areas of NDT as defectoscopy, structuroscopy, sizing, technical diagnostics of ferromagnets and products from them. There are many applied and fundamental research developments in these fields. New fields of application of the BE have also appeared, for example, to control the degree of hydrogen charging, surfacing, mechanical stresses, determining the amount of ferrite in austenitic-ferritic steels, etc. (Lomayev et al. 1984).

2.2 The Nature of the Barkhausen Effect and Its Application for Research of Ferromagnets

As is known (Wyatt 1985), despite the presence of spontaneous magnetization in ferromagnets, it is not always possible to magnetize them strongly. A hypothesis has been proposed that the ferromagnet is divided into many magnetic domains, and the direction of spontaneous magnetization varies from domain to domain. Later, in 1919, Barkhausen published a paper (Barkhausen 1919) on the discovery of the effect, which was later named after him. The essence of the phenomenon is the emission of electromagnetic noise caused by magnetization processes in ferromagnetic materials. Barkhausen himself believed that sudden noises under the influence of a smoothly time-varying applied external magnetic field, which could be heard in the microphone, were caused by the reverse movement of all magnetic domains. Later, Elmore (1938) was the first to observe the movement of domain walls in cobalt crystals under the influence of an external magnetic field, which was theoretically described by Bloch and were called the Bloch walls. However, Elmore did not associate the sudden movements of the domain walls with the noise of Barkhausen. This was done in further studies by Williams and Shockley during the investigation of magnetization processes in SiFe crystals (Williams and Shockley 1949) and substantiated by Kittel (1949).

After establishing the origin of Barkhausen noise, this effect began to be used to study the dynamics of the processes of soft ferromagnetic materials magnetization and explanation of the hysteresis of their magnetic properties. The noises themselves, due to their nature, were called Barkhausen jumps (BJ). Already the first studies in this direction (McClure and Schoeder 1976) have shown that although the Barkhausen effect is relatively easy to experimentally implement, its explanation is not simple. This is due to the stochastic character of the domain walls movement, which occurs in the form of a sequence of jumps, i.e., has an avalanche-like character with slow magnetization. The dynamics of jumps significantly depend on the material microstructure, demagnetization field, external stresses, etc. (Tebble and Newhouse 1953).

The results of further theoretical and experimental studies are presented in Gorkunov et al. (1999) and Rudiak (1970), which highlight the statistical properties of the Barkhausen's jumps, including distributions of avalanche duration and size, frequency spectra of energy, and so on. The proposed approaches to the description of known experimental data were phenomenological and were not directly related to the dynamics of domain walls. Energy spectra were described as a superposition of some elementary independent jumps with previously known spectral characteristics (Mazzetti and Montalenti 1965; Grosse-Nobis 1977; Grosse-Nobis and Lieneweg 1978). In another approach, an attempt was made to interpret the interaction of domain walls with a set of springs attached to a rigid wall, based on which the frequency spectrum of energy of such a system was calculated (Baldwin and Pickles 1972a, b).

Both reviews also present the schemes of BE experimental studies, the relationship between such schemes and the recorded signals related to the interaction of the magnetizing coil and the coil used to record the Barkhausen jumps, the influence of eddy currents, and so on.

The Barkhausen jump rate can be effectively recorded using the AE method by applying a piezoelectric transducer. It is well known that magnetic domains in the field of internal stresses have a high magnetic energy. For example, in the case of an isotropic body and one-dimensional motion of the 90° wall, we can consider the change in the magnetoelastic energy of the domain by the formula (Barteniev et al. 1981):

$$\Delta E_{\sigma_i} = \frac{3}{2}\lambda\sigma_i\Delta x, \tag{2.1}$$

where σ_i is the internal elastic stress, Δx is the interval of the jump of the domain wall, λ is magnetostriction constant.

Just because of that the phenomenon of pulsed magnetostriction during the abrupt rotation of domain boundaries in the BE becomes possible. Then, due to the change in the volume of the domains, part of the magnetoelastic energy must be released in the form of a non-constant sequence of the elastic wave pulses (magnetoelastic acoustic emission), which can be recorded by the piezoelectric transducer. Contrary to the electromagnetic recording of the Barkhausen jumps, the AE method directly carries information about the magnetoelastic energy of domain rearrangement associated with irreversible magnetostrictive phenomena in the processes of magnetization and remagnetization (Barteniev et al. 1984; Vonsovskyi and Shur 1948).

Currently, five mechanisms of the Barkhausen jumps are known: (1) irreversible movement of domain walls, which occurs at the time of transition of the potential barrier; (2) irreversible rotation of the magnetization vector of the monodomain region; (3) appearance and disappearance of the Neel peaks; (4) inversion of magnetization in single-domain parts of the ferromagnet; (5) initiation and motion of the Bloch and Neel lines in two 180° walls with inverted average magnetization (Lomayev 1977).

The study of the BJ and related MAE allows us to perform a comprehensive analysis of irreversible remagnetization processes, to determine the internal mechanical stress in ferromagnets, their damaged structure, and so on. Thus, in Boltachev et al. (1992) the Barkhausen jumps and the MAE caused by them in FeAl, FeCo and FeSi alloys were experimentally studied. Using samples of 100 × 8 × 0.2 mm the dependences of the RMS values of the BJ and MAE signals, as well as their number on the external magnetic field under remagnetization along the descending part of the hysteresis loop and the tensile stresses applied to the plate were obtained. On disks with a diameter of 26 mm and a thickness of 0.2 mm, the dependences of the parameters of the BJ signals on the mutual orientation of the axes of light magnetization and the external magnetic field were obtained. The samples were magnetized in a shielded solenoid at an external magnetic field frequency of 2 MHz. The change and the behavior of BJ and MAE depending on tensile stresses for different alloys were revealed and investigated.

The relationship between magnetic properties and the structural state of a material is widely studied. In Gorkunov et al. (1998, 1999), Mikheyev and Gorkunov (1985), Mikheyev et al. (1982) and Moskvin and Leshchenko (1983) the physical bases of magnetic structural analysis are considered, which is based on the relationship between the magnetic properties of the material with its structural state and phase composition.

Various mechanisms of the influence of changes in the structural parameters of ferromagnets on the processes of magnetization and remagnetization are presented. The possibility of using the magnetic characteristics of ferromagnetic materials to predict the structural state and strength properties of steels and alloys after various methods of hardening is discussed. In Gorkunov et al. (1999), the influence of elastic and plastic deformations and the character of the change in the domain structure on the BJ parameters in iron-based ferromagnets are considered. The influence of these changes in the domain structure on the formation of the BJ flux in materials with different signs of magnetostriction during their tension, compression, single- and multi-cycle plastic deformations is shown. The possibility of using the BJ parameters to analyze the state of the metal after the action of elastic and plastic deformations is considered. The mechanisms of the BJ formation in grains of different sizes and orientations are shown. The existence of the relationship between the grain sizes of polycrystalline ferromagnets and the size of the magnetic domains and, accordingly, the BJ parameters that determine the character of the hysteresis electromagnetic properties of materials and alloys are studied. The processes of remagnetization and the appearance of the BJ are discussed considering the irreversible displacements of the domain boundaries and domains in general, domain clusters and structural complexes.

Certain difficulties in the interpretation of the BJ are presented in Wiegman and Stege (1978) and Bertotti and Fiorillo (1981). The authors emphasize that the statistical properties of the BJ usually change along the hysteresis loop. Only by considering those parts of it where the movement of domain walls is dominant (or rather the only source of BJ) can a stationary signal be obtained and its statistical analysis performed correctly. These studies led to further experimental research, the results of which were well described by a new model of domain wall motion, known as the ABBM model (Vonsovskyi 1959; Kirenskii 1960). Though this model is also phenomenological, it allows us to describe the statistical properties of the BJ, based on simple assumptions that have experimental confirmation and allow an analytical description.

Based on additional experiments, it was found that the BJ are self-similar and are characterized by large-scale invariance and power distributions. It was suggested in Cote and Meisel (1991), Meisel and Cote (1992) and Bak et al. (1987) that the BJ is a partial case of self-ordered critical systems, which do not require fine adjustment of the control parameter, as for ordinary second-order phase transitions. In general, it should be noted that despite the significant amount of experimental data, there are very few reliable results among them even today. There are no general requirements for experimental equipment, which does not allow us to correctly compare the data obtained by different authors. Therefore, in recent years, many efforts have been made to conduct experiments in the simplest and most reproducible conditions.

During the BJ experimental studies, the induced (given) signal Φ' includes the contribution of both the applied field $\vec{H}$ on the receiving coil having a cross-sectional area A_p, and the material itself, i.e. $\Phi' = A_p + J'$, where A_p is the cross-sectional area of the sample, J' is magnetization. This expression is valid in the case of neglect of eddy currents (which is performed for very low frequencies of the order of a few Hz). For materials with high magnetic permeability, when A_p (the cross-sectional area of the sample) is not very small with respect to A_p, the contribution of the external field can be very small, so that the induced (given) flux will contain only information about the material properties. In the simplest experimental conditions, for example for a sample with only two domains (one domain wall), the induced flux is proportional to the velocity of the domain wall v, i.e., $\Phi' = N(2J_s d v)$, where J_s is the saturation magnetization, d is the sample thickness, N is the number of winds. In a more general case of the set of domain walls, we can assume that the parameter v determines the velocity of the active domain walls, although this will be a rather rough approximation.

It is known that the magnetostatic field in ferromagnetic material creates an opposing field, which is called the demagnetization one. It depends on the shape of the sample and the structure of the domains. This field is spatially homogeneous only in ellipsoidal specimens. Demagnetization significantly affects the dynamics of domain walls, limiting, for example, the maximum jump size (Mikheyev and Gorkunov). It should be noted that inductive measurements always record the movements of many domain walls, for which long-range interaction of domains plays a significant role. As a result, it is impossible to distinguish whether the recorded signal comes from individual jumps, or whether they are their spatial or temporal superposition. In White and Dahmen (2003) it was noted that at zero rates of change of the external magnetic field (adiabatic approximation) the signal should be considered as a sequence of avalanche-like elementary jumps without their superposition. In the case of a finite rate of the external field change, the signal is a sequence of pulses containing several avalanches.

In experiments, it is desirable to use simple forms of changes in the applied field, such as saw-tooth instead of sinusoidal, which is commonly used in hysteresis loop measurements. Note that modern research is performed on different types of materials that can be additionally subjected to, for example, heat treatment, annealing, and so on. These circumstances make it difficult to compare the results of BJ's research. Recently, many studies have been conducted on amorphous materials (Jansen et al. 1982; Bozort 1956; Annaiev 1951; O'Brien and Weissman 1994; Petta et al. 1996; Grosse-Nobis and Wagner 1977; Yamada and Saitoh 1992; Durin et al. 1996; Zheng et al. 2002; Zani and Puppin 2003). In general, not all soft magnetic materials (even with high magnetic permeability and low losses) may have insignificant BJ. These are amorphous alloys based on cobalt or permalloy with a limited number of structural defects. Significant noise is observed instead on materials that have a structural disorder that prevents the movement of domain walls. It is necessary to consider the sample thickness because it determines the size of the jumps. For example, SiFe tapes with a thickness of hundreds of microns need to be magnetized at very low frequencies to register the BJ. To study the statistical distribution of the BJ, it is necessary to establish the size and duration of individual jumps. Such measurements are complicated by background noise. To overcome this limitation, considering the fractal nature of the BJ, it is necessary to determine the resolution factor (Meisel and Cote 1992; Bertotti et al. 1994; Durin et al. 1995a). On the basis of such measurements, it is established that the distribution of the signal amplitude v can be described by the power law

$$P(v) = v^{-(1-c)} f(v/v_0), \tag{2.2}$$

where c is a coefficient proportional to the speed of the applied field, and its threshold is in the range of 5–15% of v_0. The duration T and the power S (integral of the amplitude within the given time interval) are determined unambiguously. Like the signal amplitude, the distributions T and S have a power character:

$$P(T) = T^{\alpha} g(T/T_0), \tag{2.3}$$

$$P(S) = S^{-\tau} f(S/S_0), \tag{2.4}$$

where T_0 and S_0 are the critical parameters of the cut, and α and τ are the critical values.

The results of experimental works, in particular (Urbach et al. 1995a; Perkovic et al. 1995), show that for different ferromagnetic materials, there are at least two types or classes of the values of critical parameters.

Many papers have been devoted to the investigation of the frequency spectra of the energy distribution of the Barkhausen signals (Mazzetti and Montalenti 1965; Kirenskii 1960; Durin et al. 1996; Bertotti and Montorsi 1990; Durin 1997; Petta et al. 1998a; Vergne et al. 1981). This is related to the practical need to reduce noise in the studied objects made of ferromagnetic materials. The record of stationary or non-stationary Barkhausen signals significantly affects the spectral characteristics. Traditional spectrum calculations using the fast Fourier transform algorithm can be applied to stationary data, otherwise unpredictable results can be obtained. Variable magnetic permeability adds low-frequency components to the spectrum that are not related to the BJ. Therefore, it is desirable to consider only two types of spectral estimates: the first one is caused by the movement of a single-domain wall at a constant speed (Grosse-Nobis 1977; Grosse-Nobis and Wagner 1977; Porteseil and Vergne 1979); the second estimate for small deviations from the coercive force (Wiegman and Stege 1978; Bertotti and Fiorillo 1981). Summarizing the performed studies of the BJ spectral distribution, it can be noted that at high frequencies the spectral function $F(\omega)$ is proportional to $1/\omega^{\theta}$, where parameter $\theta = 1.7 - 2$ can be normalized to the magnetization rate of J. That is, the spectra normalized to the mean flux $F(\omega)/SJ$ will coincide at high frequencies.

At lower frequencies, the spectra have extremes at frequencies approximately proportional to $J^{1/2}$. The peak amplitude is proportional to the magnetic permeability of 1/2 power.

At frequencies below the peak, the shape of the spectrum is determined by the dependence ω^{ψ}, $\psi \sim 0.6$ or $\psi \sim 1$. These general correlations describe the experimental results only qualitatively. In general, there are at least four different types of frequency distribution forms. The level of applied mechanical stresses has an influence on the spectral distribution parameters (Mehta et al. 2002). In Lieneweg and Grosse-Nobis (1972), Rautioaho et al. (1986) and Petta et al. (1998b), it was found that the asymptotic attenuation of the spectrum is determined not by the pulse shape, but exclusively by the distribution of the signal durations and the correlation between amplitude and duration.

The influence of hydrogen on the mechanical properties of metals has been studied quite well (Fowler et al. 1960; Fukai 1993; Kolachev 1985; Galaktionova 1967). The following sources of its diffusion into metal have been established: casting, pressure treatment of products, welding, heat treatment, acid etching, electroplating, stress corrosion, aggressive working environments, etc. Hydrogen can also penetrate into the metal from the air during storage of products in warehouses or during their operation. The presence of hydrogen in the metal changes its crystal lattice parameter, electrical resistance, magnetic, plastic, strength, structural, and other properties.

The method of vacuum extraction (vacuum heating and vacuum melting) is the most common method of controlling the hydrogen content in the metal (Volkov et al. 1987). However, it is very time-consuming and requires sampling of metal from specific products or structural elements.

There are also non-destructive methods for estimating the hydrogen content in the metal, which are based on various physical principles: determining the electron work function, impedance measurement, hysteresis loop parameters, and so on. However, these methods proved to be practically unsuitable due to high labor intensiveness, and the need to use expensive and complex equipment.

In Nechai and Moskvin (1975), an attempt has been made to use the BJ to study the effect of hydrogen on the magnetic structure of the metal under the condition of cyclic remagnetization of the sample. Hydrogen has been shown to increase the level of magnetic noise in metals with positive magnetostriction and decrease it with negative ones.

The authors of Migirenko et al. (1973) studied the effect of electrolytic hydrogen charging and mechanical loading on the BJ manifestation. The obtained results allowed us to conclude that the hydrogen charging of samples to significant concentrations (from 1 mm/100 g and more) by its action on the BJ is like the action of mechanical stresses in the elastic region of deformation. The authors hypothesized that the effect of significant concentrations of hydrogen and external elastic stresses on the BJ is associated with the influence on the magnetic structure, namely the change in its magnetization due to magnetostriction. Summarizing the above and considering the results of the review (Lomayev et al. 1984), it can be stated that the effect of hydrogen on the BJ is now studied experimentally insufficiently.

2.3 Some Theoretical Approaches to the Explanation of BE

The BE theory should explain the statistical behavior of signals caused by BJ such as frequency spectrum, statistical features of jumps, pulse shapes (Durin and Zapperi 2006). In particular, the theoretical description should predict the values of the distribution parameters and relate the properties of Barkhausen noise (BN) to the microstructure of the material. This specifies the importance of establishing the influence of microscopic inverse processes that accompany magnetization with the BN generated by them it is necessary to explain the origin of self-modeling processes and quantify the parameters of experimentally registered distributions; substantiate the origin of the fracture surface frequencies; explain the influence of microstructural parameters (lattice structure, anisotropy, spin interaction, etc.).

It is very difficult to develop a general theory that would answer all these questions. However, experimental evidence of the self-similar behavior of Barkhausen jump suggests that even relatively simple models may represent a wide range of experimental data. This is due to the fact that for phenomena that demonstrate critical behavior such fundamental relations as symmetry and conservation laws, should be decisive. Details of the quantitative behavior of ferromagnetic systems should not have a significant effect on them. Therefore, in the review of existing BE models we will pay attention to the fact how this or that theory answers the above questions. First, we consider theoretical models that describe the general properties of ferromagnetic materials and are based on the determination of the free energy of ferromagnetic systems.

Energy approach. Ferromagnetic material can be thought of as an ensemble of localized magnetic moments or spins that interact with each other and with an external magnetic field $\vec{H}$. The macroscopic magnetic properties of this material, such as Barkhausen noise or the hysteresis loop, are caused by the rotation of individual spins and can be described by microscopic theory. To build such a theory, it is necessary first to consider the interactions that determine the dynamics of local magnetization and the corresponding components of energy. In particular, the energy of the ferromagnetic material can be written as a sum

$$E = E_{\mathrm{ex}} + E_{\mathrm{m}} + E_{\mathrm{an}} + E_{\mathrm{dis}}, \tag{2.5}$$

where E_{ex} is the exchange energy; E_{m} is the magnetostatic energy; E_{an} is the energy of magnetic anisotropy; E_{dis} is the energy of disorder.

The exchange energy, which characterizes the forces of short range, trying to organize the spins, gives the most important contribution here. For a set of spins $\vec{s}(\vec{r}_i)$, the exchange energy can be written as follows

$$E_{\mathrm{ex}} = \sum_{ij} J\left(\left|\vec{r}_i - \vec{r}_j\right|\right)\vec{s}(\vec{r}_i) \cdot \vec{s}\left(\vec{r}_j\right), \tag{2.6}$$

where function $J(x)$ attenuates quickly for large x, and the summation is performed for all pairs of atoms.

Expression (2.6) can be approximated for the continuum as follows: replace the group of spins $\vec{s}(\vec{r}_i)$ with a continuous field $\vec{M}(\vec{r})$, and the exchange energy will then be

$$E_{\text{ex}} = A \int \mathrm{d}^3 r \sum_{\alpha=1}^{3} \left(\vec{\nabla} M_\alpha(\vec{r}) \right)^2, \tag{2.7}$$

where A is the exchange interaction constant.

The magnetostatic term in formula (2.5) is caused by the interaction between the spins and the external field, as well as the dipole–dipole interaction between different spins. For a uniformly magnetized sample, the contribution of the external field to the magnetostatic energy is equal to

$$E_{\text{m}} = -\frac{\mu_0}{8\pi} V \vec{M} \cdot \vec{H}, \tag{2.8}$$

where $\vec{M}$ is magnetization; V is the volume.

The energy, caused by the demagnetization field $\vec{H}_{\text{dm}}$, is defined as the energy of the field created by the local magnetization. To calculate $\vec{H}_{\text{dm}}$ it is necessary to determine the magnetic charges associated with the rupture of the normal component of the magnetization vector. For a surface, separating two regions with magnetizations $\vec{M}_1$ and $\vec{M}_2$, the surface charge density will be

$$\sigma = \vec{n} \left(\vec{M}_1 - \vec{M}_2 \right), \tag{2.9}$$

where $\vec{n}$ is the vector normal to the surface.

For example, for a sample surface where the magnetization changes abruptly from M_{s} to zero, the charge density will be $M_{\text{s}} \cos\theta$, where θ is the angle between the magnetization direction and the normal $\vec{n}$. In terms of magnetic surface charges, the demagnetization field can be written as follows

$$\vec{H}_{\text{dm}} = -\vec{\nabla} \int \frac{\mathrm{d}S' \sigma}{|\vec{r} - \vec{r}'|}, \tag{2.10}$$

where the integration is carried out on the surface that separates two areas with different magnetization.

For a uniformly magnetized ellipsoid $\vec{H}_{\text{dm}}$ is constant and proportional to the magnetization vector

$$\vec{H}_{\text{dm}} = -k\vec{M}, \tag{2.11}$$

where k is the demagnetization coefficient depending on the geometry of the body.

In this case, we obtain the total magnetostatic energy by replacing $\vec{H}$ by $\vec{H}+\vec{H}_{\mathrm{dm}}$ in Eq. (2.8). In the general case, the demagnetization field is not constant, and relation (2.8) should be replaced by the following expression

$$E_{\mathrm{m}} = -\frac{\mu_0}{8\pi}\int \mathrm{d}^3 r \vec{M}\cdot\left(\vec{H}+\vec{H}_{\mathrm{dm}}\right). \tag{2.12}$$

Here the field $\vec{H}_{\mathrm{dm}}$ is determined by dependence (2.10). This field takes into account the contribution caused by the volumetric change in magnetization.

The volumetric energy of the dipole interaction is written as

$$E_{\mathrm{m}} = -\frac{\mu_0}{8\pi}\int \mathrm{d}^3 r \mathrm{d}^3 r' \times \sum_{\alpha,\beta=1}^{3}\left(\frac{\delta_{\alpha\beta}}{|\vec{r}-\vec{r}'|} - \frac{3\left(r_\alpha - r_\alpha'\right)\left(r_\beta - r_\beta'\right)}{|\vec{r}-\vec{r}'|^5}\right)$$
$$M_\alpha(\vec{r})M_\beta\left(\vec{r}'\right). \tag{2.13}$$

Relation (2.13) can be written in abbreviated form by entering the density of bulk magnetic charges $\rho(\vec{r}) = \vec{\nabla}\cdot\vec{M}(\vec{r})$:

$$E_{\mathrm{m}} = -\frac{\mu_0}{8\pi}\int \frac{\mathrm{d}^3 r \mathrm{d}^3 r' \rho(\vec{r})\rho\left(\vec{r}'\right)}{|\vec{r}-\vec{r}'|}. \tag{2.14}$$

The magnetization of ferromagnets has prevailing directions that correspond to the crystallographic axes of the material. It is easier to magnetize the samples along these axes. This circumstance is seen in formula (2.5) due to the presence of magnetocrystalline anisotropy energy

$$E_{\mathrm{an}} = \int \mathrm{d}^3 r \sum_{\alpha,\beta} K_{\alpha\beta} M_\alpha M_\beta, \tag{2.15}$$

where M_α is α-component of the magnetization vector; $K_{\alpha\beta}$ is the symmetric tensor that describes the material anisotropy.

In the simplest case of a uniaxial crystal, dependence (2.15) will take the form

$$E_{\mathrm{an}} = \int \mathrm{d}^3 r K_0\left(\vec{M}\cdot\vec{e}\right)^2 = \int \mathrm{d}^3 r K_0 M^2 \sin^2(\varphi), \tag{2.16}$$

where φ is the angle between the axis $\vec{e}$ of light magnetization and the magnetization vector; K_0 is the constant of magnetic anisotropy in the uniaxial case.

The change in magnetization inside a ferromagnetic specimen can cause the deformation of the lattice structure, causing a phenomenon known as the magnetostriction effect. When an external mechanical stress is applied to the sample, on the contrary, it can cause a change in the magnetic structure. To describe this phenomenon, magnetoelastic energy is introduced into formula (2.5), which is defined as follows

$$E_{\text{an}} = \int \mathrm{d}^3 r \sum_{\alpha,\beta,\gamma,\delta} \lambda_{\alpha\beta\gamma\delta} \sigma_{\alpha\beta} M_\gamma M_\delta, \tag{2.17}$$

where $\sigma_{\alpha\beta}$ is the tensor of mechanical stresses; $\lambda_{\alpha\beta\gamma\delta}$ is magnetoelastic tensor.

For a crystal with isotropic magnetostriction in the case of uniaxial stress σ the anisotropy energy takes the form of relation (2.16), in which K_0 must be replaced by $K_0 + 3/2\lambda\sigma$, where λ is the magnetostriction constant in the uniaxial case, σ is the uniaxial stress.

Homogeneous systems in which all interactions are globally defined and independent on position (coordinates) are considered above. In general, ferromagnetic materials have different sources of inhomogeneities. These structural inhomogeneities are crucial for explaining the Barkhausen noise fluctuations. For this reason, in well-ordered systems, such noise is significantly lower.

There are several types of inhomogeneities that contribute to the free energy of ferromagnetic material. In crystals such inhomogeneities are caused by vacancies, dislocations or non-magnetic impurities. In polycrystalline materials, the presence of grain boundaries and changes in the directions of anisotropy in different grains should be added to the above mentioned. In amorphous alloys, internal stresses and the random arrangement of atoms have an important influence on the disorder. The presence of randomly distributed non-magnetic inclusions increases the contribution of the magnetostatic component owing to the magnetic charges that are formed at the boundaries of the inclusions (Néel 1946). This contribution can be expressed as a local fluctuation of dipole and exchange interactions. It should depend on the volume fraction of non-magnetic inclusions v. The average value of the interaction $g(\vec{r})$ will be of order $\overline{g} \cong (1 - v)g_0$, and the fluctuations—$\langle (g - \langle g \rangle)^2 \rangle \cong v g_0^2$, where g_0 is the interaction in the absence of inclusions. This type of disorder is called random link. In polycrystals, each grain has its own crystallographic anisotropy. The direction of the anisotropy axes will change in space, and in the case of uniaxial anisotropy the energy will be $E_{\text{an}} = \int \mathrm{d}^3 r K_0 \left(\vec{M} \cdot \vec{e} \right)^2$, where $\vec{e}$ is a random function of the grain-dependent position.

Internal stresses play an important role in materials with magnetostrictive properties. The random distribution of internal stresses will give the energy determined by relation (2.16), where the anisotropy constant will be $K(\vec{r}) \propto K_0 + (3/2)\lambda\sigma(\vec{r})$. Internal stresses have different origins, and their distribution can sometimes be written explicitly, this is possible for the random distribution of parallel dislocations. Stresses at point $\vec{r}$, caused by the dislocation at the center of coordinates are determined as $\sigma_{\alpha\beta}\vec{b}\mu C_{\alpha\beta}(\theta)/r$, where $\vec{b}$ is the Burgers vector, μ is the shear modulus, and $C_{\alpha\beta}$ is the function that establishes the angular distribution depending on the dislocation type (Hirth and Lothe 1992).

Formally, the stress distribution function can be written as

$$P\left(\tilde{\sigma}_{\alpha\beta}\right) = \int \mathrm{d}^{2N} r D(\vec{r}_1, \ldots, \vec{r}_N) \delta \left(\sum_k \sigma_{\alpha\beta}(\vec{r} - \vec{r}_k) \right), \tag{2.18}$$

where D is the distribution of dislocations.

The distribution of internal stresses is estimated for the random short-correlated location of dislocations. It is established that for low stresses the distribution $P(\sigma)$ is Gaussian, $\langle\Delta\sigma^2\rangle \propto \rho$, where ρ is the density of dislocations. For greater stresses, the distribution has a power character with a proportionality factor σ^{-3} (Groma and Bakó 1998).

Summarizing the above said, the total energy (2.5) of the ferromagnetic body in the uniaxial case can be written as

$$E = \sum_{\alpha=1}^{3} \int \mathrm{d}^3 r \left[A\left(\vec{\nabla} M_\alpha\right)^2 + K(M_\alpha e_\alpha)^2 - \left(H_\alpha + H_{\text{dem}}^{(\alpha)}\right) M_\alpha \right], \tag{2.19}$$

where A, K, e_α depend on the $\vec{r}$ location; $H_{\text{dem}}^{(\alpha)}$ is α-component of the demagnetization field.

Energy (2.19) can be used to calculate the equilibrium properties of ferromagnetic materials. However, this is an extremely difficult task. To describe the Barkhausen effect, it is sufficient to determine the temporal variation of magnetization of the material under the influence of the external magnetic field. The corresponding equation follows from dependence (2.19) and has the form

$$\frac{\partial \vec{M}}{\partial t} = \gamma \vec{M} \times \vec{H}_{\text{eff}}, \tag{2.20}$$

where γ is the ratio of charge to electron mass, and $\vec{H}_{\text{eff}} \equiv -\frac{\delta E}{\delta \vec{M}}$.

Equation (2.20) does not consider the mechanisms of energy dissipation and therefore gives an indefinite precession of the magnetization vector. Uncertainty can be avoided by introducing such a dissipation in a phenomenological way. This allows us to obtain, the Landau–Lifshitz and Hilbert equations. Numerical integration of this equation for different types of microstructures and boundary conditions is the task of the theory of micromagnetism. Attempts to describe the Barkhausen effect in polycrystals proceeding from this equation and using the Monte Carlo method were made in Gonzalez et al. (1997), Chubykalo et al. (1998a, b), Néel (1942, 1943) and Durin and Zapperi (2000).

The corresponding model had the form of a chain of N magnetic moments, the orientation of which was determined by the angles φ_i and θ_i. The total energy of the system looked like

$$E = \frac{1}{2} \sum_i \left(\sin^2 \alpha_i + 2H \cos\theta_i + m \sin^2\theta_i \sin^2\varphi_i - 2a \cos\beta_{i,i+1}\right), \tag{2.21}$$

where α_i is the angle between the local (random) axis of light magnetization and the external magnetic field $\vec{H}$; $H = \left|\vec{H}\right|$; a is a ratio of exchange energy constants and anisotropy energy; m corresponds to the magnetoelastic energy; β_{ij} is the angle between the magnetic moments i and j.

The change in magnetization in this model is determined by the Monte Carlo algorithm. When the external field grows slowly, the magnetic moments rotate suddenly, which corresponds to the Barkhausen jumps. The results obtained for the distribution of jumps for a certain range of parameters are qualitatively like experimental data, although quantitative differences are significant. In general, this model describes well spin rotations, magnetic anisotropy, but does not take into account the important fundamental aspects such as three-dimensionality, the character of the interaction (only the interaction between the nearest neighbors is considered), in particular dipole, which should attenuate as r^{-3}, and the effect of demagnetization. Because of this, the mentioned model cannot give a quantitative description of the Barkhausen effect. It is necessary to develop three-dimensional microscopic models that consider both near and far interaction. The latter is associated with significant mathematical and computational difficulties.

To avoid difficulties typical of micromagnetic models, it is necessary to formulate such approaches that would reflect the most sensitive aspects of the Barkhausen effect, namely—the randomness (random nature) of the magnetic system. This can be done leaving in formula (2.5) only a homogeneous magnetostatic energy E_{m}, and the rest of its terms can be replaced by a random magnetization function m:

$$E = F(m) - mH_{\mathrm{eff}}, \tag{2.22}$$

where $H_{\mathrm{eff}} = H + H_{\mathrm{dem}}$.

To understand the origin of expression (2.22), consider the movement of one rigid domain wall that divides the sample into two domains. In this case, the magnetization is proportional to the position of the wall x: $m = M_{\mathrm{s}}(2x/L - 1)$, where M_{s} is saturation magnetization, L is the width of the sample. The sample is magnetized when the applied magnetic field pushes the domain wall in a random field with potential $F(m)$. To formulate the problem, it is necessary to specify the statistical properties of the function $F(m)$, such as its distribution and correlation parameters. Neel was the first who propose such a model with a random potential to study the origin of the hysteresis loop in the Rayleigh part (Néel 1942, 1943). In his model, the random function had the form of the sum of parabolas with random curvature. Barkhausen noise based on this class of models can be explained as follows. When the effective field increases, the domain wall remains stationary as long as the condition $H_{\mathrm{eff}} < W(m)$ is met, where $W(m) \equiv -\frac{\mathrm{d}F}{\mathrm{d}m}$. When H_{eff} reaches the value of the local maximum of the function $W(m)$, the domain wall performs a series of successive jumps until the condition $H_{\mathrm{eff}} < W(m)$ is met again. The dimensions S of the Barkhausen jumps are the changes in the magnetization Δm that occur as a result of these jumps in the random field $W(m)$.

The model with a random field in the form proposed by Neel does not allow us to correctly describe the Barkhausen noise, because the random field is significantly uncorrelated. As a result, the distribution of the jumps magnitudes is exponential, while experiments indicate the power form of the distribution (Durin and Zapperi

2000; Spasojevic et al. 1996; Durin et al. 1995b; Bertotti 1983; Dhar et al. 1992; McMichael et al. 1993; Durin and Zapperi 2002; Ignatchenko and Rodichev 1960).

To obtain the power form of the distribution of the jump's values, we must assume that the random field is far correlated. This approach was first proposed in Bertotti (1986, 1987), which used the Brownian distribution of the field. On this basis, a well-known ABBM model was formulated (Alessandro et al. 1990a, b). It gives a good description of Barkhausen noise statistics. In this model, the magnetization varies according to the equation of motion with strong damping (attenuation)

$$\frac{\mathrm{d}m}{\mathrm{d}t} = ct - km + W(m), \tag{2.23}$$

where the external field increases with a constant velocity $H = ct$; k is the demagnetization coefficient, and the damping coefficient is equal to 1.

The random field is given in the form of the Brownian process, where the correlations grow as

$$\left\langle \left(W(m) - W\left(m'\right)\right)^2 \right\rangle = D\left|m - m'\right|. \tag{2.24}$$

Taking the derivative in relation (2.23) and putting $v \equiv \frac{\mathrm{d}m}{\mathrm{d}t}$, $f(m) \equiv \frac{\mathrm{d}W}{\mathrm{d}m}$, we get

$$\frac{\mathrm{d}v}{\mathrm{d}t} = c - kv + vf(m), \tag{2.25}$$

where $f(m)$ is the uncorrelated random field with dispersion D.

Dependence (2.25) can be written only in terms of variables v and m:

$$\frac{\mathrm{d}v}{\mathrm{d}m} = \frac{c}{v} - k + f(m). \tag{2.26}$$

Then it has the form of the Langevine's equation for random walks in a field with a limiting potential $U(v) = kv - c\ln(v)$. Asymptotically, the statistical distribution v is determined by the Boltzmann distribution:

$$P(v, m \to \infty) \sim \exp(-U(v)/D) = v^{c/D}\exp(-kv/D). \tag{2.27}$$

In the time domain, this distribution looks like (Alessandro et al. 1990a)

$$P(v) \equiv P(v, t \to \infty) = \frac{k^{c/D} v^{c/D-1} \exp(-kv/D)}{D^{c/D}\Gamma(c/D)}. \tag{2.28}$$

Based on Eq. (2.28) we obtain that the average velocity of the domain wall movement is $\langle v \rangle = c/k$.

For $c/D < 1$, the velocity distribution has a power character with an upper threshold, the value of which can be estimated for $k \to 0$. In this mode, the domain wall moves avalanche, the size of the jumps and their duration are also distributed according to the power law.

In the case of $c/D > 1$ the movement is smoother, with fluctuations that decrease with increasing c/D. In the case of $k \to 0$, based on Bray (2000), the following distributions of jump duration and values can be obtained

$$P(T) = T^{-\alpha} g(T/T_0); \quad P(S) = S^{-\tau} f(S/S_0).$$

Here f and g are known functions,

$$\alpha = 2 - c/D; \quad \tau = 3/2 - c/2D, \tag{2.29}$$

and the values of S_0 and T_0 can be found from the condition $c \to 0$; $S_0 \sim k^{-2}$; $T_0 \sim k^{-1}$.

Mathematical modeling of domain wall movement. Minimization of the exchange energy in the ferromagnetic material at sufficiently low temperatures and the absence of an external magnetic field leads to its uniform magnetization, oriented along the axis of the lightest magnetization. The latter reduces the energy of magnetic anisotropy. However, in most cases, this significantly increases the magnetostatic energy, which is caused by the violation of the magnetization vector continuity at the sample edges. Therefore, the formation of a domain structure that covers the entire body is energetically beneficial. Under such conditions, magnetization occurs mainly through the movement of domain walls, which must be investigated to understand the statistical properties of the Barkhausen noise (Dhar et al. 1992; Cizeau et al. 1997; Zapperi et al. 1998; Narayan 1996; Bahiana et al. 1999; de Queiroz and Bahiana 2001).

In the case of a 180° domain wall that separates two parts of the body with opposite directions of magnetization (e.g., along the Oz axis), its location can be described by the function $h(\vec{r}, t)$. Magnetostatic energy within the field $\vec{H}$, applied along the Oz axis, will be

$$E_{\mathrm{m}} = -2\mu_0 (H + H_{\mathrm{dm}}) M_{\mathrm{s}} \int \mathrm{d}^2 r h(\vec{r}, t), \tag{2.30}$$

where H_{dm} is the demagnetization field, which in the general case is a complex function $h(\vec{r}, t)$, and depends on the body shape.

In the simplest case, H_{dm} term will be proportional to the total magnetization. For magnetization from an invariable field, we will have

$$H_{\mathrm{dm}} = -\frac{k M_{\mathrm{s}}}{V} \int \mathrm{d}^2 r h(\vec{r}, t), \tag{2.31}$$

where factor k considers the domain structure and shape of the sample; V is the volume.

In general, the balance between the energy of magnetocrystalline anisotropy and exchange energy determines the width of the domain wall and its surface energy. The total energy of the domain wall due to these two terms is equal to

$$E_{\mathrm{dw}} = \gamma_{\mathrm{w}} \int \mathrm{d}^2 r \sqrt{1 + |\nabla h(\vec{r}, t)|^2}, \tag{2.32}$$

where $\gamma_{\mathrm{w}} \cong 2\sqrt{AK_0}$ is the surface energy of the domain wall.

For small gradients, expression (2.32) will take the form

$$E_{\mathrm{dw}} = \gamma_{\mathrm{w}} S_{\mathrm{dm}} + \frac{\gamma_{\mathrm{w}}}{2} \int \mathrm{d}^2 r |\nabla h(\vec{r}, t)|^2, \tag{2.33}$$

where S_{dm} is the area of the undeformed domain wall.

The local warping of the domain wall associated with the rupture of the normal component of the magnetization can be related to the density of fictitious surface magnetic charges:

$$\sigma(\vec{r}) = 2M_{\mathrm{s}} \cos\theta \approx 2M_{\mathrm{s}} \frac{\partial h(\vec{r}, t)}{\partial x}, \tag{2.34}$$

where θ is the angle between the vector normal to the surface and magnetization.

The energy associated with this charge distribution will be determined from the equality

$$E_{\mathrm{d}} = \int \mathrm{d}^2 r \mathrm{d}^2 r' \frac{\mu_0 M_{\mathrm{s}}^2}{2\pi |\vec{r} - \vec{r}'|} \frac{\partial h(\vec{r}, t)}{\partial z} \frac{\partial h(\vec{r}', t)}{\partial z'}, \tag{2.35}$$

which after integration in parts will take the form

$$E_{\mathrm{d}} = \int \mathrm{d}^2 r \mathrm{d}^2 r' K(\vec{r} - \vec{r}')\left[h(\vec{r}, t) - h(\vec{r}', t)\right]^2, \tag{2.36}$$

where the non-local kernel is determined by the formula

$$K(\vec{r} - \vec{r}') = \frac{\mu_0 M_{\mathrm{s}}^2}{|\vec{r} - \vec{r}'|^3} \left(1 - \frac{3(z - z')^2}{|\vec{r} - \vec{r}'|^2}\right), \tag{2.37}$$

and its Fourier transform is defined by the relation

$$K(p, q) = \frac{\mu_0 M_{\mathrm{s}}^2}{4\pi^2} \frac{p^2}{\sqrt{p^2 + q^2}}. \tag{2.38}$$

Here p and q are the components of Fourier transform along the z and y axes, respectively.

If we take into account that the magnetization vector may deviate slightly from the axis of the lightest magnetization, creating additional bulk charges, then the kernel (2.38) will take the form

$$\tilde{K}(p,q) \sim \frac{1}{\sqrt{Q}} \frac{p^2}{\sqrt{p^2 + Qq^2}}, \tag{2.39}$$

where $Q \equiv 1 + 2\mu_0 M_s^2/K$ is a material-dependent constant.

Violation of the order, as it is mentioned above, can take the form of non-magnetic inclusions, dislocations, or residual stresses that cause deformation and fixation of the domain wall. It can be modeled by introducing a random potential $V(\vec{r}, t)$, the derivative of which determines the force field $\eta(\vec{r}, t)$ acting on the domain wall. In the partial case of pointwise non-magnetic inclusions, the random force will be

$$\eta(\vec{r}, h) = -\sum_i f_p(\vec{r} - \vec{r}_i, h - h_i), \tag{2.40}$$

where (r_i, h_i) are coordinates of the attaching center; $f_p(x)$ is the fixation force of the domain wall, the area of application of which is proportional to the width of the domain wall $\delta_w \sim (K/A)^{1/2}$.

For the case when the distance between the centers of attachment of the domain walls is less than the averaging scale, the distribution will take the form of δ-correlated Gaussian noise

$$\langle \eta(\vec{r}, h)\eta(\vec{r}', h') \rangle = \delta^2(\vec{r} - \vec{r}')R(h - h'), \tag{2.41}$$

where $R(x)$ sharply attenuates for large x.

The specific form of the function $R(x)$ in the case of both random bonds and a random field does not significantly affect the laws of similarity of avalanche-like processes distributions that characterize BN and are associated with the movement of domain walls (Narayan and Fisher 1993).

Another possible source of domain wall stoppage is related to changes in its energy γ_w due to, for example, fluctuations in the direction of anisotropy (Néel 1946). In this case energy is the function of the wall orientation and can be written as $\gamma(\vec{r}, h) = \gamma_w + \eta(\vec{r}, h)$. Substituting this expression into relation (2.33) allows us to obtain in the simplest variant a correction for the anisotropy fluctuation.

In most cases, the domain wall motion occurs under conditions of strong damping. Then the equation of motion can be written in the form

$$\Gamma \frac{\partial h(\vec{r}, t)}{\partial t} = -\frac{\delta E(\{h(\vec{r}, t)\})}{\delta h(\vec{r}, t)}, \tag{2.42}$$

where $E(\{h(\vec{r}, t)\})$ is the total energy functional; Γ is the effective ductility.

In relation (2.42) temperature effects are neglected, as they are not important for BE in three-volume bodies (Urbach et al. 1995b) in contrast to thin films (Lemerle et al. 1998).

Summarizing all the terms for the domain wall energy, the equation of its motion can be written as

$$\frac{\partial h(\vec{r},t)}{\partial t} = H - \bar{k}\tilde{h} + \gamma_{\mathrm{w}}\nabla^2 h(\vec{r},t) + \int \mathrm{d}^2 r' K(\vec{r}-\vec{r}')\big(h(\vec{r}') - h(\vec{r})\big) + \eta(\vec{r},t), \tag{2.43}$$

where the kernel of the dipole interaction is given by Eq. (2.37), the effective demagnetization coefficient $\bar{k} = 4\mu_0 k M_{\mathrm{s}}^2/V$, $\tilde{h} = \int \mathrm{d}^2 r' h(\vec{r}',t)$.

If the demagnetization coefficient is insignificant, the domain wall will move only if the applied field exceeds the critical field H_{c}. That is, under the condition $H > H_{\mathrm{c}}$ the domain wall will move with an average velocity v which can be approximated by the dependence

$$v \sim (H - H_{\mathrm{c}})^{\beta}\theta(H_{\mathrm{c}} - H), \tag{2.44}$$

where $\theta(x)$ is a power function.

The critical behavior of the domain wall motion associated with the transfer of the fixing point was studied by the method of renormalization group (Narayan and Fisher 1993; Nattermann et al. 1992; Leschhorn et al. 2001; Chauve and Wiese 2001). These results indicate that for large-scale transfer, the critical exponents acquire field-average values (Cizeau et al. 1997; Zapperi et al. 1998), which is due to the linear dependence of the kernel in Eq. (2.37). In general, if we consider the interface, for which the kernel of the interaction of moments in the space can be represented by the dependence $K(q) = A_{\mathrm{K}}|q|^{\mu}$, the upper critical value will be $d_{\mathrm{c}} = 2\mu$, and the exponent will depend on μ. Here d is the internal dimension of the domain wall, i.e., the movement of the two-dimensional interface in a three-dimensional medium is considered. In general, we should expect the behavior of the wall depending on the medium field with the influence of surface tension stresses γ_{ω} in the wall and the critical behavior for $\mu = 2$.

For small changes in the applied field, the dimensions S of the avalanche-like motion of the domain walls are distributed according to the law

$$P(S) \sim S^{-\tau} f(S/S_0), \tag{2.45}$$

where $S_0 \sim (H - H_{\mathrm{c}})^{-1/\sigma}$, σ is some material constant.

In the case when the magnetic field is applied gradually and rather slowly, the distribution of the jump values can be obtained from dependence (2.45), integrating the latter by H:

$$p_{\mathrm{int}}(S) = \int_0^{H_c} \mathrm{d}H S^{-\tau} f\left(S(H - H_c)^{1/\sigma}\right). \tag{2.46}$$

The mean field theory, which describes well the motion of the domain wall during its separation and assumes the discretization of the equation of motion and the relationship of all fixing points with the mean position of the domain wall, leads to the following dependence (Fisher 1985)

$$\frac{\mathrm{d}h_i}{\mathrm{d}t} = ct - k\overline{h} + J\left(\overline{h} - h_i\right) + \eta_i(h), \tag{2.47}$$

where J is the effective pair interaction.

Summation by i of both parts of formula (2.47) allows us to obtain the dependence for the total magnetization m

$$\frac{\mathrm{d}m}{\mathrm{d}t} = \tilde{c}t - km + \sum_{i=1}^{N} \eta_i(h), \tag{2.48}$$

like Eq. (2.23). To make the similarity more obvious, the $\sum_i \eta_i$ component must be replaced by an effective field that blocks the motion of the domain wall $W(m)$ and is described by a Brownian correlation. In this case, the domain wall jumps between two configurations, which leads to the change in W of the form

$$W\left(m'\right) - W(m) = \sum_{i=1}^{n} \Delta\eta_i, \tag{2.49}$$

where the summation is carried out according to the positions through which the domain wall moved.

In the medium field theory, this number is proportional to the avalanche size $S = \left|m' - m\right|$. Assuming that the value of $\Delta\eta_i$ is uncorrelated and has randomly distributed signs, we can obtain a Brownian blocking field for which

$$\left\langle \left|W\left(m'\right) - W(m)\right|^2 \right\rangle = D\left|m' - m\right|, \tag{2.50}$$

where D quantitatively determines the fluctuation of W.

Thus, the Brownian blocking field quite effectively describes the disorder that results from the collective motion of flexible domain walls.

The shape of the energy spectrum. Explaining the shape of the Barkhausen noise energy spectrum was considered one of the main tasks of theoretical models. In the first attempts to build such models, the energy spectrum was considered as a result of the superposition of elementary independent events without taking into account the magnetization microprocesses (Mazzetti 1962; Arques 1968). The next step in this direction was made in the ABBM model (Vonsovskyi 1959; Kirenskii 1960; Alessandro et al. 1990a, b).

From Eq. (2.23) we can obtain the Fokker–Planck equation, which describes the evolution of the probability distribution $P(v, t|v_0)$ for the velocity v at time t, if at the beginning ($t = 0$) it was equal to (Alessandro et al. 1990a):

$$\frac{\partial P}{\partial t} = \frac{\partial}{\partial v}\left((kv - c)P + D\frac{\partial vP}{\partial v}\right). \tag{2.51}$$

The correlation function for such a rate is determined by the dependence

$$\begin{aligned} G(t) &\equiv \langle (v(t) - c/k)(v(0) - c/k)\rangle \\ &= \int \mathrm{d}v\mathrm{d}v_0(v - c/k)P(v, t|v_0)P(v_0), \end{aligned} \tag{2.52}$$

where distribution $P(v_0)$ is given by relation (2.28).

Differentiating Eq. (2.52) and using dependence (2.51), we can find the evolution equation for correlations $G(t)$:

$$\frac{\mathrm{d}G}{\mathrm{d}t} = -kG. \tag{2.53}$$

The solution of Eq. (2.53) has the form

$$G(t) = c/k\exp(-kt). \tag{2.54}$$

In this case, the energy spectrum will be

$$F(\omega) = \frac{2c}{\omega^2 + k^2}. \tag{2.55}$$

Based on this simple result, some conclusions can be drawn. First, for high frequencies the spectrum attenuates as ω^{-2}; second, the spectrum is cut at frequencies lower than $\omega_0 = 1/k$; third, the amplitude of the spectrum increases with increasing rate of change of the field c. These properties qualitatively describe the shape of experimentally measured energy spectra, but this dependence does not give a quantitative coincidence. In particular, the "tail" part of the spectrum often attenuates not as an exponent with parameter 2, and the cutoff frequency depends on k more complex than $1/k$.

To reproduce the low-frequency behavior of the spectrum in the ABBM model, the parameter ξ^*, that is the correlation length in the braking field was introduced (Alessandro et al. 1990a). This allowed us to obtain a formula for the spectral distribution in the form

$$F(\omega) \propto \frac{\omega^2}{(\omega^2 + k^2)(\omega^2 + \tau_c^{-2})}, \quad \tau_c \propto \xi^*, \tag{2.56}$$

which agrees well with the experiment.

It should also be noted that many attempts have been made to combine the Barkhausen noise energy spectrum with distributions that describe the avalanche-like motion of domain walls (Spasojevic et al. 1996; Bertotti 1983; Zapperi et al. 1998; Narayan 1996; Dahmen and Sethna 1996). However, this combination proved to be unsatisfactory. Although the exponent describes the "tail" part of the spectrum distribution quite well, the theory does not allow a complete description of the whole shape of the spectrum, which would coincide satisfactorily with the experiment. A new attempt was made in Kuntz and Sethna (2000). Combining analytical methods and computer modeling, the authors have established a scale ratio that describes the high-frequency region of the spectrum and agrees well with experimental data (Durin and Zapperi 2002). In Kuntz and Sethna (2000) it was assumed that at the critical point the avalanche average size S of duration T was determined by dependence $S(T) \sim T^{1/\sigma v z}$, and accordingly

$$v(T, t) = T^{1/\sigma v z} f_{\text{sh}}(t/T), \tag{2.57}$$

where v is the signal voltage; t is time; f_{sh} is a large-scale function that is possibly universal.

The next assumption is that the distribution $P(v|S)$ is the probability that the voltage reaches the value of v for an avalanche of size S. This probability is determined as follows:

$$P(v|S) = v^{-1} f_v\left(v S^{\sigma v z - 1}\right), \tag{2.58}$$

where f_v is another universal function.

Using dependence (2.58), one can see that the functional dependence of the energy $E = \langle v^2 \rangle$ of avalanche of size S has the form $E(S) \sim S^{2-\sigma v z}$. If to consider that the correlation function for voltage is expressed by the formula

$$G(t) \equiv \int \mathrm{d}t' \langle v(t + t') v(t) \rangle, \tag{2.59}$$

then $G(t)$ can obtain such a dependence for

$$G(t|S) = S^{2-\sigma v z} f_G\left(t S^{-\sigma v z}\right). \tag{2.60}$$

Hence the spectral distribution will be

$$F(\omega|S) = \int_0^\infty \mathrm{d}t \cos(\omega t) G(t|S) = S^2 f_{\text{energy}}\left(\omega^{-1/\sigma v z} S\right). \tag{2.61}$$

Here f_g and f_{energy} are, as above, some universal functions.

To obtain the frequency spectrum of the Barkhausen noise energy, it necessary to average relation (2.61) by the avalanche sizes. The assumption that $P(S) \sim S^{-\tau}$, gives an incorrect result (Spasojevic et al. 1996; Bertotti 1983; Dahmen and Sethna 1996), according to which $F(\omega) \sim \omega^{-(3-\tau)/\sigma v z}$. The reason for this is that the function $f_{energy}(x) \sim 1/x$ is for large x. Therefore, for $\tau < 2$ the main contribution to the averaged spectral distribution is given by the terms corresponding to the upper cutoff frequency in the avalanche size distribution. It follows that

$$F(\omega) \sim \omega^{-1/\sigma v z}. \tag{2.62}$$

If to use the values of the exponent for the model of the mean field: $\sigma v z = 1.2$, then $F(\omega) \sim \omega^{-2}$. The result, expressed by correlation (2.62), agrees well with many experimental data and is also confirmed by computer modeling of the domain wall motion.

2.4 Modeling of MAE Signals Caused by Barkhausen Jumps

Emission of elastic waves caused by BE–MAE (Barteniev et al. 1981) is associated with magnetostrictive deformations in the ferromagnet. The latter occur in some local area of the body, namely where sudden changes in the position of the domain walls are observed (Sánchez et al. 2004; Shibata and Ono 1981; Buttle et al. 1987b). In contrast to studies of the Barkhausen effect, magnetoelastic acoustic emission has been studied in less detail. The phenomenological theory of the MAE was developed in Kolmogorov (1972), Turner et al. (1969), Kameda and Ranjan (1987), Kwan et al. (1984a, b), Yudin and Lopatin (1987), Natsyk and Nechyporenko (1984) and Glukhov and Kolmogorov (1988). It considers the movement of domain boundaries and the processes of initiation and propagation of remagnetization kernels. The known experimental studies should be noted (Vonsovskyi and Shur 1948; Kolmogorov 1972; Turner et al. 1969; Kameda and Ranjan 1987; Kwan et al. 1984a, b; Glukhov and Kolmogorov 1988; McClure et al. 1974; Ono and Shibata 1980; Roman et al. 1983; Gorkunov et al. 1986a, 1987; Lomayev et al. 1981; Bezymiannyi 1981; Glukhov et al. 1985; Ranjan et al. 1986, 1987a, b; Shibata et al. 1986; Guyot et al. 1987; Buttle et al. 1987a). Most tests were performed on polycrystalline nickel, armco-iron, siliceous iron steel samples (Mazzetti and Montalenti 1965; Vonsovskyi and Shur 1948; Turner et al. 1969; Kameda and Ranjan 1987; Glukhov et al. 1985; Ranjan et al. 1986; Shibata et al. 1986; Buttle et al. 1987a), iron-nickel alloys (Mazzetti and Montalenti 1965; Kameda and Ranjan 1987; Lomayev et al. 1981; Bezymiannyi 1981; Ranjan et al. 1987b; Buttle et al. 1987a). Resonant primary acoustic transducers were used in the studies, the dynamic remagnetization mode was chosen with a frequency of 20–100 Hz, and the samples were most often rod-shaped.

In experiments, the counting speed and amplitude of the MAE signal were recorded, depending on the amplitude of the remagnetization field, external mechanical stresses, annealing temperature, hardness, size of inclusions, and distance between them (Kameda and Ranjan 1987; Buttle et al. 1987a). These studies have shown that the magnitude of the MAE signal increases with decreasing external tensile stresses and with increasing grain size (except for nickel, in which the MAE signal decreases with increasing grain size). The counting speed in many cases for polycrystalline materials, on the contrary, increases with increasing external tensile forces. In addition, experimental studies show that MAE occurs mainly during the jump-like motion of the domain walls that separate domains where the magnetization vectors are not directed in the opposite way (Rudiak 1970; Boltachev et al. 1992; Sánchez et al. 2004).

Many scientists have studied the regularities of changes in the MAE signals when changing the modes of heat treatment of structural steels. In Gorkunov et al. (1987) the effect of heat treatment on MAE in medium- and high-carbon low-alloy structural steels was studied in order to use this method to control their structural condition and strength characteristics. It is shown that MAE signals depend on the value of internal stresses, as well as on structural changes and phase transformations that occur in such steels under heat treatment. The magnetic method of quality control under high-temperature treatment of structural and simple carbon steels is also described in Gorkunov et al. (1986b), Tsarkova et al. (1981), Mikheyev et al. (1977, 1981), Bida et al. (1991, 1994), Mikheyev and Gorkunov (1981), Degtiariov and Kametskii (1977), Kuznetsov et al. (1972) and Mikhailov and Shcherbinin (1992).

The Barkhausen jump rate was effectively recorded by the AE method using a piezoelectric transducer. It is well known that magnetic domains in the field of internal stresses have a high magnetic energy. For example, in the case of an isotropic body and one-dimensional movement of the 90° boundary, the change in the magnetoelastic energy of the domain should be considered (Barteniev et al. 1981):

$$\Delta E_{\sigma_i} = \frac{3}{2}\lambda\sigma_i\Delta x, \tag{2.63}$$

where σ_i is the internal elastic stress; Δx is the domain boundary jump interval; λ is a magnetostriction constant.

Just because of this, the phenomenon of pulsed dynamic magnetostriction during the jump-like rotation of the domain boundaries is possible. Then, by changing the volume of the domains, part of the magnetoelastic energy ΔE_{σ_i} must be released in the form of a non-constant sequence of the elastic wave pulses (magnetoelastic acoustic emission), which can be recorded by the piezoelectric transducer. In contrast to the electromagnetic recording of Barkhausen jumps, the AE method directly carries information about the magnetoelastic energy of the domain rearrangement associated with irreversible magnetostrictive phenomena in the processes of magnetization and remagnetization (Sánchez et al. 2004).

2.5 General Correlations of the Theory of Magnetoelasticity

The dynamics of Barkhausen jumps, and hence the magnetoelastic acoustic emission, is related to the value of mechanical stresses acting in the area of location of the domain walls. On the other hand, based on the equations of magnetoelasticity, it is known that mechanical stresses in a ferromagnetic body create not only external loads but also a magnetic field (Panasyuk 1988).

The system of equations for determining these stresses in a ferromagnetic body consists of static Maxwell's equations $\mathrm{div}\vec{B} = 0$, $\mathrm{rot}\vec{H} = 0$ in the absence of electric field, free charges, and currents, and the basic equations of the deformable bodies mechanics. The total magnetic force acting on the volume element of the body is the sum of the volume force and the volume pair of forces, respectively:

$$f_k = \mu_0 M_j H_{kj}; \quad l_k = \mu_0 \varepsilon_{ijk} M_j H_i. \tag{2.64}$$

The following notations are used above: ε_{ijk} is the obliquely symmetric Kronecker symbol; $\mu_0 = 4\pi \times 10^{-7}\,\mathrm{H/A^2}$ is the magnetic permeability of vacuum; $\vec{H}$ is the vector of magnetic field strength; $\vec{B}$ is the vector of magnetic induction; $\vec{M}$ is the vector of magnetization, as well as of the surface force that characterizes the magnetic interaction between the element and the surrounding magnetized material.

The surface force is related to the mechanical stress σ_{ij}, forming a complete stress tensor $\sigma'_{ij} = \sigma_{ij} + \sigma''_{ij}$ (σ''_{ij} is the tensor of internal stresses). To consider the surface magnetization, the Maxwell magnetic tensor σ^{M} is introduced as

$$\mathrm{div}\,\sigma^{\mathrm{M}} \equiv \vec{f}, \quad \sigma^{\mathrm{M}}_{ij} B_i H_i - 1/2\mu_0 H_k H_k \delta_{ij}. \tag{2.65}$$

Assuming that the deformations are small, according to Pao and Yeh (1973) the general equations can be linearized by representing all magnetic quantities as the sum of two components:

$$B_i = B_{0i} + b_i, \quad H_i = H_{0i} + h_i, \quad M_i = M_{0i} + m_i. \tag{2.66}$$

Here, the values denoted by the index 0 are the magnetic characteristics in a non-deformable body, and the values denoted by lowercase letters are corrections that take into account additional changes in magnetic induction and magnetization. These summands occur due to body's deformation.

The final complete system of equations and boundary conditions of magnetoelasticity in the linearized case ($\left|H_{ij} u_{i,j}\right| << |h_i|$, where u_i are components of the displacement vector), consists of the equations and boundary conditions for an undeformed magnetic body and the equations and boundary conditions for a deformed state.

The equations for an undeformed body are written as

$$\begin{aligned}&\operatorname{div} B_0 = 0, \quad \operatorname{rot} H_0 = 0, \quad B_{0i} = \mu,\\ &H_{0i} = \mu_0(1+\chi)H_{0i}, \quad M_{0i} = \chi H_{0i},\end{aligned} \tag{2.67}$$

where χ is the magnetic susceptibility of the material.

Boundary conditions for an undeformed body can be written as

$$\varepsilon_{ijk} n_{0k}[H_{0k}] = 0; \quad n_{0j}\left[B_{0j}\right] = 0. \tag{2.68}$$

Here, square brackets indicate a jump in the magnitude on the edge surface.

The equations of magnetoelasticity for the disturbed state consist of the equilibrium equations:

$$\sigma'_{ji,j} + \sigma^M_{ji,j} = 0, \tag{2.69}$$

$$\operatorname{div} \vec{h} = 0, \quad \operatorname{rot} \vec{h} = 0 \tag{2.70}$$

and the equations of state:

$$\begin{aligned}&b_i = \mu h_i, \quad m_i = \chi h_i,\\ &\sigma'_{ij} = \sigma_{ij} + \sigma''_{ij}, \quad \sigma_{ij} = \lambda\delta_{ij}u_{k,k} + G\left(u_{i,j} + u_{j,i}\right),\\ &\sigma''_{ij} = \mu_0\chi\left(H_{0i}H_{0j} + H_{0i}h_j + H_{0j}h_i\right),\\ &\sigma^M_{ij} = \mu\left(H_{0i}H_{0j} + H_{0i}h_j + H_{0j}h_i\right) - 1/2\mu_0\delta_{ij}(H_{0k}H_{0k} + 2H_{0k}h_k),\end{aligned} \tag{2.71}$$

where μ is the magnetic permeability of the environment; $\mu = \mu_0(1+\chi)$; λ is the Lamé constant; G is the shear modulus.

Boundary conditions are defined by the equalities

$$\begin{aligned}&n_{0i}[b_i] - n_{0m}u_{m,i}[B_{0i}] = 0, \quad e_{ijk}\left\{n_{0j}[h_k] - n_{0m}u_{m,j}[H_{0k}]\right\} = 0,\\ &n_{0j}\left[\sigma'_{ji} + \sigma^M_{ji}\right] = 0, \quad [u_i] = 0.\end{aligned} \tag{2.72}$$

The equilibrium equation in displacements will look like

$$\nabla^2 u_i + \frac{1}{1-2\nu}\frac{\partial}{\partial x_i}\left(\nabla H_{0i}\left(\vec{H}_0 + \vec{h}\right) + \vec{H}_0\nabla h_i\right) = 0. \tag{2.73}$$

Crack-like defects, which are the most dangerous in terms of loss of load-bearing capacity of structural elements, create stresses in the ferromagnetic material, which affect the dynamics of the Barkhausen jumps and, accordingly, the MAE. As follows from the system of Eqs. (2.67)–(2.73) of magnetoelasticity, the applied magnetic field will create mechanical stresses in the ferromagnetic material. Stress intensity factor (SIF) describes the concentration of these stresses at the crack-like defects. To determine it, in particular at a disk-shaped crack, in Shindo (1978) in a cylindrical coordinate system with the center O coinciding with the center of the crack and the axis Oz, perpendicular to the plane of location of such a defect, the corresponding boundary conditions were formulated:

$$\begin{aligned}
&\sigma_{zz}(r,0) = \frac{\chi^2}{\mu_\mathrm{r}}\left(\frac{B_0^2}{2\mu_0\mu_\mathrm{r}} + B_0\varphi_z\right), \quad r < r_0; \quad u_z(r,0) = 0, \quad r > r_0;\\
&\sigma_{rz}(r,0) = 0, \quad r < \infty;\\
&h_\mathrm{r} = \frac{\chi B_0}{\mu}\frac{\partial u_z}{\partial r}, \quad h_z^{(e)} - (1+\chi)h_z = 0, \quad r < r_0;\\
&\Phi(r,0) = 0,
\end{aligned} \tag{2.74}$$

where Φ is the magnetostatic potential that satisfies the Laplace equation, and:

$$h_\mathrm{r} = \frac{\partial \Phi}{\partial r}, \quad h_z = \frac{\partial \Phi}{\partial z}$$

After solving the problem formulated in this way, the SIF value is obtained.

Based on the above said, instead of traditional acoustic emission non-destructive testing methods in the case of ferromagnetic bodies with crack-like defects, the "loading" can be carried out by applying an external magnetic field created, for example, by a solenoid. Concentration of mechanical stresses generated by this field at the crack-like defects, which is characterized by the SIF value, will affect the intensity of the Barkhausen jumps in these areas.

References

Alessandro B, Beatrice C, Bertotti G, Montorsi A (1990a) Domain wall dynamics and Barkhausen effect in metallic ferromagnetic materials. I. Theory. J Appl Phys 68(11):2901–2908

Alessandro B, Beatrice C, Bertotti G, Montorsi A (1990b) Domain wall dynamics and Barkhausen effect in metallic ferromagnetic materials. II. Experiments. J Appl Phys 68(11):2908–2915

Annaiev RG (1951) Magneto-elektricheskiie yavleniia v ferromagnitnykh metallakh (Magneto-electric phenomena in ferromagnetic metals). Izdatelstvo Akademii Nauk TSSR (Publishing House of the Academy of Sciences of TSSR), Ashgabat

Arques P (1968) Sur certains problémes statistiques liés a l'effet de Barkhausen. J Phys 29:369–384

Bahiana M, Koiller B, de Queiroz SLA, Denardin JC, Sommer RL (1999) Domain size effect in Barkhausen noise. Phys Rev E 59:3884–3887

Bak P, Tang C, Wiesenfeld K (1987) Self-organized criticality: an explanation of the 1/f noise. Phys Rev Lett 59:381–384

Baldwin JA, Pickles GM (1972a) Autocorrelation function of Barkhausen noise in simple materials. J Appl Phys 43(8):1263–1268

Baldwin JA, Pickles GM (1972b) Power spectrum of Barkhausen noise in simple materials. J Appl Phys 43(12):4746–4749

Barkhausen H (1919) Zwei mit Hilfe der neuen Verstärker entdeckte Erseheinugen. Phys Z 20:401–403

Barteniev OA, Khamitov VA, Gorkunov ES (1981) Akusticheskaia emissiia pri dinamicheskoi magnitostriktsii v effektie Barkhausena. Signoelektriki i piezoelektriki (Acoustic emission during dynamic magnetostriction in the Barkhausen effect. Ferroelectrics and piezoelectrics). Publishing House of Kalinin State University

Barteniev OA, Khamitov VA, Gorkunov ES (1984) Sposob nablyudeniia skachka Barkhausena po izlucheniiu zvuka (A method of observing the Barkhausen jump in sound radiation). Prib Tekh Eksp (Instrum Exp Techn) 1:140–187

Bertotti G (1983) Space-time correlation properties of the magnetization noise and magnetic losses. Noise Phys Syst 1/f Noise 2:355–358

Bertotti G (1986) Statistical interpretation of magnetization processes and eddy current losses in ferromagnetic materials. In: Gorzkowski W, Lachowicz H, Szymczak H (eds) Proceedings of the 3rd international conference on physics of magnetic materials, Singapore, pp 489–508

Bertotti G (1987) Dynamics of magnetic domain walls and Barkhausen noise in metallic ferromagnetic systems. In: Balucani U, Lovesey SW, Rasetti MG, Tognetti V (eds) Magnetic excitations and fluctuations II. Springer

Bertotti G, Fiorillo F (1981) Barkhausen noise and domain structure dynamics in Si–Fe at different points of the magnetization curve. J Magn Magn Mater 23:136–148

Bertotti G, Montorsi A (1990) Dependence of Barkhausen noise on grain size in ferromagnetic materials. J Magn Magn Mater 86(1):214–216

Bertotti G, Durin G, Magni A (1994) Scaling aspects of domain wall dynamics and Barkhausen effect in ferromagnetic materials. J Appl Phys 75(12):5490–5492

Bezymiannyi YG (1981) Issledovaniie vozmozhnostei methoda magnetoakusticheskikh shumov dla kontrolia ustalosti nikelia. Effekt Barkhausena i yego ispolzovaniie v technike (Investigation of the possibilities of the magnetoacoustic noise method for nickel fatigue control. The Barkhausen effect and its use in technology). Kalinin

Bida GV, Tsarkova TN, Kostin BN, Sazhyka EY (1991) Ispolzovaniie relaksatsionnykh magnitnykh svoistv dlia nerazrushayushchego kontrolia zakalionnykh i otpushchennykh stalei (The use of relaxation magnetic properties for non-destructive testing of hardened and tempered steels). Defektoskopiya (Defectoscopy) 12:39–43

Bida GV, Tsarkova TN, Sazhyka EY (1994) O vybore rezhyma peremagnichivaniia pri nerazrushayushchem kontrolie kachestva otpushchennykh detaliei po vtorichnoi ostatochnoi magnitnoi induktsii (On the choice of the remagnetization mode for non-destructive quality control of tempered parts by secondary residual magnetic induction). Defektoskopiya (Defectoscopy) 3:26–31

Boltachev VD, Golovshchikova IV, Yernakov AY, Dragoshanskyi YN (1992) Effekt Barkhausena i magnitoakusticheskaia emissiia v splavakh FeAl, FeCo i FeSi (Barkhausen effect and magnetoacoustic emission in FeAl, FeCo, and FeSi alloys). Fiz Met Metalloved (Phys Met Mater Sci) 12:59–67

Bozort R (1956) Ferromagnetizm (Ferromagnetism). Foreign Literature Publishing

Bray AJ (2000) Random walks in logarithmic and power-law potentials, nonuniversal persistence, and vortex dynamics in the two-dimensional XY model. Phys Rev E 62:103–112

Buttle DJ, Yakubovies JP, Briggs JAD (1987a) Magnetoacoustic and Barkhausen emission from domain. Wall interaction with precipitates in Incoloy 904. Philos Mag A 55(6):735–756

Buttle DJ, Scruby CB, Yakubovics JP, Briggs JAD (1987b) Magnetoacoustic and Barkhausen emission: their dependence on dislocation in iron. Philos Mag 55(6):717–734

Chauve P, Wiese PLD (2001) Renormalization of pinned elastic systems: how does it work beyond one loop. Phys Rev Lett 86:1785–1788

Chubykalo OA, Gonzalez J, Gonzalez JM (1998a) Avalanches as propagating domain walls in a micromagnetic model. Phys D 113:382–386

Chubykalo OA, Gonzalez JM, Gonzalez J (1998b) Barkhausen jump distributions in a micromagnetic model. J Magn Magn Mater 184:257–261

Cizeau P, Zapperi S, Durin G, Stanley HE (1997) Dynamics of a ferromagnetic domain wall and the Barkhausen effect. Phys Rev Lett 79(10):4669–4672

Cote PJ, Meisel LV (1991) Self-organized criticality and the Barkhausen effect. Phys Rev Lett 67:1334–1337

Dahmen K, Sethna JP (1996) Hysteresis, avalanches, and disorder-induced critical scaling: a renormalization-group approach. Phys Rev B 53:14872–14905
de Queiroz SLA, Bahiana M (2001) Finite driving rates in interface models of Barkhausen noise. Phys Rev E 64:66127
Degtiariov MV, Kametskii V (1977) Struktura, mekhanichekiie i magnitnyie svoistva stali 30KhGSA posle kompleksnogo vozdeistviia kholodnoi plasticheskoi deformatsii i termichwskoi obrabotki (Structure, mechanical and magnetic properties of 30ChGSA steel after combined exposure to cold plastic deformation and heat treatment). Defektoskopiya (Defectoscopy) 7:15–20
Derzhavnyi Standart Ukrainy. DSTU 2865-94 (State Standard of Ukraine 2865-94) (1995) Kontrol neruinivnyi. Terminy ta vyznachennia (Nondestructive testing. Terms and definition). Derzhstandart of Ukraine (Publishing House of State Standard of Ukraine), Kyiv
Dhar A, Jadandish C, Atherton DI (1992) Using the Barkhausen effect to determine the easy axis if magnetization in steels. Mater Eval 10:1139–1141
Durin G (1997) The Barkhausen effect: new perspectives for an old problems. In: Claeys C, Simoe E (eds) Noise in physical systems and 1/f fluctuations. World Scientific, Singapore, pp 577–582
Durin G, Zapperi S (2000) Scaling exponents for Barkhausen avalanches in polycrystalline and amorphous ferromagnets. Phys Rev Lett 84:4075–4078
Durin G, Zapperi S (2002) On the power spectrum of magnetization noise. J Magn Magn Mater 242–245:1085–1088
Durin G, Zapperi S (2006) The Barkhausen effect. Physics 13:461–479
Durin G, Magni A, Bertotti G (1995a) Fractals, scaling and the question of self-organized criticality in magnetization processes. Fractals 3(2):351
Durin G, Magni A, Bertotti G (1995b) Fractal properties of the Barkhausen effect. J Magn Magn Mater 140–144:1835–1836
Durin G, Magni A, Bertotti G (1996) Measurements of the Barkhausen effect in FeCoB alloys. J Magn Magn Mater 299–301
Elmore W (1938) The magnetic structure of cobalt. Phys Rev J 53:757–764
Fisher DS (1985) Sliding charge-density waves as a dynamic critical phenomenon. Phys Rev B 31(8):1396–1427
Fowler CA, Fryer EM, Trevers D (1960) Observation of domains in iron whiskers under high fields. J Appl Phys 31(12):2267–2272
Fukai Y (1993) The metal-hydrogen system-basic bulk properties. Springer-Verlag
Galaktionova NA (1967) Vodorod v metallakh (Hydrogen in metals). Metallurgizdat Publishing House
Glukhov NA, Kolmogorov VN (1988) Sviaz parametrov akusticheskikh shumov v peremagnichivaiemykh konstruktsionnykh materialakh (Relation between acoustic noise parameters in remagnetizing structural materials). Defektoskopiya (Defectoscopy) 2:26–29
Glukhov NA, Kolmogorov VN, Miletskii BI (1985) Issledovaniie akusticheskikh shumov v peremagnichiveyemykh konstruktsionnykh staliakh (Study of acoustic noise in remagnetizing structural materials). Defektoskopiya (Defectoscopy) 9:36–40
Gonzalez J, Chubykalo O, Gonzalez J (1997) Quantitative analysis of the collective behavior in a micromagnetic model. Phys Rev B 55(3):921–930
Gorkunov ES, Barteniev OA, Khamitov VA (1986a) Magnitouprugaia akusticheskaia emissiia v monokristallakh kremnistogo zheleza (Magnetoelastic acoustic emission in silicon iron single crystals). Izv Vuzov Ser Fiz (Proc Univ Ser Phys) 62–66
Gorkunov ES, Somova VM, Buldakova NB (1986b) Ustoichicost sostoyaniia ostatochnoi namagnichennosti razlichno termicheski obrabotannykh stalei k vozdeistviyu postoyannykh razmnagnichivayushchikh polei (Resistance of the state of residual magnetization of differently heat-treated steels to the effects of constant demagnetizing fields). Defektoskopiya (Defectoscopy) 9:23–31

Gorkunov ES, Khamitov VA, Barteniev OA (1987) Magnitouprugaia akusticheskaia emissiia v termicheski obrabotannykh konstruktsionnykh stalyakh (Magnetoelastic acoustic emission in heat-treated structural steels). Defektoskopiya (Defectoscopy) 3:3–9

Gorkunov ES, Deagoshansky YN, Rodionova SS (1998) Vliyaniie struktury stalei na protsessy peremagnichivaniia v slabykh i silnykh magnitnykh poliakh i resheniia zadachi magnitnoi struktoroskopii izdelii iz etikh stalei (Influence of steel structure on magnetization remagnetization processes in weak and strong magnetic fields and solution of the problem of magnetic structuroscopy of products made of these steels). Defektoskopiya (Defectoscopy) 6:42–59

Gorkunov ES, Drogoshanskyi YuN, Mikhovskii M (1999) Effekt Barkhausena i yego ispolzovaniie v strukturoskopii ferromagnitnykh materiallov. Obzory 1–3 (Barkhausen effect and its application in structuroscopy of ferromagnetic materials. Overviews 1–3). Defektoskopiya (Defectoscopy) 6(3–23), 7(3–32), 8(3–25)

Greshnikov VA, Drobot YuB (1976) Akusticheskaya emissiya. Primineniie dlia ispytanii materialov i izdelii (Acoustic emission. Application to testing of materials and products). Publishing House of Standards

Groma I, Bakó B (1998) Probability distribution of internal stresses in parallel straight dislocation systems. Phys Rev B 58(9):2969–2974

Grosse-Nobis W (1977) Frequency spectrum of the Barkhausen noise of a moving 180 domain wall. J Magn Magn Mater 4:247–253

Grosse-Nobis W, Lieneweg U (1978) Spatial extent of the flux due to local irreversible changes of the magnetic polarization in ferromagnetic strips and rods. J Magn Magn Mater 9:343–348

Grosse-Nobis W, Wagner W (1977) Fluctuations of the driving field in a tape wound core of Metglas 2826. J Magn Magn Mater 6:77–79

Guyot M, Merceron T, Cagan T (1987) Acoustic emission along the hysteresis loops of various ferro- and ferrimagnets. J Appl Phys 63(8):3955–3957

Hirth J, Lothe J (1992) Theory of dislocations. Krieger Publishing Company

Ignatchenko VA, Rodichev AM (1960) O raspredelenii skachkov Barkhausena po velichine (On the distribution of Barkhausen jumps in magnitude). In: Magnitnaya struktura ferromagnetikov (Magnetic structure of ferromagnets). Academy of Sciences of the USSR Publishing House

Jansen K, Grosse-Nobis W, Klebrink H (1982) Propagation properties of irreversible magnetizing processes in amorphous and crystalline material. J Magn Magn Mater 26(1):267–269

Kameda J, Ranjan R (1987) Nondestructive evaluation of steels using acoustic and magnetic Barkhausen signals. I. Effect of carbide precipitation and hardness. Acta Metall 35(7):1515–1526

Kirenskii LV (ed) (1960) Magnitnaia struktura ferromagnetikov: materialy Vsesoyuznogo soveshchaniya, Krasnoyarsk, 10–16 iyunia 1958 (Magnetic structure of ferromagnets: proceedings of all-union meeting, Krasnoyarsk, 10–16 June, 1958). Novosibirsk

Kittel C (1949) Physical theory of ferromagnetic domains. Rev Mod Phys 21:541–583

Kolachev BA (1985) Vodorodnaia khrupkost metallov (Hydrogen embrittlement of metals). Metallurgiya Publishing House

Kolmogorov VN (1972) Nekotoryie issledovaniya akusticheskoi emissii v ferromegnitnykh materialakh (Some studies of acoustic emission in ferromagnetic materials). In: Tezisy doklada Vsesoyuznogo n/t seminara (Abstracts of all-union scientific and technical seminar). Khabarovsk

Kuntz MC, Sethna JP (2000) Noise in disordered systems: the power spectrum and dynamic exponents in avalanche models. Phys Rev B 62:11699–11708

Kuznetsov VG, Somova VM, Bashkirov YuP (1972) Magnitnyie, elektricheskiie i mekhanicheskiie svoistva stalei 45KhN i 45NMFA poslie razlichnykh termicheskikh obrabotok (Magnetic, electrical and mechanical properties of 45ChN and 45ChMFA steels after various heat treatments). Defektoskopiya (Defectoscopy) 5:13–20

Kwan MM, Ono K, Shibata M (1984a) Magnetomechanical acoustic emission of ferromagnetic materials at low magnetization levels (type I behavior). J Acoust Emiss 3(3):144–156

Kwan MM, Ono K, Shibata M (1984b) Magnetomechanical acoustic emission of ferromagnetic materials at high magnetization levels (type II behavior). J Acoust Emiss 3(4):190–203

Lemerle S, Ferre J, Chappert C, Mathet V, Giamarchi T, Doussal PL (1998) Domain wall creep in ultrathin films. Phys Rev Lett 80(6):849–852

Leschhorn H, Nattermann T, Stepanowm S, Tang LH (2001) Driven interface depinning in a disordered medium. Ann Phys 6:1–34

Lieneweg U, Grosse-Nobis W (1972) Distribution of size and duration of Barkhausen pulses and energy spectrum of Barkhausen noise investigated on 81% nickel-iron after heat treatment. Int J Magn 3:11–16

Lomayev GV (1977) Metod magnitnykh shumov v nerazrushayushchem kontrole ferromagnetikov (Method of magnetic noise in nondestructive testing of ferromagnets). Defektoskopiya (Defectoscopy) 4:75–94

Lomayev GV, Komarov VA, Rubtsov BI (1981) Eksperimentalnoie issledovaniie akusticheskogo proyavleniia effekta Barkhausena v konstruktsionnykh staliakh (Experimental study of the acoustic manifestation of the Barkhausen effect in structural steels). Kalinin

Lomayev GV, Malyshev VS, Degtiariov AP (1984) Obzor primeneniya metoda Barkhousena v nerazrushayushchem kontrole (Overview of the application of the Barkhausen method in nondestructive testing). Defektoskopiya (Defectoscopy) 3:54–71

Mazzetti P (1962) Study of non-independent random pulse trains with application to the Barkhausen noise. Nuovo Cim 25(9):1322–1342

Mazzetti P, Montalenti G (1965) The theory of the power spectrum of the Barkhausen noise (1964). In: Proceedings of the international conference on magnetism, Nottingham, London, pp 701–706

McClure JC, Schoeder K (1976) The magnetic Barkhausen effect. Crit Rev Solid State Mater Sci 6:45–83

McClure JC, Battacharya S Jr, Schreder K (1974) Correlation of Barkhausen effect type measurements with acoustic emission in fatigue crack growth studies. IEEE Trans Magn 10(3):913–915

McMichael RD, Swartzendruber LJ, Bennet LH (1993) Langevin approach to hysteresis and Barkhausen noise jump modelling in steel. J Appl Phys 73(12):5848–5850

Mehta A, Mills A, Dahmen K, Sethna J (2002) Universal pulse shape scaling function and exponents: a critical test for avalanche models applied to Barkhausen noise. Phys Rev E 65:136–139

Meisel LV, Cote PJ (1992) Power laws, flicker noise, and the Barkhausen effect. Phys Rev B 46:822–828

Migirenko GS, Moskvin VN, Nechai EP (1973) Primineniie metoda magnitnykh shumov dlia issledovaniia navodorozhyvaniia stali. Metody opredeleniia i issledovaniia sostoiania gazov v metallakh (Applications of the magnetic noise method for the study of hydrogenation of steel. Methods of determination and study of the state of gases in metals). Part I. Publishing House of Moscow House of Scientific and Technical Propaganda

Mikhailov SN, Shcherbinin VE (1992) Fizicheskiie osnovy magnetograficheskoi defektoskopii (Physical basis of magnetographic flaw detection). Nauka Publishing House

Mikheyev MN, Gorkunov ES. Sviaz magnitnykh svoistv so strukturnym sostoyaniiem veshchestva—fizicheskaia osnova magnitnogo strukturnogo analiza (obzor) (Relationship of magnetic properties with the structural state of matter—the physical basis for magnetic structural analysis (review)). Defektoskopiya (Defectoscopy) 8:5–22

Mikheyev MN, Gorkunov ES (1981) O vozmozhnykh prichinakh razlichiia protsessov peremagnichivaniia v slabykh i srednikh magnitnykh poliakh termicheski obrabotannykh konstruktsionnykh stalei (On the possible reasons for the difference in remagnetization processes in weak and medium magnetic fields of thermally treated structural steels). Fiz-Khim Mekh Mater (Physicochem Mech Mater) 51(4):749–755

Mikheyev MN, Gorkunov ES (1985) Magnitnyie metody nerazrushayushchego kontrolia strukturnogo sostoyaniia i prochnostnykh kharakteristik termicheski obrabotannykh izdelii (Magnetic methods for non-destructive testing of the structural state and strength characteristics of heat-treated products). Defektoskopiya (Defectoscopy) 3:2–21

Mikheyev MN, Gorkunov ES, Dinayev FN (1977) Nerazrushayushchii magnitnyi kontrol zakalionnykh i otpushchennykh izdelii iz nizkolegirovannykh konstruktsionnykh i prostykh uglerodistykh stalei (Non-destructive magnetic testing of hardened and tempered products from low-alloy structural and simple carbon steels). Defektoskopiya (Defectoscopy) 6:7–12

Mikheyev MN, Bida GV, Tsarkova TN, Kostin BN (1981) Issledovaniie rezhymov peremagnichivaniia pri kontrole kachestva zakalionnykh i otpushchennykh izdelii po velichinie magnitnoi induktsii (Investigation of remagnetization modes during quality control of quenched and tempered products by the value of residual magnetic induction). Defektoskopiya (Defectoscopy) 8:69–79

Mikheyev MN, Gorkunov ES, Somova VM, Kutiin AB (1982) Vzaimosviaz magnitnykh i mekhanicheskikh svoistv so strukturnym sostoyaniyem zakalionnykh i otpushchennykh izdelii (Interrelation of magnetic and mechanical properties with the structural state of quenched and tempered products). Defektoskopiya (Defectoscopy) 9:66–75

Moskvin VN, Leshchenko AI (1983) Elektromagnitnyi strukturoskop, osnovan na effektie Barkhausena (Electromagnetic structuroscope based on the Barkhausen effect). Defektoskopiya (Defectoscopy) 3:35–40

Narayan O (1996) Self-similar Barkhausen noise in magnetic domain wall motion. Phys Rev Lett 77:3855–3858

Narayan O, Fisher DS (1993) Threshold critical dynamics of driven interfaces in random media. Phys Rev B 48:7030–7042

Natsyk VD, Nechyporenko IN (1984) Akusticheskoie izlucheniie zarodyshei pri fazovykh perekhodakh v magnetikakh. Akustichekaia emissiia materialov i konstruktsii (Acoustic radiation of nuclei during phase transitions in magnets. Acoustic emission of materials and structures), Part 1. Rostov-on-Don

Nattermann T, Stepanow S, Tang LH, Leschhorn H (1992) Dynamics of interface depinning in a disordered medium. J Phys 2:1483–1488

Nazarchuk Z, Skalskyi V, Serhiyenko O (2017) Acoustic emission. Methodology and application. Springer International Publishing AG

Nechai EP, Moskvin VN (1975) Ob ispolzovanii effekta Barkhausena dla kontrolia stepeni navodorazhyvaniia metallov (On the use of the Barkhausen effect to control the degree of hydrogenation of metals). Tezisy dokladov II Vsesoyuznoi mezhvuzovskoi konferentsii po el./mag. metodam kontrolia kachestva materialov i izdelii (Abstracts II all-union interuniversity conference by email/mag. methods of quality control mater. and products). Part II, Riga

Néel L (1942) Theorie des lois d'aimantation de Lord Rayleigh. I: le deplacements dun paroi isolee. Can J Phys 12:1–20

Néel L (1943) Theorie des lois d'aimantation de Lord Rayleigh. II: multiples domaines et champ coercitif. Can J Phys 13:18–30

Néel L (1946) Bases d'une nouvelle theorie generale du champ coercitif. Ann Univ Grenoble 22:299–343

O'Brien K, Weissman MB (1994) Statistical characterization of Barkhausen noise. Phys Rev E 50(8):3446–3452

Ono K, Shibata M (1980) Magnetomechanical acoustic emission of iron and steels. Mater Eval 38:55–61

Panasyuk VV (ed) (1988) Mechanika ruinuvannia i mitsnist materialiv. T. 1 (Fracture mechanics and strength of materials. Vol. 1). Naukova Dumka Publishing House

Pao Y-H, Yeh C-S (1973) A linear theory for soft ferromagnetic elastic solids. Int J Eng Sci 11:415–416

Perkovic O, Dahmen K, Sethna JP (1995) Avalanches, Barkhausen noise, and plain criticality. Phys Rev Lett 75(10):4528–4531

Petta JR, Weissman MB, O'Brien K (1996) Multiple magnetization paths in Barkhausen noise. Phys Rev E 54(5):1029–1031

Petta JR, Weissman MB, Durin G (1998a) Barkhausen pulse structure in an amorphous ferromagnet: characterization by high order spectra. Phys Rev E 57(6):6363–6369

Petta JR, Weissman MB, Dorin G (1998b) Determination of Barkhausen signal scaling from high order spectral analysis. IEEE Trans Magn 34:1171–1173

Porteseil JL, Vergne R (1979) Irreversible processes in Bloch wall motion. J Appl Phys 50(B3):2131–2133

Ranjan R, Jiles DC, Rastogi PK (1986) Magnetoacoustic emission, magnetization and Barkhausen effect in decarburized steels. IEEE Trans Magn 22(5):511–513

Ranjan R, Buck O, Thompson RB (1987a) A study on the effect of dislocation on the magnetic properties of nickel using magnetic NDT methods. J Appl Phys 61(8):3196–3198

Ranjan R, Jiles DC, Buck O, Thompson RB (1987b) Grain size measurement using magnetic and acoustic Barkhausen noise. J Appl Phys 61(8):3199–3201

Rautioaho R, Karialainen P, Moillanen M (1986) Coercivity and power spectrum of Barkhausen noise in structural steels. J Magn Magn Mater 61:183–192

Roman J, Maharshak S, Amir J (1983) Magnetomechanical acoustic emission: a non-destructive characterization technique of precipitation hardness steels. J Acoust Emiss 2(1/2):64–66

Rudiak VM (1970) Effekt Barkhausena (Barkhausen effect). Usp Fiz Nauk (Adv Phys Sci) 111(3):429–462

Sánchez RL, Pumarega MIL, Armeite M (2004) Barkhausen effect and acoustic emission in a metallic glass—preliminary results. Rev Quant Nondestruct Eval 23:1328–1335

Shibata M, Ono K (1981) Magnetomechanical acoustic emission—a new method of nondestructive stress measurement. NDT Int 14(5):227–234

Shibata M, Kobajashi E, Ono K (1986) The detection of longitudinal rail force via magnetomechanical acoustic emission. J Acoust Emiss 4(4):93–100

Shindo Y (1978) Magnetoelastic interaction of a soft ferromagnetic elastic solid with a penny-shaped crack in a constant axial magnetic field. Trans ASME J Appl Mech 45:291–296

Spasojevic D, Bukvic S, Milosevic S, Stanley HE (1996) Barkhausen noise: elementary signals, power laws, and scaling relations. Phys Rev E 77(9):2531

Tebble RS, Newhouse VL (1953) The Barkhausen effect in single crystals. Proc Phys Soc Lond 66(8):633–641

Tsarkova TP, Bida GV, Mikheyev MN, Gorkunov ES (1981) O magnitnom metode kontrolia kachestva vysokotemperaturnogo otpuska konstruktsionnykh i prostykh stalei (On the magnetic method for quality control of high-temperature tempering of structural and simple carbon steels). Defektoskopiya (Defectoscopy) 3:14–17

Turner PA, Stockbridge CD, Theuerer HC (1969) Magnetic domain nucleation and propagation in fine wires. J Appl Phys 40(4):1864–1869

Urbach JS, Madison RC, Markert JT (1995a) Interface depinning, self-organized criticality, and the Barkhausen effect. Phys Rev Lett 75(2):276–279

Urbach JS, Madison RC, Markert JT (1995b) Reproducible noise in a macroscopic system: magnetic avalanches in Perminvar. Phys Rev Lett 75(10):4694–4697

Vergne R, Cotillard JC, Porteseil JL (1981) Quelques aspects statistiques des processus d'aimaintation dans les corps ferromagnétiques. Cas du déplacement d'une seule paroi de Bloch à 180∘ dans un millieu monocristallin aléatoirement perturbé. Rev Phys Appl 16(2):449–476

Volkov VV, Kumeishyn VF, Cherniakhovskyi MY, Lovastov YuP, Lepskyi SM (1987) Ob akusticheskoi emissii peremagnichivayemykh ferromagnetikov (On acoustic emission of remagnetized ferromagnets). Defektoskopiya (Defectoscopy) 4:21–28

Vonsovskyi SB (ed) (1959) Magnitnaia struktura ferromagnetikov (Magnetic structure of ferromagnets). Moscow

Vonsovskyi SV, Shur YS (1948) Ferromagnetizm (Ferromagnetism). Publishing House of Technical & Theoretical Literature

White RY, Dahmen KA (2003) Driving rate effects on crackling noise. Phys Rev Lett 91(12):5702

Wiegman NJ, Stege R (1978) Barkhausen effect in magnetic thin films: general behavior and stationarity along the hysteresis loop. Appl Phys 16:167–174

Williams HJ, Shockley W (1949) A simple domain structure in an iron crystal showing a direct correlation with the magnetization. Phys Rev J 75:178–183

Wyatt R (1985) Kvantovaia teoriia magnetizma (Quantum theory of magnetism). Mir Publishing House

Yamada K, Saitoh T (1992) Observation of the Barkhausen effect in ferromagnetic amorphous ribbons by sensitive pulsed magnetometer. J Magn Magn Mater 104–107:341–342

Yudin AA, Lopatin MV (1987) K teorii magnitnoi akusticheskoi emissii (To the theory of magnetic acoustic emission). In: Dep. № 3158–B87. VINITI (All-union institute of scientific and technical information). Moscow

Zani M, Puppin E (2003) Negative Barkhausen jumps in amorphous ribbons of $Fe_{63}B_{14}Si_8Ni_{15}$. J Appl Phys 92(12):5901–5904

Zapperi S, Cizeau P, Durin G, Stanley HE (1998) Dynamics of a ferromagnetic domain wall: avalanches, depinning transition and the Barkhausen effect. Phys Rev B 58(12):6353–6366

Zheng G-P, Li M, Zhang J (2002) Dynamics of Barkhausen jumps in disordered ferromagnets. J Appl Phys 92(4):883–888

Chapter 3
Models of MAE and Interaction of Magnetic Field with Cracks

As shown in the previous chapter, the emission of elastic waves caused by the Barkhausen effect, i.e., magnetoelastic acoustic emission, in a ferromagnetism occurs in local areas of the body and is caused by jump-like movements of domain walls under the influence of the external magnetic field. A study of this effect was conducted by the authors (Sánchez et al. 2004; Buttle et al. 1987a; Turner et al. 1969; Skal's'kyi et al. 2009) and developed in Buttle et al. (1987a, b), Kolmogorov (1972), Kameda and Ranjan (1987a), Kwan et al. (1984a, b), Yudin and Lopatin (1987), Natsik and Nechyporenko (1984), Glukhov and Kolmogorov (1988), McClure et al. (1974), Ono and Shibata (1980), Roman et al. (1983), Gorkunov et al. (1986a, b, 1987), Lomayev et al. (1981), Bezymiannyi (1981), Glukhov et al. (1985), Ranjan et al. (1986, 1987a, b), Shibata et al. (1986), Guyot et al. (1987), Brechko et al. (2004) and Krinchik (1960). However, the obtained theoretical and experimental results do not include models of the interaction of the external magnetic field with the cracks presented in the ferromagnet. Therefore, they do not make it possible to relate material damage with such micro- or macrodefects to MAE parameters. The research presented in this chapter is aimed at a partial solution to this problem.

3.1 Quantitative Evaluation of a Single Volume Jump of 90° Domain Wall

Modeling of signals caused by domain wall jumps. To estimate the components of the displacement vector and hence the magnetoelastic acoustic emission signals caused by domain wall jumps, assume that a certain body volume, namely the one where the jumps occur, is a volume source of transformational (without external mechanical stresses) deformations caused by magnetostriction (Aki and Richards 1983).

V. Skalskyi and Z. Nazarchuk, *Magnetoelastic Acoustic Emission*, Springer-AAS Acoustics Series, https://doi.org/10.1007/978-981-99-4032-5_3

The elastic displacements in the body caused by the sudden appearance of such a volume source can be estimated using the approach developed by Eshelby (1957). First, we need to imagine separating the material of the source area by cutting it out on the surface Σ, which covers this volume. At this stage, assume that the material "extracted" from the body retains its original shape due to the stresses that on the surface Σ have the same values as the stresses that act on the surface Σ from the rest of the body before cutting. Next, assume that the cut out material (inclusion) undergoes transformational deformation of the value of $\Delta\varepsilon_{rs}$. This deformation occurs without changing the stresses inside the inclusion. It is the deformation that characterizes the MAE source.

In the next step, we apply additional surface stresses that will restore the volume of inclusion to the original value. To do this, the stress, $-\Delta\tau_{pq} = -c_{pqrs}\Delta\varepsilon_{rs}$, must be added throughout the inclusion and additional stresses $-c_{pqrs}\Delta\varepsilon_{rs}v_q$ to its surface Σ, where c_{pqrs} are elastic constants and v_q is the external normal to Σ. Since $\Delta\tau_{pq}$ is a static field, then according to the equilibrium equation we have $\Delta\tau_{pq,q} = 0$. The stresses in the body are considered to be constant because they retain the original shape due to the stresses applied to the inner surface Σ, which have the same values as the stresses created in the body by the inclusion before it was cut out.

Next, put the inclusion in the cut out (which has the same shape) and glue the material along the cut. The stresses on Σ^- in this case will be greater by the value $-c_{pqrs}\Delta\varepsilon_{rs}v_q$ than on Σ^+. This creates a jump of stresses $c_{pqrs}\Delta\varepsilon_{rs}v_q$ in the direction to the external normal. These stresses act on the inclusion to preserve its shape. They are caused by the surface forces that are external with respect to the volume source.

Then, remove the surface forces applied to Σ. Since the stresses must be continuous on Σ, it is necessary to add their imaginary jump of the value $-c_{pqrs}\Delta\varepsilon_{rs}v_q$. In this case, the elastic field created in the external environment by the above sequence of steps will be the same as the field caused by the imaginary stress jump $[T_{\mathrm{p}}] = -c_{pqrs}\Delta\varepsilon_{rs}v_q$ on the surface Σ that surrounds the volume source.

The presented approach can be generalized to the dynamic problem of excitation of magnetoelastic acoustic emission, since for an arbitrary moment of time the transformation deformation $\Delta\varepsilon_{rs}$ can be determined for undeformed material, and $\Delta\tau_{pq,q} = 0$. Finally, for the components of the displacement vector, using Betti theorem, we can obtain the dependence (Aki and Richards 1983)

$$u_n(\vec{x}, t) = \int_{-\infty}^{\infty} \mathrm{d}\tau \iint_{\Sigma} c_{pqrs}\Delta\varepsilon_{rs}v_q G_{np}(\vec{x}, t-\tau; \xi, 0)\mathrm{d}\Sigma(\xi), \tag{3.1}$$

where $G_{np}(\vec{x}, t-\tau; \xi, 0)$ is the Green's function.

If the subintegral expression and also its ξ derivatives are continuous, then applying the Ostrogradsky–Gauss theorem, we obtain

$$u_n(\vec{x}, t) = \int\limits_{-\infty}^{\infty} \mathrm{d}\tau \iiint\limits_V \frac{\partial}{\partial \xi}\left\{c_{pqrs}\varepsilon_{rs}G_{np}(\vec{x}, t-\tau; \xi, 0)\right\}\mathrm{d}V(\xi). \tag{3.2}$$

Here V is the volume of inclusion, i.e., the area occupied by the MAE emission source.

Using the dependence $\partial\left(c_{pqrs}\Delta\varepsilon_{rs}\right)/\partial\xi_q = \Delta\tau_{pq,q} = 0$, relation (3.2) can be rewritten as follows:

$$u_n(\vec{x}, t) = \iiint\limits_V c_{pqrs}\varepsilon_{rs} * \frac{\partial G_{np}}{\partial \xi_q}\mathrm{d}V. \tag{3.3}$$

Here symbol $*$ means time convolution.

Denote $c_{pqrs}\Delta\varepsilon_{rs} = \frac{\mathrm{d}M}{\mathrm{d}V}$, where $\mathrm{d}M/\mathrm{d}V$ is tensor of the seismic moment density (Aki and Richards 1983). Then for the components of the displacement vector we find

$$u_n(\vec{x}, t) = \iiint\limits_V \frac{\mathrm{d}M_{pq}}{\mathrm{d}V} * \frac{\partial G_{np}}{\partial \xi_q}\mathrm{d}V. \tag{3.4}$$

In the case of a long-wave approximation, the volume V can be considered to be a "system of force pairs" with a seismic moment tensor equal to the integral of the seismic moment density tensor concentrated at a point (for example, in the center V), i.e.,

$$M_{pq} = \iiint\limits_V c_{pqrs}\varepsilon_{rs}\mathrm{d}V. \tag{3.5}$$

For example, if there is a transformational increase in the volume of a spherical region of radius a, the components of the strain tensor will be $\varepsilon_{11} = \varepsilon_{22} = \varepsilon_{33} = \Delta\theta/3$, where $\Delta\theta$ is the relative change in the volume, and the other components of the strain tensor are zero. For a homogeneous isotropic body based on (3.5) we have $M_{ij} = 4/3\pi a^3\Delta_p\delta_{ij}$, where $\Delta_p = [\lambda + 2/3\mu]\Delta\theta$ (λ and μ are Lamé constants). Thus, a spherical source with a transformational increase in volume is equivalent to three mutually perpendicular dipoles.

In Kishi (1985), in the same long-wave approximation the dependence was obtained for the components of the Green's function

$$G_{np,q} = \frac{\alpha_n\alpha_p\alpha_q}{4\pi\rho c_1^3 r}\delta'\left(t - \frac{r}{c_1}\right), \tag{3.6}$$

where α_i are the directing cosines, ρ is the density of the environment, $\delta(\cdot)$ is the Dirac delta function, r is the distance from the source to the observation point, and c_1 is the velocity of the longitudinal wave.

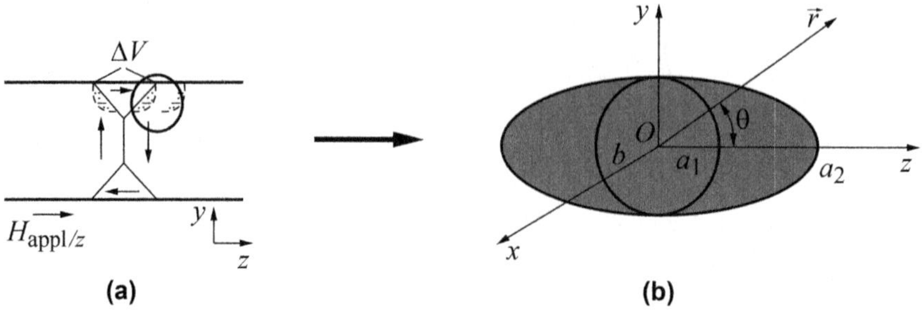

Fig. 3.1 Scheme of MAE emission area (**a**) and volume emission source (**b**)

At the next stage of construction of the approximate model of elastic waves emission, caused by the MAE, on the basis of Sánchez et al. (2004) we assume that the main contribution to the acoustic emission signals is done by the location of the domain walls shown at Fig. 3.1a (area surrounded by a circle), i.e., for the case when the magnetic field is applied along the Oz axis (Skal's'kyi et al. 2009).

Suppose that the region, where the domain structure is rearranged and which due to the magnetostrictive effect is a source of acoustic emission, initially has a spheroidal shape with semi-axes a_1 and $b(a_1 > b)$ (Glukhov and Kolmogorov 1988). The magnetostrictive change of the remagnetization region itself is symmetric with respect to the center—point O in Fig. 3.1b. As a result of this change the region will be elongated along the Oz axis, so that the major semi axis of the ellipsoid will be equal to a_2 (see Fig. 3.1b). For the case of such an increase in the spheroidal region of transformational deformations due to the magnetostriction effect, the components of the seismic moment tensor according to Eshelby (1957) and Shibata (1984) will be

$$M_{xx} = M_{yy} = \lambda \Delta V \varepsilon_{zz}; \quad M_{zz} = (\lambda + 2\mu)\Delta V \varepsilon_{zz}, \tag{3.7}$$

and change of its volume $\Delta V = 4\pi b\left(a_2^2 - a_1^2\right)/3$.

Note that according to Rudiak (1986) the component of the strain tensor ε_{zz} is related to the magnetization J by the dependence

$$\varepsilon_{zz} \sim \lambda_s J / J_s, \tag{3.8}$$

where λ_s is a magnetostriction constant, J is the magnetization, and J_s is the saturation magnetization.

For the isotropic case $\lambda_s = (2\lambda_{[100]} + 3\lambda_{[111]})/5$, where $\lambda_{[ijk]}$ are magnetostriction constants along the corresponding crystallographic axes. For polycrystalline iron, in particular, $\lambda_s = -3.48 \times 10^{-6}$.

Then, on the basis of relations (3.4)–(3.7) to estimate the components u_r of the displacement vector in the polar coordinate system r, θ (angle θ is calculated from the xOz plane, see Fig. 3.1b) we obtain the following dependence:

$$u_r \approx \frac{\lambda + 2\mu\cos^2\theta}{4\pi\rho c_1^3 r}\varepsilon_{zz}\frac{d}{dt}[\Delta V(t - r/c_1)]. \tag{3.9}$$

Component u_r corresponds to the propagation of the longitudinal elastic wave caused by the change in the ferromagnet domain structure due to the Barkhausen effect. The amplitude values A of the magnetoelastic acoustic emission signals can be estimated using dependence (3.9), assuming that they are proportional to the maximum values of the components of the displacement vector. Besides, using dependence (3.8), for A we obtain the following relation:

$$A \sim C\dot{V}\lambda_s\frac{J}{J_s}(\lambda + 2\mu\cos^2\theta)/r, \tag{3.10}$$

where $\dot{V}$ is the time derivative of change of the remagnetization region volume and C is the coefficient of proportionality between mechanical and electric values.

Note that according to Glukhov and Kolmogorov (1988), the remagnetization processes caused by the jump-like movement of the domain walls occur under the condition that the applied magnetic field reaches a certain critical value

$$H_k \approx \frac{\lambda_s\sigma_{int}}{\mu_0 J_s}, \tag{3.11}$$

where μ_0 is the magnetic permeability of the vacuum and σ_{int} is the average value of fluctuations of internal mechanical stresses

Thus, when the magnetic field strength reaches its critical value of H_k, domain wall jumps will occur, and the amplitude values of the signals caused by these jumps can be estimated by dependence (3.10). It follows that the amplitude values of the MAE signal are proportional to the value of the transformation deformations (multiplier $\lambda_s I/I_s$) and to the rate of change $\dot{V}$ of the volume of the remagnetization region. This result was experimentally confirmed in Sánchez et al. (2004), Buttle et al. (1987a) and Shibata and Ono (1981), in which a similar dependence was experimentally established for amplitude values on the basis of the registered MAE signals analysis.

Technique of experimental research and discussion of results. Two representatives of ferromagnetic materials were selected for experimental research, commercially pure (99.7%) nickel and 30 steel (see Appendix). When choosing the type of ferromagnet we considered their different physical structure, the wide presentation of studies of magnetic properties, and the practical application of the results to create appropriate NDT methods. The mechanical properties of the selected materials are given in Table 3.1.

In the experiments lamellar samples with dimensions of $1100 \times 45 \times 0.2$ mm^3 were used. These were magnetized in a solenoid with a diameter of 330 mm with a winding length of 1000 mm, having 1600 coils of copper wire with a diameter of 1.8 mm. The solenoid magnetic field strength distribution was previously calculated, as described in Chap. 1.

Table 3.1 Mechanical properties of ferromagnets

Material	σ_{02} (MPa)	σ_b (MPa)	E (MPa)	δ (%)	ψ (%)
Nickel	180–210	530	196×10^3	36	78
30 steel	340	400–580	204×10^3	21	50

A ferromagnet sample with resistance strain gauges glued to it was placed in diamagnetic guides inside a solenoid in which a magnetic field with a voltage of up to 16.0 kA/m could be formed, causing magnetostriction in the metal. The resistance strain gauges were divided into two groups, in each of which they were connected in series and linked to the measuring amplifier input. Together with the input constant resisters of the amplifier a bridge circuit with two active elements was formed in such a way (Fig. 3.2). The amplifier was powered by bipolar direct current from the power supply unit, and the output was connected to an oscilloscope or recording equipment. The magnetic field of the solenoid was formed by a direct current in the range 0–30 A.

To measure the weak signals of the bridge circuit unbalance against the background of cophased noise with a voltage of U_{cc} a measuring amplifier was developed, the scheme of which is shown in Fig. 3.3. Its characteristic feature is a small "zero drift", which is important when measuring with a bridge circuit.

The voltage was supplied to the input of the measuring amplifier from the output of the bridge resistance strain gauge transducer, which had two variable arms. Therefore, it had a differential input, high rate of gain and input resistance, low zero level at the output, and a large (more than 80 dB) attenuation of the inphase signal K_{ais}. When

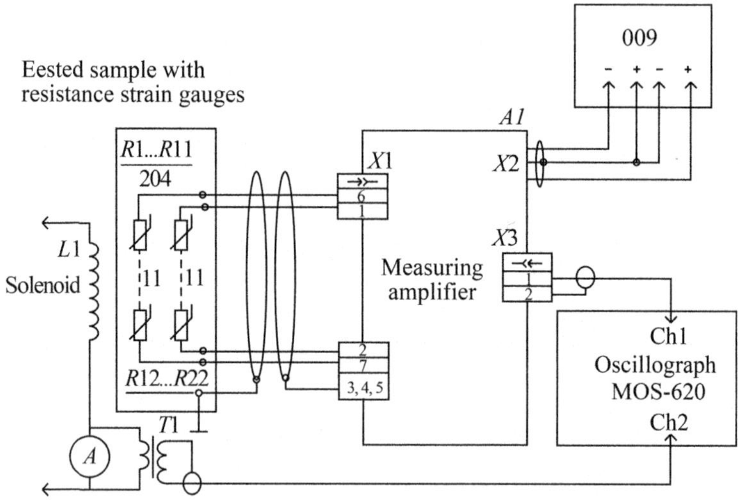

Fig. 3.2 Scheme of an experimental stand for MAE measuring during Barkhausen jumps and magnetostriction in ferromagnets

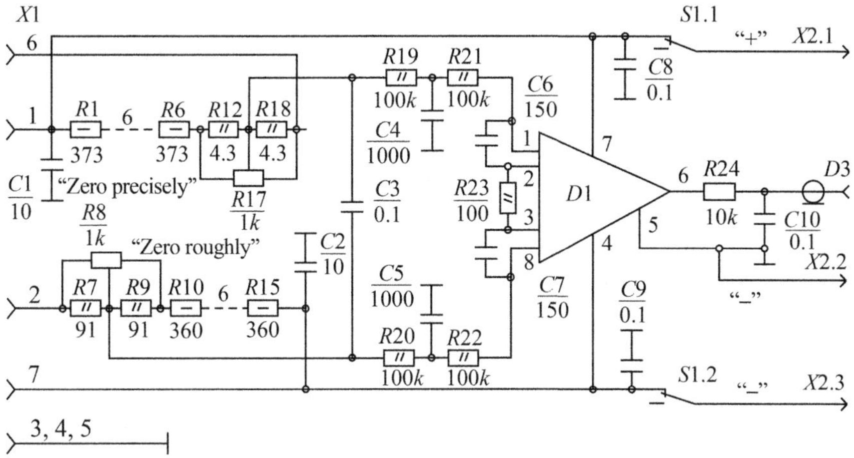

Fig. 3.3 Scheme of the measuring amplifier

all four arms of the bridge circuit are equal in resistance value, the voltage in the measuring diagonal will be $E_1 = E_2 = U_{out}/2$. If due to the magnetostriction there is a change in the resistance of two measuring arms by $\pm\,\Delta R$, then $E_1 \neq E_2$, and the polarity of E_2 relative to E_1 depends on ΔR: $E_1 - E_2 = U_{out}\Delta R/2R$. In our case, each active arm of the bridge circuit was assembled from 10 to 15 serially connected resistance strain gauges with a total resistance $R = 2000$–$3000\ \Omega$.

The voltage at the input of the amplifier during measurements increases to μV units with the absolute elongation of the test sample by 5–10 μm and with changes in resistance in the measuring arms on an average of 0.0245 Ω. The total gain of the measuring amplifier, constructed on the AD625 chip from analog device, can reach 60 dB. Therefore, the voltage U_{out} at its output is tens of mV, which is enough for its reliable registration during elongation or reduction of the sample.

If to take into account that the relative sensitivity of the resistance strain gauge is $\varepsilon = \Delta l \times R(\Delta R \times l)^{-1} = 2.15$, then the absolute magnetostrictive elongation is $\Delta l = U_{out} \times 4.3l/(U_{sup} \times K_u)$. In these dependences K_u is the gain factor of the amplifier, Δl is the change in length (linear magnetostriction), l is the length of the sample, R is the resistance of the bridge measuring circuit, ΔR is the change of resistance, and U_{sup} is the supply voltage of the bridge circuit.

Considering that $U_{sup} = 24$ V, as well as the background noise, we choose $K_u = 495$. Then

$$\Delta l = 0.36195 U_{out}. \tag{3.12}$$

A two-element T-shaped low-frequency RC filter is included to the scheme to increase the signal stability in the conditions of high-frequency noise between the input of the measuring amplifier and the output of the strain-resistive bridge measuring circuit.

The general error of estimation of the magnetostriction value consists of the errors of: transformation of the measuring bridge scheme; amplification of the measuring amplifier; estimation of the signal value by measuring equipment; and methodological error of the above formula for Δl. In total, it did not exceed 10%.

Magnetoelastic acoustic emission caused by the Barkhausen jumps was recorded by a highly sensitive (coefficient of the elastic waves conversion into electrical signals not less than 1.6×10^9 V/m) transducer 3, which had a non-uniformity $\pm$ 7 dB of the conversion factor in the operating frequency band of 0.2–0.6 MHz. The MAE electrical signals were supplied from the converter output to the pre-amplifier (gain factor is 34 dB) and to the device that further amplified and processed them, and later to the two-channel oscilloscope (Fig. 3.3).

The results of magnetostriction measurements for samples made of nickel and 30 steel are shown in Figs. 3.4 and 3.5, respectively.

As can be seen from the figures, the elongation Δl (linear magnetostriction) of the nickel plate was 10.0×10^{-3} mm; the change in volume ΔV was 9.0×10^{-2} mm^3. For the steel sample, these values were $\Delta l = 1.6 \times 10^{-3}$ mm; $\Delta V = 1.44 \times 10^{-2}$ mm^3, respectively.

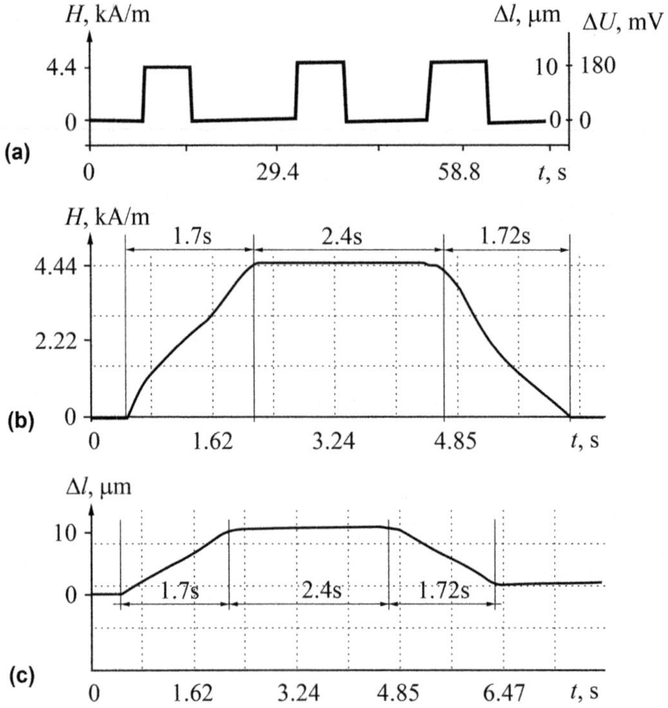

Fig. 3.4 Type of the magnetization pulses to check the consistency of operation of the bridge measuring circuit ("zero drift") when measuring the magnetostriction of the nickel sample (**a**) and the time change of the magnetic field strength of the solenoid (**b**) and linear magnetostriction (**c**)

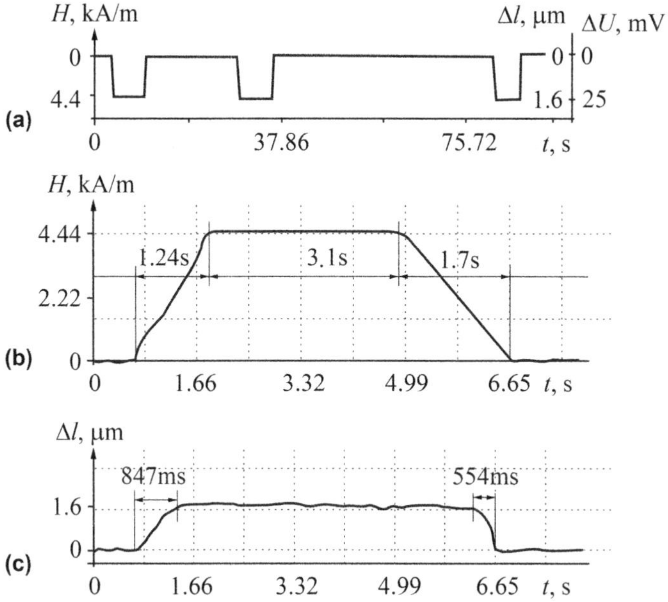

Fig. 3.5 Type of the magnetization pulses to check the consistency of operation of the bridge measuring circuit ("zero drift") when measuring magnetostriction of the 30 steel sample (**a**) and time change of magnetic field strength of solenoid (**b**) and linear magnetostriction (**c**)

Results of registration of the magnetoelastic acoustic emission signals are shown at Fig. 3.6. As it follows from Fig. 3.6 MAE maxima during remagnetization (as a result of Barkhausen jumps) are manifested in the ascending and descending parts of the sinusoid, which corresponds to the steep section of the dynamic hysteresis loop. Their amplitudes increase with increasing the external magnetic field $\vec{H}$ value.

Figure 3.7 shows the MAE signals at points A and B, which correspond to both of these areas at a magnetic field strength of $H_A = 2$ kA/m and $H_B = 1.2$ kA/m. The frequency distribution of these signals is shown in Fig. 3.8.

Thus, the experimental data required for quantitative estimation of the Barkhausen jumps according to MAE parameters were obtained. For the studied ferromagnetic materials the calculations according to the dependence (3.9) give: the maximum

Fig. 3.6 MAE signals during remagnetization of the nickel sample: current in the solenoid is 7.4 A, remagnetization frequency—50 Hz

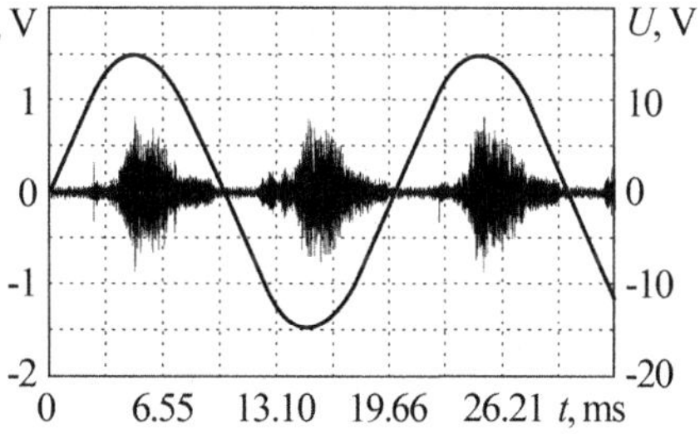

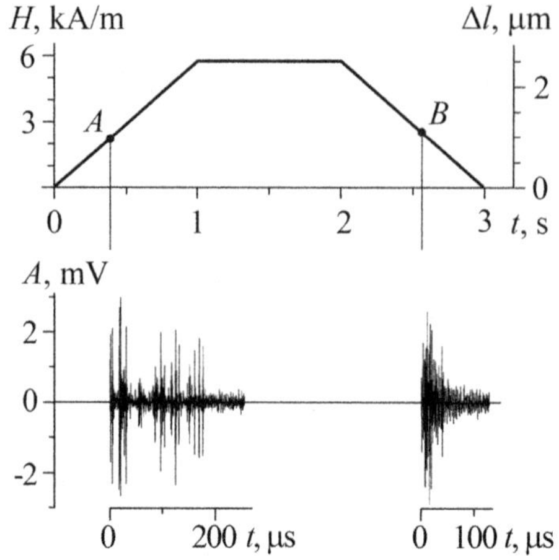

Fig. 3.7 Type of MAE signals in the ascending and descending sections of the magnetization pulse

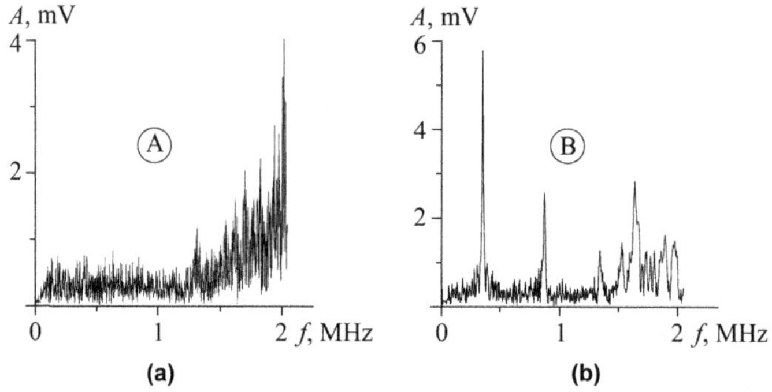

Fig. 3.8 Spectral characteristics of MAE signals at points A (**a**) and B (**b**) of the remagnetization diagram

values of displacements for nickel are $u_r = 1.28 \times 10^{-12}$ m and for 30 steel—$u_r = 3.47 \times 10^{-14}$ m. The corresponding experimental data for nickel are $(1-4) \times 10^{-12}$ m. In the calculations according to formula (3.9) on the basis of the experimental results it was assumed for nickel: $\rho = 8900$ kg/m^3; $E = 210$ GPa; $\upsilon = 0.3$; $\Delta l = 10^{-6}$ m; $\varepsilon_{zz} = 9.1 \times 10^{-7}$; $\dot{V} = 1.0 \times 10^{-3}$ m/s; $r = 0.1 \times 10^{-3}$ m; and for 30 steel: $\rho = 7800$ kg/m^3; $E = 210$ GPa; $\upsilon = 0.28$; $\Delta l = 1.6 \times 10^{-7}$ m; $\varepsilon_{zz} = 1.46 \times 10^{-7}$; $\dot{V} = 1.6 \times 10^{-6}$ m/s^1; and $r = 0.1 \times 10^{-3}$ m.

Thus, on the basis of the conducted researches the model of quantitative estimation of parameters of the Barkhausen jumps according to the MAE parameters was formulated. The experimental tests allowed us to obtain the necessary data to estimate the magnetostriction and amplitudes of the MAE in nickel and steel.

3.2 Magnetoacoustic Diagnostics of Thin-Walled Ferromagnets with Plane Cracks

It is known that defects emit AE signals, in particular crack-like defects, which propagate under the influence of forces applied to the structural element (Nazarchuk and Skalskyi 2009). But if the stress level is not high enough, the defects will not propagate and therefore will not cause emittance, thus remaining undetected by acoustic emission NDT. More efforts should be applied to detect them. In many cases, this is purposeless or even dangerous. For ferromagnetic materials, there is a method of local loading of the structural element with a magnetic field created by an external source. This magnetic field will create a mechanical stresses concentration in the vicinity of the defect, the level of which will be determined by the stress intensity factor (Parton and Morozov 1985; Bagdasarian and Hasanyan 2000; Shindo 1983; Liang et al. 2002; Andreikiv et al. 2008), causing the Barkhausen jumps in the vicinity of such a defect and acoustic signals emission due to it (Shibata 1984; Kameda and Ranjan 1987b; Harris and Pott 1984). Consider the scheme of such method for testing of a thin-walled ferromagnetic structural element.

Considering that the magnetic field in the material disappears quickly with distance, it is possible to model an element of a metal construction with a plane surface, a ferromagnetic half-space with a magnetic dipole being located above its surface (Fig. 3.9) (Andreikiv et al. 2011).

Suppose that the given structural element contains many cracks scattered throughout its volume in such a way that the minimum distance between two adjacent cracks is not less than twice the size of the larger one.

Determining the stress–strain state near any crack on the basis of this assumption it can be considered isolated from the adjacent ones. Choose for diagnostics the most dangerous crack in terms of its orientation, i.e., one whose plane is located perpendicular to the force lines created by the magnetic field dipole (Fig. 3.9). Assuming that the crack is at a sufficient distance from the surface of the half-space and is close to the circular one, it is possible to determine the SIF from the solution of the analog of the Sak's problem under the action of an external magnetic field.

The expression for SIF in this case is known from the literature and is written as (Brechko et al. 2004; Aki and Richards 1983)

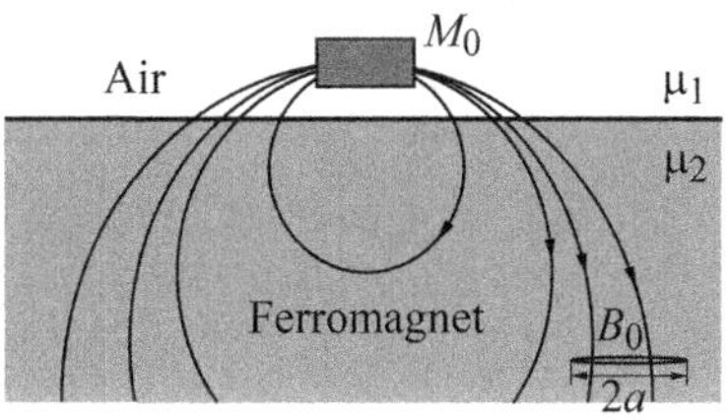

Fig. 3.9 Scheme of magnetoacoustic diagnostics of a ferromagnetic plate

$$K_{\mathrm{I}} = -\frac{\chi \cdot (\chi - 2) \cdot b_{\mathrm{c}}^2 \cdot \left[(1+\chi)^2 + [2 \cdot (1-\nu) + (5-6\nu) \cdot \chi] \cdot \chi \cdot b_{\mathrm{c}}^2\right]}{2 \cdot (1+\chi)^2 \cdot \left[2 \cdot (1+\chi)^2 - \chi^2 \cdot b_{\mathrm{c}}^2 \cdot [2\nu - 1 + 2 \cdot (1-\nu) \cdot \chi]\right]} \cdot G \cdot \sqrt{a}, \tag{3.13}$$

where $b_{\mathrm{c}}^2 = \frac{B_0^2}{G_{\mu_0}}$, B_0 is the absolute value of the magnetic induction vector of the applied field; ν is the Poisson's ratio; G is the shear modulus; $\mu_0 = 4\pi \times 10^{-7}$ (N/A^2) is magnetic constant; χ is magnetic permeability of the environment; and a is the crack radius.

Thus, to estimate the SIF value, it is necessary to determine the value of B_0 as a function of the spatial coordinates x, y, z in the ferromagnetic region. To do this, consider the problem of determining the magnetic field in a magnetoconducting half-space. There is a magnetic dipole with a magnetic moment M_0 at a certain height h above the boundary of a half-space (Fig. 3.10). To find the distribution of induction B_0 in a ferromagnetic half-space, it is necessary to solve the corresponding problem of magnetostatics. In Shaposhnikov (1980), an effective approach is proposed to solve this problem, on the basis of which the following relation for the magnetic potential in a ferromagnetic half-space is obtained:

$$\varphi_2(x, y, z) = \frac{2\mu_1}{\mu_1 + \mu_2} \frac{M_0 \cos\theta}{x^2 + (y-h)^2 + z^2}. \tag{3.14}$$

Here θ is the angle between the vector of the magnetic moment of the dipole and the radius vector of the observation point A (Fig. 3.10).

Then, according to the known dependence of magnetostatics for the magnetic induction vector B_0 we have

$$\vec{B}_0 = -\mu_i \cdot \operatorname{grad} \varphi_i, \tag{3.15}$$

where i is the number of the subregion ($i = 1$ is air, and $i = 2$ is ferromagnet).

Therefore, based on relations (3.14) and (3.15), the expression for the projection of the vector $\vec{B}_0$ on the Oy axis at an arbitrary point A with coordinates (x, y, z) of the ferromagnetic region (Fig. 3.10) will have the form

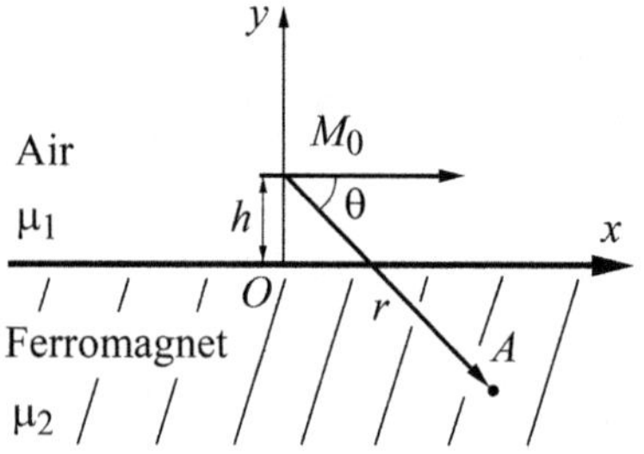

Fig. 3.10 Draft of a magnetic scanner and coordinate systems

$$B_y(x, y, z) = 6M_0 \frac{\mu_1 \mu_2}{\mu_1 + \mu_2} \frac{x(y - h)}{\left[x^2 + (y - h)^2 + z^2\right]^{5/2}}. \tag{3.16}$$

Substituting the expression for B_y to the relation for SIF (3.13), we obtain the dependence of the stress intensity factor K_I as a function of the spatial coordinates of the disk-shaped crack located in the ferromagnetic body.

The proposed approach was tested on the example of calculation of SIF arising in a ferromagnetic plate under the magnetic field action generated by an electromagnetic scanner placed above it. Because of the small size of the scanner, it can be modeled with a magnetic dipole with a certain value of the magnetic dipole moment M_0.

Without reducing the generality, it was assumed that the defect (a disk-shaped crack) is in the plane $z = 0$ (Fig. 3.10). Then in the polar coordinate system $Or\theta$ (Fig. 3.10) expression (3.16) takes the form

$$B_y(r, \theta) = -3M_0 \frac{\mu_1 \mu_2}{\mu_1 + \mu_2} \frac{\sin 2\theta}{r^3}. \tag{3.17}$$

Investigating function (3.17) for the extremum points, it is found that the value of B_y runs into maximum values in the directions $\theta_{\max} = \arccos(1/\sqrt{5})$ and $\theta_{\max} = \pi - \arccos(1/\sqrt{5})$ (about 63.4° and 116.6°, respectively, Fig. 3.11).

Figure 3.12 shows plots of the dependence of SIF K_I of a disk-shaped crack ($a = 5 \times 10^{-4}$ m) on the depth d of its location in the ferromagnetic plate for different values of angle θ.

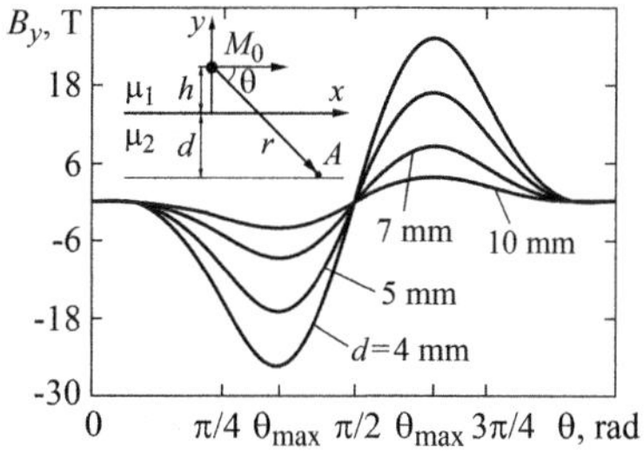

Fig. 3.11 Distribution of the B_y value on lines parallel to the surface of the ferromagnetic half-plane, for different depth d of their location

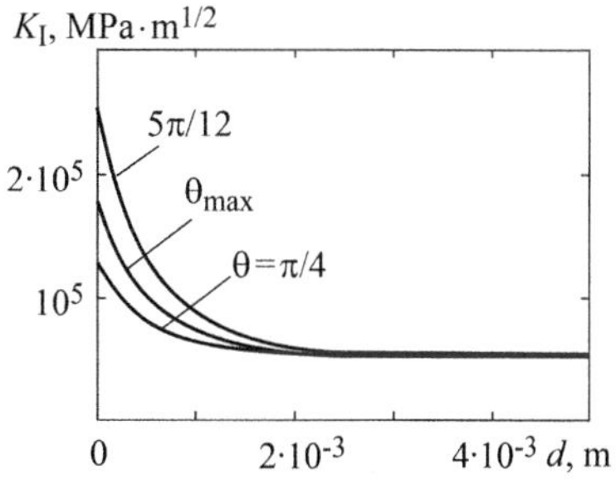

Fig. 3.12 Change the value of SIF K_I of a disk-shaped crack with its depth d for different angles θ

To construct these dependences it is assumed: $M_0 = 4.0$ A m^2; $\chi = 104$; $\mu_1 = \mu_0$; $\mu_2 = \mu_1(\chi + 1)$; $G = 62.7 \times 10^9$ Pa; $\nu = 0.3$; and $h = 3 \times 10^{-3}$ m. As can be seen from Fig. 3.12 the K_{I} function reaches the maximum values in the direction of the largest values of magnetic field B_y induction.

Based on the above presented, we see that the applied magnetic field in the ferromagnetic material will create a concentration of mechanical stresses at the crack tip, the intensity coefficient of which can be determined by dependences (3.13) and (3.16). Since the speed of counting the signals of magnetoelastic acoustic emission is related to the value of SIF, according to the scheme of Fig. 3.9, a new approach of non-destructive testing is proposed, which is as follows. The AE signals caused by a magnetic field from an external source are recorded using the transducers located on the surface of a ferromagnetic body with crack-like defects. The concentration of the field of the corresponding mechanical stresses is the maximum in the vicinity of the defect, contributing to sudden irreversible changes in the domain structure in this area. Having determined the value of SIF near the defect by the MAE signals, it is possible to estimate its size by relation (3.13), and accordingly to assess the risk of fracture of the entire structural element with a crack.

3.3 Estimation of the Stress Intensity Factors, Caused by the Magnetic Field Near Cracks in Ferromagnets

When investigating the boundary equilibrium state of cracked bodies we consider, as a rule, the internal stress–strain state caused by external mechanical loading. However, in some materials, the internal stress–strain state can be created not only by external mechanical forces, but also by an external magnetic field. Such materials, as already mentioned, include ferromagnets.

A number of works have been devoted to the study of the stress–strain state in the vicinity of cracks in ferromagnets, in particular (Nazarchuk and Skalskyi 2009; Parton and Morozov 1985; Bagdasarian and Hasanyan 2000; Shindo 1983; Liang et al. 2002; Andreykiv and Lysak 1989), where the authors managed to obtain solutions to some boundary-value problems of magnetoelasticity for bodies with cracks of simple geometry. For bodies of more complex configuration, what is more common in practice, solving the corresponding problems of magnetoelasticity meets significant mathematical difficulties.

To solve this problem, a method of equivalent congruences (comparisons) is proposed, which allows us to determine approximately, but with sufficient accuracy, the SIF of a body with a crack of arbitrary configuration under an external magnetic field (magnetoelastic problem), if for such a configuration the SIF is known, caused by external mechanical load only (elastic problem) (Andreikiv et al. 2008).

Formulation of the method. Consider a ferromagnet with a crack of some complex configuration b_1 under the action of an external magnetic field with a magnetic induction vector $\vec{B}_0$ directed perpendicular to the crack plane (Fig. 3.13). The problem is to determine the SIF for this case.

For bodies with cracks loaded by external forces, the following representation is often used (Andreykiv 1982; Panasyuk 1988a, b; Shindo 1978):

$$K_{\rm I} = \sigma_{\rm nom}\alpha, \tag{3.18}$$

where $\sigma_{\rm nom}$ is the force part of the SIF, which has the stress units (MPa); α is the geometric part of the SIF, which has units ($\sqrt{\rm m}$).

From the analysis of elastic problems solutions for bodies with cracks (Panasyuk 1988a, b) it follows that for two cases of the same body under the same load, but with different crack geometries, the force parts of SIF $\sigma_{\rm nom}$ will be the same, while α will be different. Similarly it can be assumed that relation (3.18) will be valid for the case when a body with a crack of configuration b_1 is subjected to the action of an external magnetic field with a magnetic induction vector $\vec{B}_0$ directed perpendicular to the crack plane. To prove this, consider two main problems of fracture mechanics: the analog of the Griffiths problem (an infinite plate with a rectilinear crack of length $2l$) and the analog of the Sak's problem (an infinite body with a circumferential crack of radius a). For these cases, according to the results of Shindo (1978) and Achenbach and Harris (1979), the values of $\sigma_{\rm nom}$ and α will be equal to

$$\begin{aligned}
\sigma_{\rm nom}^{\rm (mG)} &= \frac{\chi(\chi-2)b_{\rm c}^2\left[(1+\chi)^2+[2(1-\nu)+(5-6\nu)\chi]\chi\cdot b_{\rm c}^2\right]\cdot G}{2\sqrt{\pi}(1+\chi)^2\left[2(1+\chi)^2-\chi^2 b_{\rm c}^2[2\nu-1+2(1-\nu)\chi]\right]},\\
\alpha^{\rm (mG)} &= \sqrt{\pi\cdot l};\\
\sigma_{\rm nom}^{\rm (mS)} &= \frac{\chi(\chi-2)b_{\rm c}^2\left[(1+\chi)^2+[2(1-\nu)+(5-6\nu)\chi]\chi\cdot b_{\rm c}^2\right]\cdot G}{2\sqrt{\pi}(1+\chi)^2\left[2(1+\chi)^2-\chi^2 b_{\rm c}^2[2\nu-1+2(1-\nu)\chi]\right]},\\
\alpha^{\rm (mS)} &= 2\sqrt{\frac{a}{\pi}}.
\end{aligned} \tag{3.19}$$

Here $b_{\rm c}^2 = \frac{B_0^2}{G\cdot\mu_0}$, G is the shear modulus, and χ is the magnetic permeability of the environment.

As can be seen from relations (3.19), an equality holds

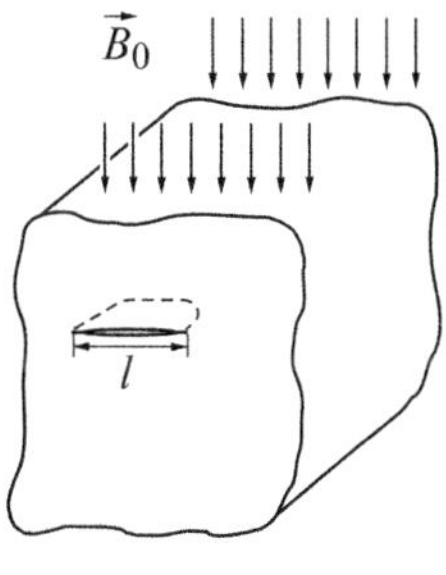

Fig. 3.13 Scheme of the crack plane orientation in the ferromagnet to the direction of the magnetic field

$$\sigma_{\text{nom}}^{(\text{mG})} = \sigma_{\text{nom}}^{(\text{mS})}, \tag{3.20}$$

while $\alpha^{(\text{mG})}$ and $\alpha^{(\text{mS})}$ are equal to the geometric parts of the force cases of the Griffiths $\alpha^{(\text{G})}$ and Sak's $\alpha^{(\text{S})}$ problems, respectively (Panasyuk 1988b). We can assume that equality (3.20) will also hold for more complex crack configurations. Then we will have

$$\frac{K_{\text{I}(1)}}{K_{\text{I}(2)}} = \frac{K_{\text{I}(1)}^{(m)}}{K_{\text{I}(2)}^{(m)}}. \tag{3.21}$$

Here $K_{\text{I}(1)}$ is SIF for a body of some configuration M under force load p with a crack of configuration b_1; $K_{\text{I}(2)}$ is SIF for a crack configuration b_2; $K_{\text{I}(1)}^{(m)}$ is SIF for a body of configuration M with a crack of configuration b_1 under magnetic load $\vec{B}_0$; and $K_{\text{I}(2)}^{(m)}$ is similar SIF for a crack of configuration b_2. It is assumed that the crack configuration b_2 is quite simple, and $K_{\text{I}(2)}$, $K_{\text{I}(2)}^{(m)}$ are well known. At present, sufficiently powerful mathematical methods have been developed (Panasyuk 1988a, b) for the determination of $K_{\text{I}(1)}$. Based on this, you can write the following formula to find $K_{\text{I}(1)}^{(m)}$:

$$K_{\text{I}(1)}^{(m)} = K_{\text{I}(2)}^{(m)} \frac{\alpha_{(1)}\left(a_1^{(1)}, \ldots, a_i^{(1)}\right)}{\alpha_{(2)}\left(a_1^{(2)}, \ldots, a_i^{(2)}\right)}, \tag{3.22}$$

where $\alpha_{(1)}, \alpha_{(2)}$ are the SIF geometric parts for bodies under power load with cracks of configuration b_1 and b_2, respectively; $a_i^{(1)}$ are some dimensionless parameters that completely determine the geometry of the body with a crack of configuration b_1; and parameters $a_i^{(2)}$ completely determine the geometry of a body with a crack configuration b_2.

Thus, using formula (3.22) we can approximately find the SIF for a body with a crack of complex configuration b_1 under the action of an external magnetic field, having a magnetoelastic solution for the SIF in a cracked body of simple configuration b_2 and corresponding solutions for the SIF of similar elastic problems under the action of external mechanical load only.

Circular crack in an infinite body. To verify the method described above, consider the problem of determining the SIF of a circular crack in space under the action of an external magnetic field perpendicular to the crack plane. Without reducing the generality of the proposed method, determine the SIF $K_{\text{I}(1)}^{(m)}$ only on the outer edge of the crack of radius b (Fig. 3.14).

Solution of such a problem under the action at infinity of only uniformly distributed mechanical load of intensity p perpendicular to the crack plane is known in the literature. For example, in Panasyuk (1988a) the following expression is proposed for the stress intensity factor $K_{\text{I}(1)}$:

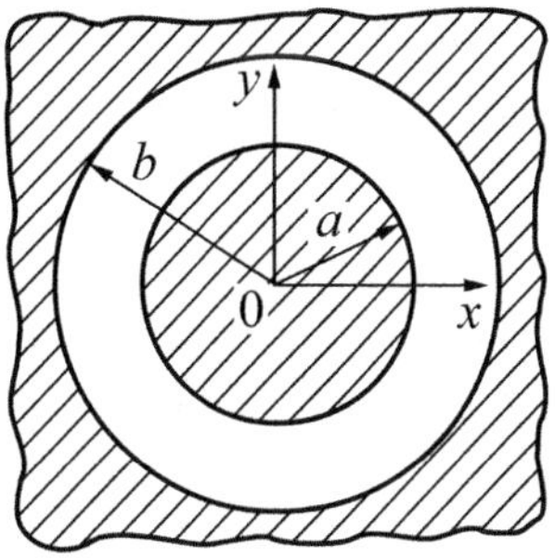

Fig. 3.14 Infinite body with a circular crack

$$K_{I(1)}(\varepsilon) = 2p\sqrt{\frac{b}{\pi}} \cdot \left[\sqrt{1-\varepsilon^2} - \frac{2}{\pi} \cdot \left(0.637 \cdot \varepsilon - 0.526 \cdot \varepsilon^2 - 0.124 \cdot \varepsilon^3\right)\right], \tag{3.23}$$

where $0 < (\varepsilon = a/b) < 1$.

As an auxiliary problem consider the Sak's problem for a space with a disk-shaped crack of radius b (Fig. 3.15).

The solution for the SIF of this problem is as follows:

$$K_{I(2)} = 2p\sqrt{\frac{b}{\pi}}. \tag{3.24}$$

In Parton and Morozov (1985), a magnetoelastic solution to this problem is obtained under action on the body of an external magnetic field with a magnetic induction vector $\mathbf{B}_0$, perpendicular to the crack plane

$$K_{I(2)}^{(m)} = \frac{2}{\pi} \frac{\left[2 \cdot (1+\chi)^2 + [2 \cdot (1-\nu) + (5-6\nu) \cdot \chi] \cdot \chi \cdot b_c^2\right] \cdot P_{h_0}\sqrt{b}}{\left[2 \cdot (1+\chi)^2 + [1 - 2\nu - 2 \cdot (1-\nu) \cdot \chi] \cdot \chi^2 \cdot b_c^2\right]}, \tag{3.25}$$

where $b_c^2 = \frac{B_0^2}{G \cdot \mu_0}$, $P_{h_0} = -\frac{\chi(\chi-2) \cdot B_0^2}{2 \cdot (1+\chi)^2 \cdot \mu_0}$.

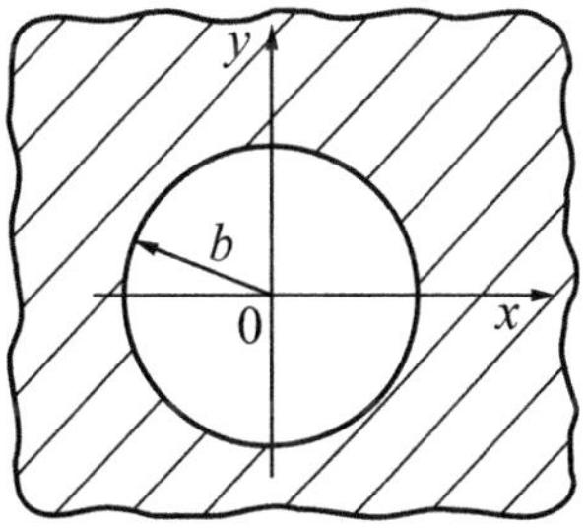

Fig. 3.15 Infinite body with a disk-shaped crack

Then, according to the method of equivalent congruences (comparisons), using relation (3.22) and solution of the mechanical problem for the circular crack (3.23), for $K_{\mathrm{I}(1)}^{(m)}$ we obtain the following approximate relation:

$$K_{\mathrm{I}(1)}^{(m)} = \frac{2}{\pi} \frac{\left[2\cdot(1+\chi)^2 + [2\cdot(1-\nu)+(5-6\nu)\cdot\chi]\cdot\chi\cdot b_{\mathrm{c}}^2\right]\cdot P_{\mathrm{h}_0}\sqrt{b}}{\left[2\cdot(1+\chi)^2 + [1-2\nu-2\cdot(1-\nu)\cdot\chi]\cdot\chi^2\cdot b_{\mathrm{c}}^2\right]} \times \left[\sqrt{1-\varepsilon^2} - \frac{2}{\pi}\cdot\left(0.637\cdot\varepsilon - 0.526\cdot\varepsilon^2 - 0.124\cdot\varepsilon^3\right)\right], \quad (0<\varepsilon<1). \tag{3.26}$$

In Shindo (1983) an exact magnetoelastic solution of this problem is found for $K_{\mathrm{I}(1)}^{(m)}$. In Fig. 3.16 the SIFs obtained in Shindo (1983) and calculated by formula (3.26) are compared.

A good agreement between the SIF, calculated by the approximate method of equivalent comparisons, and a precise solution indicates the effectiveness of the proposed approach and the possibility of its application for technical diagnostics.

Elliptical crack in an infinite body. Consider the next example of the use of the proposed approach to determination of the SIF for elliptical crack in an infinite body under action of magnetic field with a magnetic induction vector $\vec{B}_0$ perpendicular to the crack plane (Fig. 3.17). Solution of this problem is important for engineering practice, since an ellipse can describe approximately many convex contours.

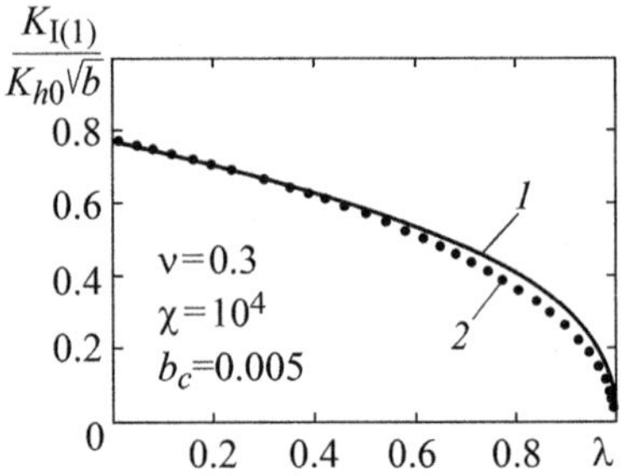

Fig. 3.16 Comparison of the SIF of the circular crack calculated by formula (3.26) (curve *1*) and calculation by Shindo (1983) (curve *2*)

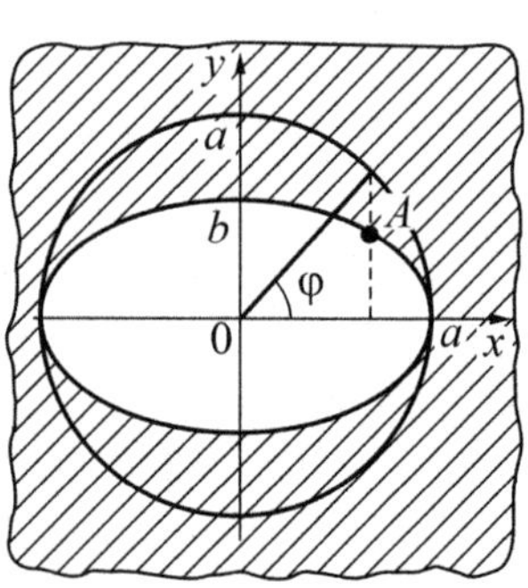

Fig. 3.17 Infinite body with an elliptical crack

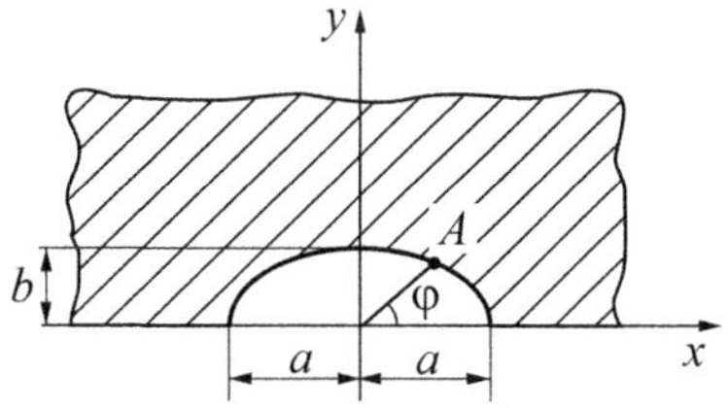

Fig. 3.18 Infinite body with a surface semi-elliptical crack

The solution of the similar problem in the force load, when at infinity uniformly distributed stresses of intensity p are applied perpendicular to the crack plane, is known in the literature and is written as (Achenbach and Harris 1979)

$$K_{\mathrm{I}(1)}(A) = 2p\sqrt{\frac{b}{\pi}} \cdot \frac{\pi}{2} \cdot \left[\sin^2\varphi + (b/a)^2\cos^2\varphi\right]^{1/4} E^{-1}(k), \tag{3.27}$$

where $E(k)$ is a complete elliptic integral of the second kind with modulus k, $k^2 = 1 - (b/a)^2$, $a \geq b, 0 \leq \varphi \leq 2\pi$.

Having chosen the Sak's problem as the auxiliary problem, on the basis of relation (3.22) for $K_{\mathrm{I}(1)}^{(m)}$ in case of elliptical crack we obtain the following expression:

$$K_{\mathrm{I}(1)}^{(m)}(A) = K_{\mathrm{I}(2)}^{(m)}(A) \cdot \frac{\pi}{2}\left[\sin^2\varphi + \left(\frac{b}{a}\right)^2\cos^2\varphi\right]^{1/4} E^{-1}(k). \tag{3.28}$$

Semi-elliptical surface crack in an infinite body. Consider one more important example from the point of view of practical application—a half-space with a surface semi-elliptical crack under action of the magnetic field with a magnetic induction vector $\vec{B}_0$ perpendicular to the plane of the crack (Fig. 3.18).

A solution for the SIF of the corresponding mechanical problem is known in the literature (Achenbach and Harris 1979)

$$K_{\mathrm{I}(1)}(A) = 2p\sqrt{\frac{b}{\pi}} \cdot \frac{\pi}{2}\frac{(1.13 - 0.09\varepsilon)}{\sqrt{1 + 1.464\varepsilon^{1.65}}}\left[1 + 0.1(1 - \sin\varphi)^2\right] \times \left[\sin^2\varphi + \varepsilon^2\cos^2\varphi\right]^{1/4}. \tag{3.29}$$

Here $\varepsilon = b/a$, $\varepsilon \leq 1, 0 \leq \varphi \leq \pi$.

Similar to the previous example, based on relation (3.22) for $K_{\mathrm{I}(1)}^{(m)}$ in case of semi-elliptical crack we obtain the expression

$$K_{\mathrm{I}(1)}^{(m)}(A) = K_{\mathrm{I}(2)}^{(m)}(A)\frac{\pi}{2}\frac{(1.13 - 0.09 \cdot \varepsilon)}{\sqrt{1 + 1.464 \cdot \varepsilon^{1.65}}}\left[1 + 0.1(1 - \sin\varphi)^2\right] \times \left[\sin^2\varphi + \varepsilon^2\cos^2\varphi\right]^{1/4}. \tag{3.30}$$

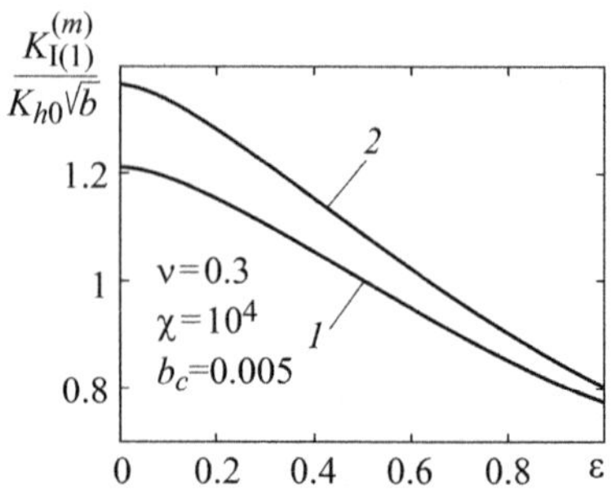

Fig. 3.19 Graphical dependences of the SIF of elliptical (curve *1*) and semi-elliptical (curve 2) cracks on the parameter ε at the point of their contours, which corresponds to $\varphi = \pi/2$

Figure 3.19 presents the SIF graphs for elliptical (curve *1*) and semi-elliptical (curve *2*) cracks, calculated by relations (3.28) and (3.30) at the point of their contours, which corresponds to the value of angle $\varphi = \pi/2$.

As the above examples show, the method of equivalent comparisons is simple and effective. It allows us to determine with sufficient accuracy the SIF in a body with a crack of any configuration under the external magnetic field action, if for such a configuration the SIF, caused only by external mechanical stress, is known.

References

Achenbach JD, Harris JG (1979) Acoustic emission from a brief crack propagation event. ASME J Appl Mech 46(1):107–112

Aki K, Richards PK (1983) Kolichestvennaia seismologiia: teoriia i metody (Quantitative seismology: theory and methods), vol 1. Mir Publishing House

Andreikiv OE, Nazarchuk ZT, Skal's'kyi VR, Rudavs'kyi DV, Serhienko OM (2008) Stress intensity factors caused by magnetic fields in ferromagnets. Mater Sci 44(3):456–460

Andreikiv OE, Skal's'kyi VR, Rudavs'kyi DV, Serhienko OM, Matviiv YY (2011) Magnetoacoustic diagnostics of thin-walled ferromagnets with plane cracks. Mater Sci 46(6):795–799

Andreykiv OY (1982) Prostransvennyie zadachi teorii treshchin (Spatial problems of the theory of cracks). Naukova Dumka Publishing House

Andreykiv OY, Lysak NV (1989) Metod akusticheskoi emissii v isledovaniiakh protsessov razrusheniia (Acoustic emission method in the study of fracture processes). Naukova Dumka Publishing House

Bagdasarian GY, Hasanyan DJ (2000) Magnetoelastic interaction between soft ferromagnetic elastic half-plane with a crack and a constant magnetic field. Int J Solids Struct 37:5371–5383

Bezymiannyi YG (1981) Issledovaniie vozmozhnostei metoda magnetoakusticheskikh shumov dlia kontrolia ustalosti nikelia. Effekt Barkhausena i yego ispolzovaniie v tekhnike (Investigation of the possibilities of the magnetoacoustic noise method for nickel fatigue control. The Barkhausen effect and its use in technology). Kalinin

Brechko TM, Skriabina NE, Spivak LV, Bramovich MY (2004) Domennaia struktura i effekt Barkhausena v amorfnom splave $Fe_{78}B_{12}Si_9Ni_1$ (Domain structure and Barkhausen effect in the $Fe_{78}B_{12}Si_9Ni_1$ amorphous alloy). Pisma Zh Tech Fiz (Lett J Tech Phys) 30(9):68–72

Buttle DJ, Scruby CB, Yakubovics JP, Briggs JAD (1987a) Magnetoacoustic and Barkhauzen emission: their dependence on dislocation in iron. Philos Mag 55(6):717–734

Buttle DJ, Yakubovies JP, Briggs JAD (1987b) Magnetoacoustic and Barkhauzen emission from domain. Wall interaction with precipitates in Incoloy 904. Phillos Mag 55(6):735–756

Eshelby JD (1957) The determination of the elastic field of an ellipsoidal inclusion and related problems. Proc R Soc A 241:379–396

Glukhov NA, Kolmogorov VN (1988) Sviaz parametrov akusticheskikh shumov v peremagnichevayemykh konstruktsionnykh materialakh (Relationship between acoustic noise parameters in remagnetizable structural materials). Defektoskopiya (Defectoscopy) 2:26–29

Glukhov NA, Kolmogorov VN, Miletski BI (1985) Issledovaniie akusticheskikh shumov v peremagnichivayemykh konstruktsionnykh materialakh (Study of acoustic noise in remagnetized structural materials). Defektoskopiya (Defectoscopy) 96:36–40

Gorkunov ES, Berteniev OA, Khamitov VA (1986a) Magnitouprugaia akusticheskaia emissiia v monokristallakh kremnistogo zheleza (Magnetoelastic acoustic emission in silicon iron single crystals). Izv Vuzov Ser Fiz (Rep Univ Ser Phys) 62–66

Gorkunov ES, Somova VM, Buldakova NB (1986b) Ustoichivost sostoianiia ostatochnoi namagnichennosti razlichno termicheski obrabotannykh stalei k vozdeistviiu postoiannykh razmagnichivaiushchikh polei (Resistance of the state of residual magnetization of differently heat-treated steels to the effects of constant demagnetizing fields). Defektoskopiya (Defectoscopy) 9:23–31

Gorkunov ES, Khamitov VA, Berteniev OA (1987) Magnitouprugaia akusticheskaia emissiia v termicheski obrabotannykh kontruktsionnykh staliakh (Magnetoelastic acoustic emission in heat-treated structural steels). Defektoskopiya (Defectoscopy) 3:3–9

Guyot M, Merceron T, Cagan T (1987) Acoustic emission along the hysteresis loops of various ferro- and ferrimagnets. J Appl Phys 63(8):3955–3957

Harris JG, Pott J (1984) Surface motion excited by acoustic emission from a buried crack. Trans ASME J Appl Mech 51(1):77–83

Kameda J, Ranjan R (1987a) Nondestructive evaluation of steels using acoustic and magnetic Barkhausen signals. I. Effect of carbide precipitation and hardness. Acta Metall 35(7):1515–1526

Kameda J, Ranjan R (1987b) Nondestructive evaluation of steels using acoustic and magnetic Barkhausen signals. II. Effect of intergranular impurity segregation. Acta Metall 35(7):1527–1531

Kishi T (1985) Acoustic emission source characterization and its application to microcracking. Z Met 76(7):512–515

Kolmogorov VN (1972) Nekotoryie issledovania akusticheskoi emissii v ferromagnitnykh materialakh (Some studies of acoustic emission in ferromagnetic materials). Tezisy dokladov vsesoyuznogo nauchno-tekhnicheskogo seminara (Theses of the all-union scientific and technical seminar), Khabarovsk

Krinchik GA (1960) Structura domennoi granitsy i dinamicheskiie svoistva ferromagnetikov. Magnitnaia struktura feromagnetikov (Domain wall structure and dynamic properties of ferromagnets. Magnetic structure of ferromagnets). In: Materialy Vsesoyuznogo soveshchaniya, Krasnoyarsk, 10–16 iyunia 1958 (Proceedings of all-union meeting, Krasnoyarsk, 10–16 June 1958). Novosibirsk

Kwan MM, Ono K, Shibata M (1984a) Magnetomechanical acoustic emission of ferromagnetic materials at low magnetization levels (type I behavior). J Acoust Emiss 3(3):144–156

Kwan MM, Ono KY, Shibata M (1984b) Magnetomechanical acoustic emission of ferromagnetic materials at high magnetization levels (type II behavior). J Acoust Emiss 3(4):190–203

Liang O, Fang D, Shen Y, Soh AK (2002) Nonlinear magnetoelastic coupling effects in a soft ferromagnetic material with a crack. Int J Solids Struct 39(10):3997–4011

Lomayev GV, Komarov VA, Rubtsov VI (1981) Eksperimentalnoie issledovaniie akusticheskogo proiavleniia effekta Barkhausena v konstruktsionnykh staliakh. Effekt Barkhausena i yego ispolzovaniie v tekhnike (Experimental study of the acoustic manifestation of the Barkhausen effect in structural steels. The Barkhausen effect and its use in technology). Kalinin

McClure JC Jr, Battacharya S, Schreder K (1974) Correlation of Barkhausen effect type measurements with acoustic emission in fatigue crack growth studies. IEEE Trans Mag 10(3):913–915

Natsik VD, Nechyporenko IN (1984) Akusticheskoie izlucheniie zarodyshei pri fazovykh perekhodakh v magnetikakh. Akustichekaia emissiia materialov i konstruktsii (Acoustic radiation of nuclei during phase transitions in magnets. Acoustic emission of materials and structures), Part 1. Rostov-on-Don

Nazarchuk ZT, Skalskyi VR (2009) Akustyko-emisiine diagnostuvannia konstruktsii. T. 1: Teoretychni osnovy metodu akustychnoi emisii (Acoustic-emission diagnostics of structural elements. Vol. 1: theoretical foundations of the acoustic emission method). Naukova Dumka Publishing House

Ono K, Shibata M (1980) Magnetomechanical acoustic emission of iron and steels. Mater Eval 38:55–61

Panasyuk VV (ed) (1988a) Mekhanika razrusheniia i prochnost meterialov (Fracture mechnaics and strength of materials). In: Savruk MP (ed) Koeffitsienty intensivnosti napriazhenii v telakh s treshchinami (Stress intensity factors in bodies with cracks), vol 2. Naukova Dumka Publishing House

Panasyuk VV (ed) (1988b) Mekhanika razrusheniia i prochnost meterialov (Fracture mechnaics and strength of materials), vol 1. Naukova Dumka Publishing House

Parton VZ, Morozov EM (1985) Mekhanika uprugoplasticheskogo razrusheniia (Mechanics of elastoplastic fracture). Nauka Publishing House

Ranjan R, Jiles DC, Rastogi PK (1986) Magnetoacoustic emission, magnetization and Barkhausen effect in decarburized steels. IEEE Trans Mag 22(5):511–513

Ranjan R, Buck O, Thompson RB (1987a) A study on the effect of dislocation on the magnetic properties of nickel using magnetic NDT methods. J Appl Phys 61(8):3196–3198

Ranjan R, Jiles DC, Buck O, Thompson RB (1987b) Grain size measurement using magnetic and acoustic Barkhauzen noise. J Appl Phys 61(8):3199–3201

Roman J, Maharshak S, Amir J (1983) Magnetomechanical acoustic emission: a non-destructive characterization technique of precipitation hardness steels. J Acoust Emiss 2(1/2):64–66

Rudiak VM (1986) Protsessy perekliucheniia v nielineinykh krystallakh (Switching processes in nonlinear crystals). Nauka Publishing House

Sánchez RL, Pumarega MIL, Armeite M (2004) Barkhausen effect and acoustic emission in a metallic glass—preliminary results. Rev Quant Nondestruct Eval 23:1328–1335

Shaposhnikov AB (1980) Teoreticheskiie osnovy elektromagnitnoi defektoskopii metallicheskikh tel (Theoretical foundations of electromagnetic flaw detection of metal bodies), vol 1. Tomsk

Shibata M (1984) A theoretical evaluation of acoustic emission signals. The rise-time effect of dynamic force. Mater Eval 42(1):107–120

Shibata M, Ono K (1981) Magnetomechanical acoustic emission—a new method of nondestructive stress measurement. NDT Int 14:227–234

Shibata M, Kobajashi E, Ono K (1986) The detection of longitudinal rail force via magnetomechanical acoustic emission. J Acoust Emiss 4(4):93–100

Shindo Y (1978) Magnetoelastic interaction of a soft ferromagnetic elastic solid with a penny-shaped crack in a constant axial magnetic field. Trans ASME J Appl Mech 45:291–296

Shindo Y (1983) Singular stresses in a soft ferromagnetic solid with a flat annular crack. Acta Mech 50:50–56

Skal's'kyi VR, Serhienko OM, Mykhal'chuk VB, Semehenivs'kyi RI (2009) Quantitative evaluation of Barkhausen jumps according to the signals of magnetoacoustic emission. Mater Sci 45(3):399–408

Turner PA, Stockbridge CD, Theuerer HC (1969) Magnetic domain nucleation and propagation in fine wires. J Appl Phys 40(4):1864–1869

Yudin AA, Lopatin MV (1987) K teorii magnitnoi akusticheskoi emisii (On the theory of magnetic acoustic emission). Dep. N 3158–B87. VINITI (Vsiesojuznyi Institut nauchno-tekhnicheskoi informacii) (All-Union Institute of Scientific and Technical Information), Moscow

Chapter 4
Models of Hydrogen Cracks Initiation as Sources of Elastic Waves Emission

To use the MAE with the aim to assess the damage of ferromagnetic materials and hence for the NDT and TD of the responsible equipment, it is necessary to build the correct physical models of elastic wave emission. They foresee solving the corresponding nonstationary dynamic problems of crack theory in displacements. Their solutions make it possible to determine the expected amplitudes of the AE electrical signals generated during the initiation or development of ferromagnetic material. It is especially important to be able to distinguish the amplitude-frequency characteristics of signals that correspond to different mechanisms of their generation by sources that appear in the material under the action of the applied load and environment. This, in turn, serves as a basis for the creation of original methodological approaches, means of AE signals selection and processing with appropriate algorithms and software.

4.1 Subcritical Crack Growth, Local on the Front

Problem formulation To build a model of AE emission of signals caused by the macrocrack growth, local on the front, in the ferromagnetic material, consider an elastic half-space weakened by Mode I plane macrocrack, bounded by a smooth contour L.

Let at time $t = 0$ in the local region of the body, where stresses (or strains) reach a certain limit value, due to the application of external tensile forces and magnetic field, which is known to cause additional mechanical stresses (Andreikiv et al. 2008), the microcrack occurs at the crack contour (Fig. 4.1). As a result of unloading the edges of a newly formed microcrack from the initial level $\sigma_0 = \sigma_{\mathrm{mec}} + \sigma_{\mathrm{mag}}$ to zero, elastic waves are emitted. They reach the surface of the object and can be registered by the primary AE converters.

V. Skalskyi and Z. Nazarchuk, *Magnetoelastic Acoustic Emission*, Springer-AAS Acoustics Series, https://doi.org/10.1007/978-981-99-4032-5_4

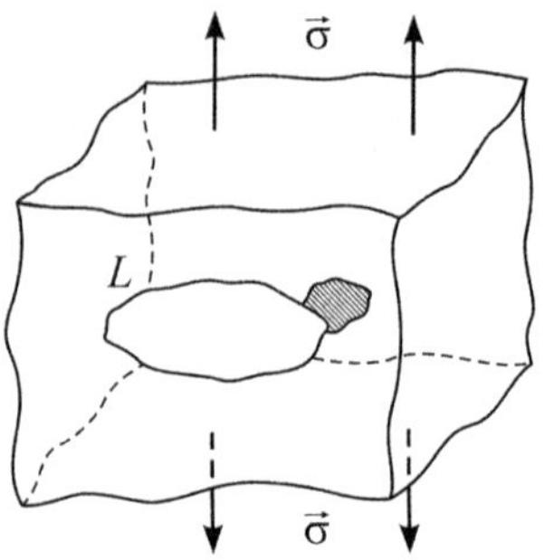

Fig. 4.1 Scheme of local growth of an internal crack

Based on the above presented and the results obtained in (Nazarchuk and Skalskyi 2009; Achenbach and Harris 1979) for the problem of semi-infinite crack growth, the displacement field caused by the formation of a disk-shaped micro-crack in the vicinity of a macrocrack for distances much larger than the radius of the latter is written as the product of displacements in an infinite body and function $\gamma_i(\alpha)$, that takes into account the influence of the free surface

$$u_i(\vec{R},t) = \gamma_i(\alpha) u_R^{(d)}(\vec{R},t). \tag{4.1}$$

Here, $u_1(\vec{R},t) = u_R(\vec{R},t)$, and $u_2(\vec{R},t) = u_\theta(\vec{R},t)$, $u_R^{(d)}(\vec{R},t)$ and $u_\theta^{(d)}(\vec{R},t)$ determine the displacement in an infinite body and are presented in Nazarchuk and Skalskyi (2009). Functions $\gamma_i(\alpha)$ are obtained by formula (Achenbach and Harris 1979)

$$\gamma_i(\alpha) = \frac{\sqrt{1 + c_1 / c_i \cos\alpha}}{\left(1 + c_R / c_i \cos\alpha\right) K\left(-\frac{\cos\alpha}{c_1}\right)}, \tag{4.2}$$

where $K(\cdot)$ is some integral, the value of which we found by numerical integration; c_R is the Rayleigh wave velocity; $\cos\alpha = \frac{\cos\theta \sin\varphi + \delta}{\sqrt{(\cos\theta \sin\varphi + \delta)^2 + \sin^2\theta}}$; angle α is counted out from the plane of a semi-infinite crack; $\delta = \Delta / R$, Δ is the distance between the center of a disk-shaped crack and the edge of a semi-infinite crack; and the angle φ is counted out from tangent line to a contour of macrocrack's front in the touching point of micro- and macrocracks.

The components u_R and u_θ of the displacement vector, as can be seen from relations (4.1), (4.2), have the same time dependence as the components $u_R^{(d)}$ and $u_\alpha^{(d)}$. However, the radiation patterns in this case will be different from those obtained for an isolated disk-shaped crack. Figure 4.2 shows such a diagram constructed for the angle $\varphi = \pi / 2$ with the maximum values of the component $u_R(R,t)$, which corresponds to the propagation of a longitudinal wave. Points indicate the experimental data (Scruby et al. 1983). As can be seen from Fig. 4.2, the angular dependence differs the most from that for an isolated disk-shaped crack if angle θ is close to π. In the

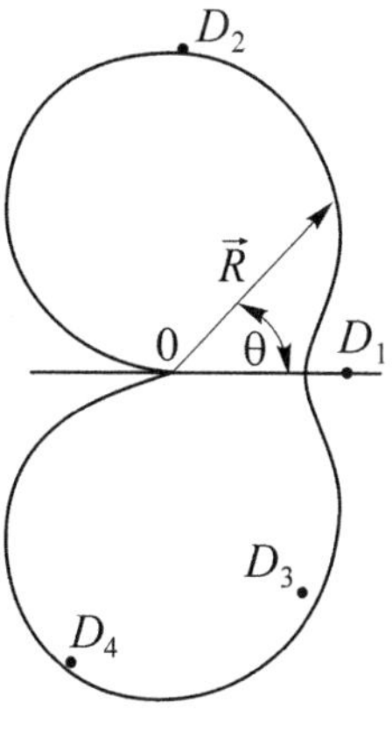

Fig. 4.2 Angular distribution of the maximum of the displacement vector modulus for longitudinal wave during formation of a disk-shaped crack near the front of internal macrocrack

case of $|\theta| < \pi/2$ diagrams are almost identical. In this case, if the primary AE converters are located in the domains of angle $|\theta| < \pi/2$, the effect of the free surface on the AE signals will be negligible.

From the dependences describing the angular distribution of radiation, we can see that the maximum values of displacements at the front of the emitted longitudinal wave are proportional to r_0^2, i.e., to the area of the newly formed defect.

Speaking about the problem of the formation of a microdefect with an area $S = \pi r_0^2$ at an internal plain macrocrack contour, assume that there is a linear relationship between the amplitudes A of the AE signals and the maximum values of displacements at the longitudinal wave front, that is,

$$A = \frac{\lambda S}{R} \Phi_1^{(d)}(\theta) \gamma_i(\alpha), \tag{4.3}$$

where $\lambda = \lambda_0 \lambda_1$, λ_0 is the coefficient of proportionality between the electrical signals at the output of the AE primary converters and the maximum values of displacements at the longitudinal wave front, $\lambda_1 = \delta_1 \sigma_0 / \left(\pi \rho c_1^2\right)$, $\sigma_0 = \sigma_{\text{mec}} + \sigma_{\text{mag}}$.

If during local extension N microdefects are formed at the internal crack front, the total area of the internal macrocrack extension will be

$$\Delta S = \sum_{k=1}^{N} b A_k, \tag{4.4}$$

where b is the proportionality factor.

Thus, as follows from dependence (4.4), the amplitudes of the AE signals during local extension of the internal plane crack are proportional to the newly formed area. This type of relationship between A and S was confirmed experimentally in (Skalskyi et al. 1999; Takahashi et al. 1981; Gerberich et al. 1975; Nazarchuk et al. 2017a).

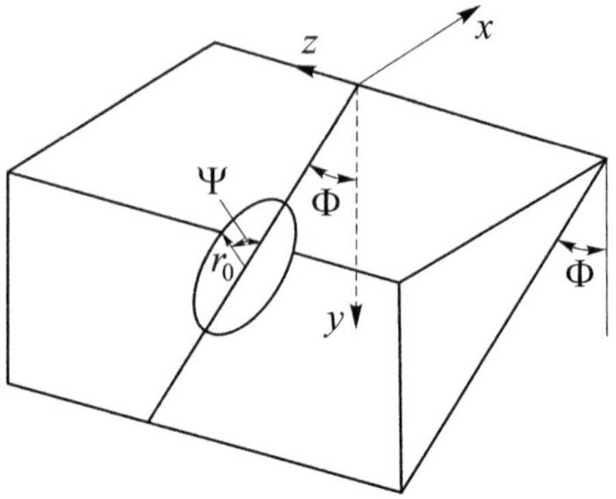

Fig. 4.3 Scheme of a disk-shaped crack located in an elastic half-space

Influence of the body boundaries. Signals are recorded by AE primary converters located on the body's surface. Therefore, an important issue of AE diagnostics is to take into account the influence of the free surface on the signal parameters.

Figure 4.3 presents the geometry of the corresponding problem. Based on the results of (Nazarchuk and Skalskyi 2009; Nazarchuk et al. 2017b; Harris and Pott 1984), the total displacements on the surface in the plane of symmetry $\psi = 0$ for the zone closest to the epicenter can be written in the form

$$\vec{U}_i(R, \alpha, t) = \left[R_x^{(1)}(\alpha)\vec{i} + R_y^{(1)}(\alpha)\vec{j} \right] u_i^{(d)}(R, \alpha, t), \tag{4.5}$$

where

$$\begin{aligned} R_x^{(1)}(\alpha) &= \frac{2\varepsilon^{-1} \sin 2(\alpha + \Phi) \sin \delta}{\varepsilon^2 \cos^2 2\delta - \sin 2\delta \sin 2(\alpha + \Phi)}, \\ R_y^{(1)}(\alpha) &= \frac{2\varepsilon^{-2} \cos(\alpha + \Phi) \cos 2\delta}{\varepsilon^2 \cos^2 2\delta - \sin 2\delta \sin 2(\alpha + \Phi)}, \end{aligned} \tag{4.6}$$

$$\begin{aligned} R_x^{(2)}(\alpha) &= \frac{2\varepsilon^{-2} \cos(\alpha + \Phi) \cos 2(\alpha + \Phi)}{\varepsilon^{-2} \cos^2 2(\alpha + \Phi) - \sin 2\gamma \sin 2(\alpha + \Phi)}, \\ R_y^{(2)}(\alpha) &= \frac{-2 \sin 2\gamma \cos(\alpha + \Phi)}{\varepsilon^{-2} \cos^2 2(\alpha + \Phi) - \sin 2\gamma \sin 2(\alpha + \Phi)}. \end{aligned} \tag{4.7}$$

Here $\vec{U}_1(R, \alpha, t)$ and $\vec{U}_2(R, \alpha, t)$ correspond to a falling longitudinal and transverse waves; $\vec{i}, \vec{j}$ are orts on the body surface; $R = d/\cos(\alpha + \Phi)$ is the distance from the crack center to the observation point, which is located at the half-space boundary; θ is the angle between the crack and the direction to the observation point, drawn from the defect center; δ is the angle given by the equation: $\cos\delta = -\varepsilon \sin(\alpha + \Phi)$; and γ is the angle given by the equation: $\varepsilon \cos\gamma = -\sin(\alpha + \Phi)$. In addition, in the case of a transverse crack, the angle $\alpha + \Phi$ must satisfy the condition $|\sin(\alpha + \Phi)| < \varepsilon$.

In accordance with the considerations presented in Harris and Pott (1984), dependence (4.5) is true for the moments of time that correspond to the front-line areas of wave propagation and condition $d >> r_0$.

Finally, taking into account formulas (4.1), (4.5)–(4.7) for the maximum values of the displacement vector caused by local growth of the internal crack, we obtain the expression

$$u^{(k)}_{\max|c_i} = \frac{\delta_i \sigma_0 \Phi_i^{(d)}(\theta) r_0^2}{\rho c_1^2 R} \gamma_i(\alpha) R^{(i)}_{(k)}(\alpha), \tag{4.8}$$

where $\sigma_0 = \sigma_{\text{mec}} + \sigma_{\text{mag}}$; $k = x,\ y$; $i = 1$ corresponds to the longitudinal and $i = 2$—to the transverse waves; and functions $\Phi_1^{(d)}(\theta)$ and $\Phi_2^{(d)}(\theta)$ determine the angular distribution of radiation for the longitudinal and transverse waves, respectively. Note that for $d \gg r_0$ the difference between θ and α is insignificant.

Figure 4.4 shows the dimensionless maximum values of the modulus of the displacement vector $U_{\max|c_1} = \left[U^{(x)^2}_{\max|c_1} + U^{(y)^2}_{\max|c_1}\right]^{1/2}$ on the surface of a half-space for the longitudinal wave depending on the dimensionless distance l/d ($l = d\text{tg}\Phi + x$) to the epicenter for some angles of crack orientation. It was assumed in the calculations that $d/r_0 = 200$. Number *1* indicates the maximum of the modulus of the displacements vector of falling wave, and number 2 is the maximum values of the modulus of the surface displacements vector, calculated by the relations (4.8). We see that the maximum values of surface oscillations are almost twice the corresponding values for the falling wave. At distances approximately equal to seven depths of the defect, the difference between the maxima of displacements caused by the falling wave and the total displacements of the surface is insignificant.

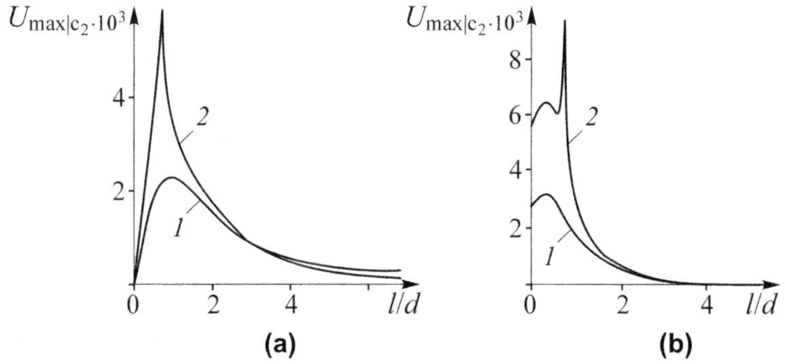

Fig. 4.4 Dependence of the dimensionless value $U_{\max|c_i}$ on the surface of a half-space on d ($d/r_0 = 200$) for the falling longitudinal wave: **a** orientation angle $\Phi = 0°$, **b** $\Phi = 75°$; *1*—incident wave, *2*—total wave

The obtained dependences between the parameters of subcritically growing cracks and the amplitudes of the AE elastic waves are included in the algorithms for processing signals of acoustic emission technical means created in the Karpenko Physico-Mechanical Institute of the NAS of Ukraine.

4.2 Modeling of Displacements Caused by Elastic Rayleigh Waves Generated by an Internal Source in a Half-Space

Formulation of the problem. The surface of the body, which is controlled during technical diagnostics, distorts the characteristics of the AE signals generated by the crack. Such changes are ambiguous and depend on the geometry and size of the body. The known results concern mainly the investigations carried out at the epicenter of emission (Shibata 1984; Sinclair 1979). Consider the typical cases that occur during AE testing.

The peculiarity of the AE recording on large-scale objects is that besides longitudinal and transverse waves, the Rayleigh waves appear in them. Depending on the location of the AE converters relative to the source epicenter, certain types of waves will dominate. Therefore, it is first necessary to determine the limits of the waves dominance, and then write down the equation for calculating the components of the displacement vector.

In order to compare the Rayleigh waves with other waves, consider the formation of an internal source in a homogeneous half-space (Andreikiv et al. 1993). Assume that at the initial moment of time at a distance z_0 from the free surface, a point source of emission is activated, in particular, due to the phenomenon of magnetostriction in the ferromagnetic material (Fig. 4.5). Introduce a system of cylindrical coordinates $Or\theta z$ so that its center (point O) is located in the epicenter at the boundary of the half-space, and the axis Oz is perpendicular to it.

Boundary conditions on the surface of the half-space are as follows:

$$\sigma_z = \tau_{rz} = 0,\ z = 0,\ \varphi(r, z, t) = \varphi_0 \delta(r)\delta(z - z_0)H(t)/2\pi r, \qquad (4.9)$$

where φ_0 is the source intensity; $\delta(...)$ is Dirac delta function; and $H(...)$ is Heaviside function.

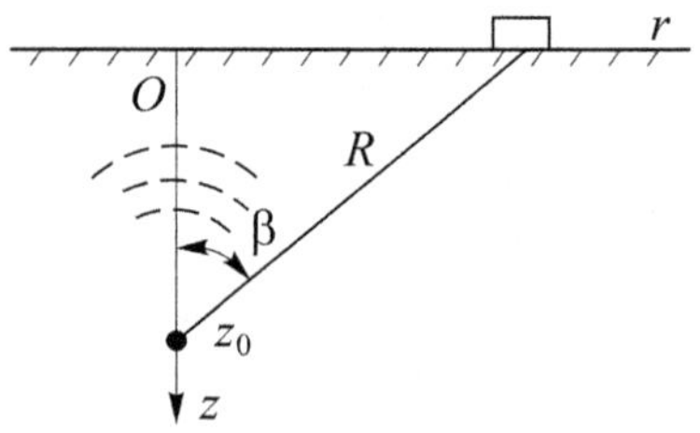

Fig. 4.5 Schematic representation of the defect (AE sources) in a half-space

The formulated problem is axisymmetric. Wave potentials φ and ϕ satisfy the system of wave equations. In addition, a term $\varphi_0\delta(r)\delta(z - z_0)H(t)/2\pi r$ appears in the right part of the equation. The initial conditions are zero.

Solution of the problem. Applying the Hankel integral transform according to coordinate r and the Laplace integral in time t with respect to such wave equations, we obtain the components of the displacement vector in the Laplace image space

$$\overline{u}_r = \int\limits_0^\infty \alpha^2\left[B\gamma_2\exp(-\gamma_2 z) - A\exp(-\gamma_2 z) - \varphi_0\exp(-\gamma_1|z - z_0|)\right]J_1(\alpha r)\mathrm{d}\alpha,$$

$$\overline{u}_z = \int\limits_0^\infty \alpha\left[-A\gamma_1\exp(-\gamma_1 z) + B\exp(-\gamma_1|z - z_0|) + \varphi_0\exp|-\gamma_1 z|/s\, J_0(\alpha r)\right]\mathrm{d}\alpha, \tag{4.10}$$

where $\gamma_l = \left(\alpha^2 + s^2/c_l^2\right)^{1/2}$, $l = 1{,}2$; $J_0(\ldots)$, and $J_1(\ldots)$ are Bessel functions of zero and first order, correspondently.

Unknown constants A and B are determined from a system of two algebraic equations derived from the dependences $u_r = -\partial\phi/\partial r - \partial\varphi/\partial z$; $u_z = \partial\phi/\partial z + \partial\varphi/\partial x + \varphi/r$ and boundary conditions (4.9) (ϕ and φ are the wave potentials). Substituting constants A and B found in this way in (4.10), we determine the movement of the free surface in space of Laplace images

$$\overline{u}_z(r, 0, s) = 2\varphi_0/\left(c_2^2 s\right)\int\limits_0^\infty \frac{\alpha\exp(-\gamma_1 z_0)\left[\alpha^2 + s^2/\left(2c_2^2\right)\right]}{R(\alpha, s)}J_0(\alpha r)\mathrm{d}\alpha, \tag{4.11}$$

where $R(\alpha, s) = \left[\alpha^2 + s^2/\left(2c_2^2\right)\right]^2 - \alpha^2\gamma_1\gamma_2$ is the Rayleigh function.

Writing the Bessel function $J_0(\ldots)$ in terms of Hankel function of the first kind $H_0^{(1)}(\ldots)$ as $J_0(x) = 0{,}5\left[H_0^{(1)}(x) - H_0^{(1)}(-x)\right]$ (Andreikiv et al. 1993) and using the relation $H_0^{(1)}(-iz) = 2K_0(z)/\pi i$, where $K_0(\ldots)$ is the modified Hankel function, after replacing the integration variable $\alpha = \eta/i$ wing (4.11), we obtain

$$\overline{u}_z(r, 0, s) = 2\varphi_0 s/\left(\pi i c_2^2\right)\int\limits_{-i\infty}^{i\infty} \frac{\eta\exp(-\gamma_1 z_0)\left[-\eta^2 + s^2/\left(2c_2^2\right)\right]}{R(\alpha, s)}K_0(\eta r)\mathrm{d}\eta. \tag{4.12}$$

Search the original of Laplace transform by the Kanyar method (Cagniard 1962). For this purpose, we modify expression (4.12) by introducing a new integration variable $p = s\eta$. Then, taking into account the condition $K(z^*) = [K(z)]^*$ (asterisk means complex conjugation), we obtain

$$\overline{u}_z(r,0,s)=\frac{8\varphi_0 s}{\pi c_2^2}\mathrm{Im}\left(\int\limits_0^{i\infty}\frac{p\left(c_2^{-2}-2p^2\right)\exp(-s\overline{\gamma}_1 z_0)K_0(spr)}{R(p)}\mathrm{d}p\right),\tag{4.13}$$

where $\overline{\gamma}_i=\left(c_i^{-2}-p^2\right)^{-1/2}$; $R(p)=\left(c_2^{-2}-2p^2\right)^2+4p^2\overline{\gamma}_1\overline{\gamma}_2$.

Introduce a real integration variable $\tau=pr+\overline{\gamma}_1 z_0$. Then the relationship

$$p(\tau)=\begin{cases}\left[r\tau+iz_0\left(\tau^2-R_0^2/c_1^2\right)^{1/2}\right]/R_0^2,\ \tau\geq R_0/c_1,\ R_0=\left(z_0^2+r^2\right)^{1/2};\\ \left[r\tau-iz_0\left(R_0^2/c_1^2-\tau^2\right)^{1/2}\right]/R_0^2,\ \tau<R_0/c_1\end{cases}\tag{4.14}$$

describes the contour in the complex plane, which consists of a segment of the real axis $[-z_0c_1/R_0,\ rc_1/R_0]$ for $\tau<R_0/c_1$ and a hyperbola for $\tau\geq R_0/c_1$. To apply the Jordan's lemma, consider a closed circuit with an imaginary half-axis $[0,\ \infty]$, a line (4.14), and a part of a circle of infinite radius connecting the imaginary half-axis with this contour. In the area bounding the closed loop, the subintegral function has no specific features. The function $K_0(x)$ with large values of x has an asymptote $\pi\exp(-x)\left[1+o\left(x^{-1}\right)\right]/(2x)^{1/2}$. Thus, based on the Jordan's lemma, the integration along the imaginary axis in relation (4.13) can be replaced by the integration along the contour (4.14) for τ, which varies from 0 to ∞. After determining original of the Laplace transform from the product of functions $\exp(-s\gamma_1 z_0)\,K_0(spr)$ after its differentiation we find the expression for calculating the displacements on the half-space surface

$$u_z(r,0,t)=\frac{-8\varphi_0}{\pi c_2^2}\frac{\mathrm{d}}{\mathrm{d}t}\mathrm{Im}\left(\int\limits_{R_0/c_1}^{t}\frac{\Phi_1(r,t;\tau)}{\Phi_2(r,t;\tau)}\mathrm{d}\tau\right),\quad t>R_0/c_1,\tag{4.15}$$

where

$$\Phi_1(r,t;\tau)=ip\overline{\gamma}_1\left(\overline{\gamma}_2^2-p^2\right),$$
$$\Phi_2(r,t;\tau)=R(\tau)(t-\tau)^{1/2}(t-\tau-2pr)\left(\tau^2-R_0^2/c_1^2\right)^{1/2}.$$

Differentiate time t and normalizing the variables $\tilde{r}=r/z_0$, $\tilde{t}=tc_1/z_0$, $\tilde{\tau}=\tau c_1/z_0$, $\tilde{R}=R_0/z_0$, $\tilde{u}_z=-u_z z_0^2/\left(8\pi\varphi_0\varepsilon^2\right)$, $\tilde{\gamma}_1=c_1\overline{\gamma}_1$, $\tilde{\gamma}_2=c_1\overline{\gamma}_2$, we obtain the final dependence for calculating the components of the displacement vector at the half-space boundary (Andreikiv et al. 1993)

$$u_z(r,0,t)=\left(\frac{1}{t-R_0}\int\limits_{R_0}^{t}\frac{F_1(r,t;\tau)}{F_2(r,t;\tau)}\mathrm{d}\tau\right)H(t-R_0)$$

$$+\left(-\int\limits_{R_o}^{t}\frac{\mathrm{Re}F_3(r,t;\tau)}{F_2(r,t;\tau)}\mathrm{d}\tau\right)\delta(t-R_0), \tag{4.16}$$

where

$$F_1(r,t;\tau)=\frac{1}{2}\left[\frac{\mathrm{Re}F_3(r,t;\tau)}{\sqrt{\tau+R_0}}-(t-R_0)\sqrt{\tau+R_0}\mathrm{Re}\left(\frac{F_3(r,t;\tau)}{t-\tau-2pr}\right)\right]$$
$$-\mathrm{Im}\left\{\frac{(\gamma_2^2-p^2)(\gamma_1^2-\gamma_2^2)-4p^2\gamma_1^2}{R(p)\sqrt{t-\tau-2pr}}\right\}\frac{\tau-R_0}{\tau+R_0}$$
$$+\mathrm{Im}\left\{4p\gamma_1F_3(t,\tau)\frac{2\gamma_1\gamma_2-(\gamma_1/\gamma_2+\gamma_2/\gamma_1)p^2-2(\gamma_2-p^2)}{R(p)}\right\}$$
$$\times\frac{\sqrt{\tau-R_0}}{\tau+R_0}--\mathrm{Re}\left\{F_3(r,t;\tau)\frac{2ir\gamma_1-\sqrt{\tau^2+R_0^2}}{2(t-\tau+2pr)}\right\}\frac{\sqrt{\tau-R_0}}{\tau+R_0};$$
$$F_2(r,t;\tau)=(t-\tau)^{1/2}\left(\tau^2-R_0^2\right)^{1/2};$$
$$F_3(r,t;\tau)=p\gamma_1(\gamma_2^2-p^2)/\left[R(p)\,(t-\tau-2pr)^{1/2}\right].$$

In relations (4.16) the sign ~ above the normalized values is omitted to simplify the notation.

It follows from relation (4.16) that in the case of a longitudinal wave we have a delta pulse. This can be explained by the fact that the source was modeled by the center of comprehensive expansion. As expected, the attenuation of the Rayleigh wave amplitudes, with increasing distance from the epicenter to the observation point, is weaker than for longitudinal or transverse waves generated by an internal source in space. At short distances from the epicenter, the Rayleigh wave is slightly visible or absent (Fig. 4.6a). During $\tilde{r}$ growth at the time of arrival of the Rayleigh wave, the shape of the elastic oscillation changes, and the trailing and leading edges of the pulse become more pronounced (Fig. 4.6b).

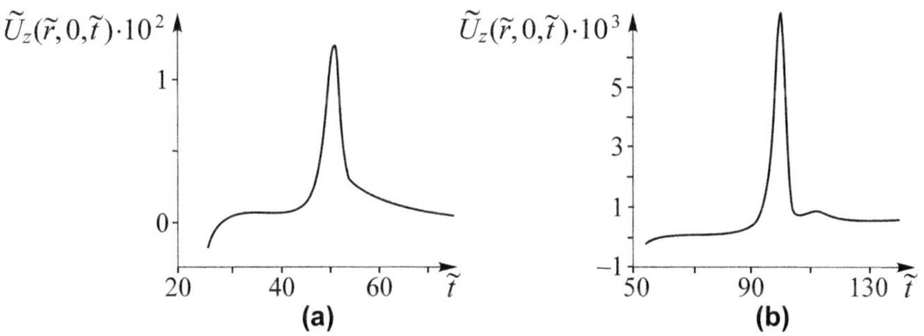

Fig. 4.6 Change of dimensionless displacement $\tilde{u}_z = -u_z z_0^2/(8\pi\varphi_0\varepsilon^2)$ in time $\tilde{\tau} = \tau c_1/z_0$ at distances $\tilde{r} = r/z_0 = 1$ (**a**) and $\tilde{r} = 25$ (**b**)

To establish the analytical dependence that will allow us to estimate the distances where the Rayleigh wave will appear, use the approach proposed in Aki and Richards (1983). According to it, the Rayleigh wave will be noticeable when the inequality is satisfied

$$|p(r/c_R) - 1/c_R| \leq 1/c_R - 1/c_2. \tag{4.17}$$

Substituting in (4.16) instead of p its value from (4.14) for $t \geq R_0/c_1$, we obtain (Andreikiv et al. 1993)

$$\begin{aligned} \mathrm{tg}\beta &\geq [2/(c_{2R} - 1)]^{1/2}, \\ c_{2R} &= c_2/c_R. \end{aligned} \tag{4.18}$$

The c_{2R} value is determined by equating the function $R(p)$ to zero and choosing the largest among the obtained solutions (Seismov 1976). For example, in the case of metals for $\upsilon = 0.3$, the Rayleigh wave will be observed at distance $r \geq 5z_0$. This conclusion coincides with the calculations that follow directly from relations (4.16).

Note that when locating defects or determining their parameters, it is necessary to take into account the type of wave being recorded. Incorrectly calculated type of wave can cause significant errors.

4.3 Calculational Model for Initiation of Fatigue Microcrack at the Hydrogenated Stress Notch Tip

As practice shows, fatigue facture is the main cause of operational damage of the majority of modern metal structures. One of the most important and difficult problems in the theory of fatigue is to determine the period of fatigue crack initiation. The solution of this problem by the known methods of the classical theory of elasticity and plasticity is associated with significant mathematical difficulties, in particular with the solutions of complex nonlinear equations in partial derivatives. Effective methods for solving such equations have not been developed yet. Therefore, the energy approach to the material fracture mechanics remains the most suitable for solving this problem. Based on it, computational models are proposed to determine the period of fatigue macrocrack initiation at the hydrogenated stress notch tip.

Experiments show that two main stages of fatigue macrocrack initiation are clearly visible. This is the stage of fatigue microcrack initiation, which usually occurs jump-like and the stage of the initiated crack growth to a macroscopic size. Therefore, the total period N_2 of initiation of the fatigue macrocrack can be determined using the relation

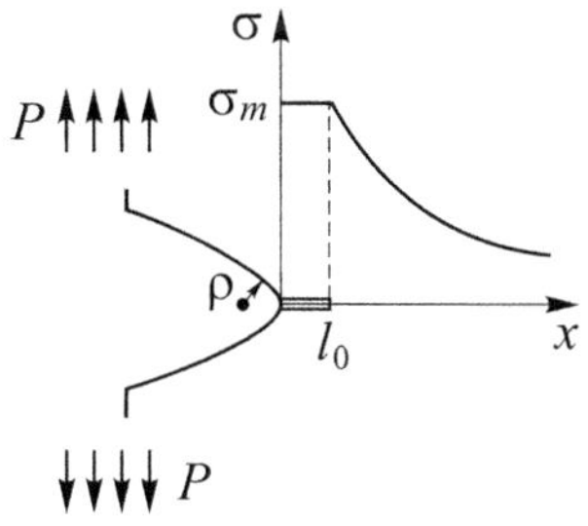

Fig. 4.7 Scheme of the stress notch tip with a microcrack

$$N_2 = N_1 + \int_{l_0}^{l_1} v^{-1}(l)\mathrm{d}l, \tag{4.19}$$

where l_0 is the elementary length of the microcrack that appears at the notch tip after N_1 load cycles of the first stage; l_1 is the minimum length of the macrocrack; and $v(l)$ is the rate of the microcrack growth to the macroscopic size.

Consider an elastic–plastic plate hydrogenated to the level of hydrogen concentration C_{H} with a stress concentrator's curvature radius ρ at the tip. Under the action of cyclic load evenly distributed in infinity with a force parameter p, perpendicular to the notch plane, an initial plastic zone of length l_1 appears at the tip of the latter. Let after N_1 load cycles, an elementary microcrack appears jump-like at the crack tip. According to the experiments (Yarema and Popovich 1985), cyclic deformation of the material elementary volume at the stress concentrator tip, due to defects in the microstructure (accumulation of microplastic deformations, formation of microvoids and a net of differently oriented microcracks, etc.) an elementary Mode I microcrack of length l_0, appears. The crack length approximately coincides with the distance from the notch tip to the point of action of the maximum tensile stresses (Fig. 4.7). Therefore, we can take the distance, which has the order of magnitude of the notch tip opening displacement δ, as the minimum microcrack of length l_0 formed at the stress notch tip (McMeeking 1977).

The model is based on the energy fracture criterion, according to which fracture occurs in any elementary microvolume of the material, if the total energy of plastic deformations scattering in it during all loading cycles $W^{(\mathrm{H})}$ (index H indicates the presence of hydrogen in the material) reaches some critical value, namely

$$W^{(\mathrm{H})} = \alpha \cdot W_c^{(\mathrm{H})}. \tag{4.20}$$

Expanding the total scattering energy of plastic deformations into static and cyclic components, we obtain

$$W_s^{(\mathrm{H})} + W_f^{(\mathrm{H})} = \alpha \cdot W_c^{(\mathrm{H})}, \tag{4.21}$$

where $W_c^{(\mathrm{H})}$ is the static energy of the hydrogenated material fracture and α is the Morrow coefficient (Troshchenko 1981).

Assuming that the length of the generated microcrack l_0 is small enough, the static component $W_s^{(\mathrm{H})}$ of the total energy of plastic deformations scattering and the fracture energy of the material can be written it terms of their densities (assuming these densities be constant within the length l_0) as

$$\begin{aligned} W_s^{(\mathrm{H})} &= \gamma_s^{(\mathrm{H})} l_0, \\ W_c^{(\mathrm{H})} &= \gamma_c^{(\mathrm{H})} l_0, \end{aligned} \tag{4.22}$$

where $\gamma_s^{(\mathrm{H})}$ is the density of the static component of the energy of plastic deformations scattering in material and $\gamma_c^{(\mathrm{H})}$ the energy density of the material fracture.

Then on the basis of (4.21), (4.22) the energy criterion (4.20) will take the form

$$\gamma_s^{(\mathrm{H})} l_0 + (N_1 - 1)\left(w_f^{(\mathrm{H})} - w_{th}^{(\mathrm{H})}\right) = \alpha \cdot \gamma_c^{(\mathrm{H})} l_0, \tag{4.23}$$

where $w_f^{(\mathrm{H})}$, $w_{th}^{(\mathrm{H})}$, is the cyclic component of the energy of plastic deformations scattering per the load cycle and its threshold value.

According to experiments (Panasyuk et al. 1982; Karpenko 1960), hydrogen charging of metal samples leads mainly to a decrease in their deformation characteristics, which for low concentrations of hydrogen are well described by the linear dependence

$$\varepsilon_c^{(\mathrm{H})} = \varepsilon_c - A_1 C_{\mathrm{H}}, \tag{4.24}$$

where A_1 is the experimental constant of the "metal–hydrogen–containing medium" system. Assuming that the concentration of hydrogen in the metallic material is not high enough, its effect on fracture is modeled by relation (4.24).

The fracture energy density of the material is estimated as the area under the tensile diagram with strengthening, which is approximated by the power dependence (Fig. 4.8)

$$\sigma = \sigma_T + B\varepsilon^n, \tag{4.25}$$

where B, n are some constants of the material.

The density of the static component of the energy of plastic deformations dissipation is determined using the relationship

$$\gamma_s^{(\mathrm{H})} = \sigma_{\max}\varepsilon_{\max}. \tag{4.26}$$

The cyclic component is determined as the hysteresis loop area, which is considered symmetric about the axis $O\varepsilon$ (Fig. 4.8), that is almost always implemented in practice regardless of the stress ratio

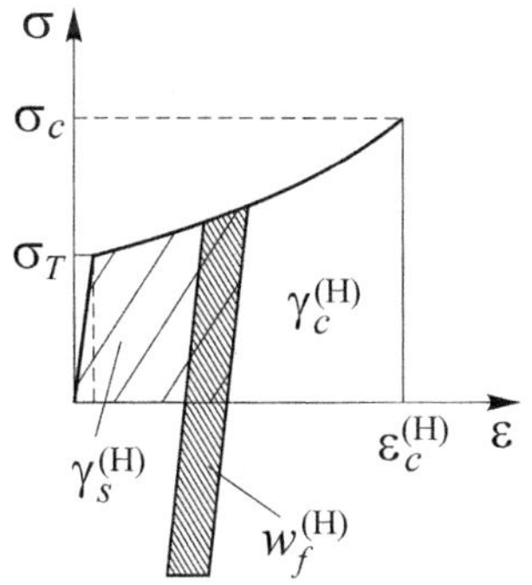

Fig. 4.8 Model representation of the cyclic tensile diagram in the pre-fracture zone

$$w_f^{(H)} = \left(\sigma_T + B\varepsilon_{\max}^n\right)\Delta\varepsilon l_0. \tag{4.27}$$

Here $\Delta\varepsilon$ is the value of the plastic deformation range in the zone of reversible plastic yield (width of the hysteresis loop) (Fig. 4.8).

The value of $\Delta\varepsilon$ is determined based on the Rice scheme in terms of the stress ratio R and deformation $\varepsilon_{\max}$. In (Panasyuk 1991) it was shown that the deformation at the notch tip can be determined with sufficient accuracy from the relation

$$\varepsilon_{\max} = \left(1 + (\rho/\rho_0)^2\right)^{-1/2}\left(K_{\mathrm{I}\max}^{\rho}/K_c\right)\varepsilon_c, \tag{4.28}$$

where $K_{\mathrm{I}\max}^{\rho}$ is the stress intensity factor at the notch tip, when $\rho \to 0$; $\rho_0 = 4K_c^2/(\pi E\sigma_T\varepsilon_c)$ is material constant.

According to the Rice scheme (Panasyuk 1988; Rice 1967), replacing the load by the load range, and doubling the yield strength of the material, we obtain the approximate formula for the deformation range $\Delta\varepsilon$

$$\Delta\varepsilon \approx 0,5\,(1 - R)^2\varepsilon_{\max}. \tag{4.29}$$

The threshold value of the energy of plastic deformations scattering $w_{th}^{(H)}$ is found from the relation

$$w_{th}^{(H)} = \sigma_T\Delta\varepsilon_{th} l_0, \tag{4.30}$$

where $\Delta\varepsilon_{th}$ is the deformation threshold range in the hydrogenated material (material constant).

The density of the static fracture energy of hydrogenated material, with account of hydrogen concentration according to relation (4.24), is calculated as the area under the total tensile diagram (Fig. 4.8)

$$\gamma_c^{(H)} = \left(\sigma_T + \frac{B}{n+1}(\varepsilon_c - A_1 C_{\mathrm{H}})\right)(\varepsilon_c - A_1 C_{\mathrm{H}}). \tag{4.31}$$

Having thus determined the energies included in the energy condition of initiation (4.23), we obtain such relationship for the period of the fatigue microcrack initiation at the stress notch tip in the hydrogenated material

$$N_1 = \frac{\alpha\left[\sigma_T + B\,(n+1)^{-1}\,(\varepsilon_c - A_1 C_{\mathrm{H}})^n\right](\varepsilon_c - A_1 C_{\mathrm{H}}) - (\sigma_T + B\varepsilon_{\max}^n)\,\varepsilon_{\max}}{(1-R)^2(\sigma_T + B\varepsilon_{\max}^n)\,\varepsilon_{\max} - 2\sigma_T\,\Delta\varepsilon_{th}}. \tag{4.32}$$

It should be noted that all the material constants in relation (4.32) are determined from the static tension testing. No constant is estimated from the fatigue experiment. The transition to the expression of cyclic loads here is carried out only with the help of the Morrow coefficient α, which allows us to determine the total energy of cyclic plastic deformations scattering during material fracture through its static fracture energy. This coefficient is calculated empirically by the dependence

$$\alpha = \left(\frac{\sigma_{ic}}{\sigma_a}\right), \tag{4.33}$$

proposed by Morrow based on many experiments on different materials (Troshchenko 1981).

Here σ_{ic} is the real ultimate strength; σ_a is the stress amplitude in the pre-fracture zone.

Expressing the value of σ_{ic} through the relative narrowing ψ, in our case we obtain

$$\alpha = \frac{\sigma_b^4}{\left(1-\psi^{(\mathrm{H})}\right)^4\left(\sigma_T + B\varepsilon_{\max}^n\right)^4}, \tag{4.34}$$

where σ_b is the ultimate strength of the material.

The obtained calculation data were compared with the experimental ones in Ostash et al. (1998) for U8A steel and 08kp steel for different geometries of stress notch in air (Fig. 4.9). This experiment was performed on disk samples at a fixed value of the deformation range $\Delta\varepsilon$ at the notch tip (stress ratio $R = 0.1$). In Fig. 4.9 points indicate the experimental data for U8A steel.

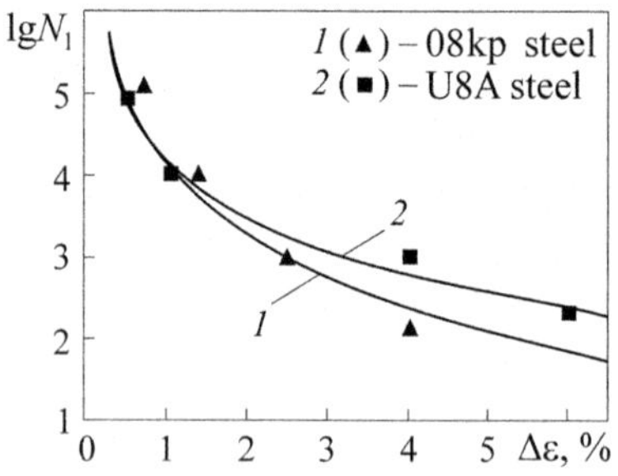

Fig. 4.9 Comparison of the period N_1 of fatigue microcrack initiation (lines 1, 2) calculated by formula (4.32) with experimental data

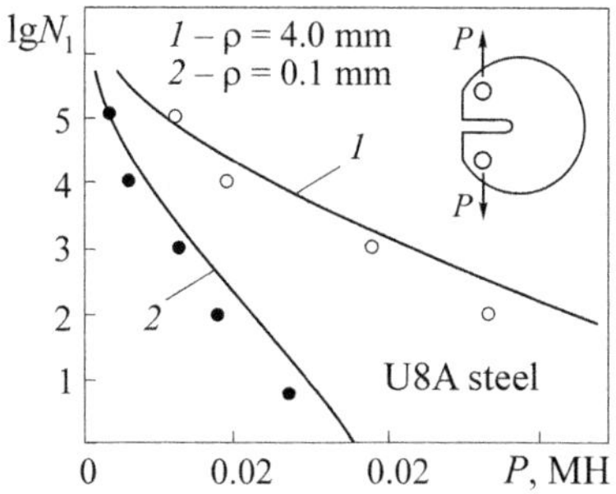

Fig. 4.10 Graphical comparison of the calculation data (lines 1, 2) with experiment (○, ●) for samples of different geometry

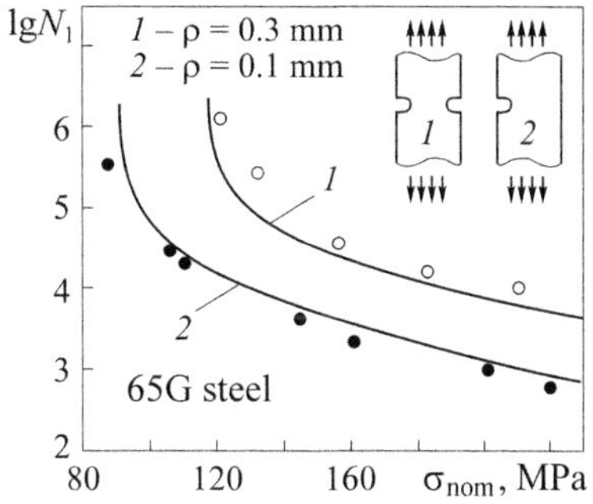

Fig. 4.11 Comparison of the calculated period of fatigue microcrack initiation (lines 1, 2) with experimental data

The experiment was carried out on disk samples for different values of the curvature radius ρ at the stress notch tip. In Fig. 4.10 the calculation is compared with the experiment for sufficiently sharp ($\rho = 0.1$ mm) and blunt ($\rho = 4.0$ mm) notches.

The calculation curves were also compared with the experimental data obtained in Yarema and Popovich (1985) for different geometries of the 65G steel samples (Fig. 4.11).

For the hydrogenated material, the calculation and experimental data were compared for Kh18N10T steel. The essence of the experiment was as follows. The round cylindrical Kh18N10T steel sample was subjected to cyclic loading, rotating in grips at some angle. Thus, the material on the sample surface was subjected to cyclic tension with the stress ratio $R = -1$. The experiment was performed in hydrogen (hydrogen pressure 35 MPa) and in helium (Fig. 4.12a). Constant A_1 of the hydrogen-metal system was evaluated under static tensile testing in hydrogen (Fig. 4.12b).

The calculation done for the hydrogenated material was compared with other experiments, for example, with data for 30KhGSNA and EI-643 steels (Karpenko 1985) (Fig. 4.13).

Mechanical characteristics of these steels are given in Table 4.1. As one can see, the calculated data are in good agreement with the results of all the experiments presented here.

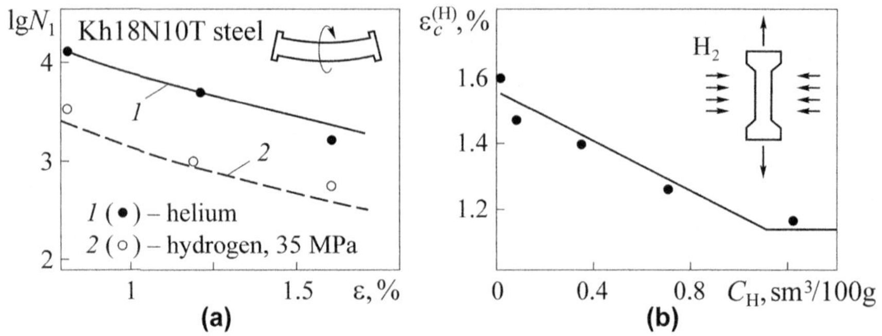

Fig. 4.12 Graphical representation of the $N_1 \sim \varepsilon_{max}$ dependence calculated by formula (4.32) for hydrogenated material (line) in comparison with the experimental data (○, ●) (**a**) and the dependence of the critical deformation $\varepsilon_c^{(H)}$ on hydrogen concentration C_H (**b**)

Fig. 4.13 Comparison of calculation (lines) and experimental data for hydrogenated material

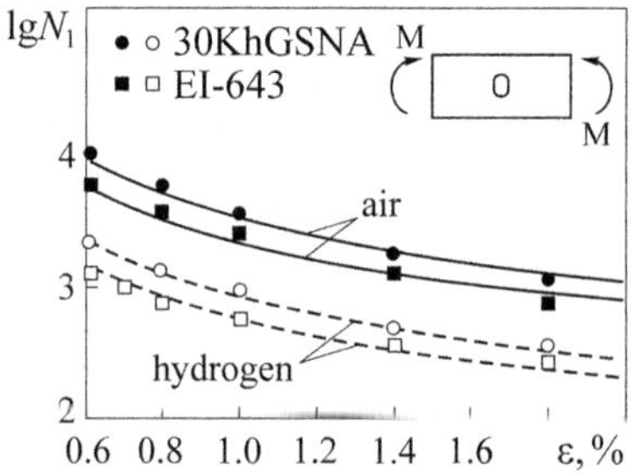

Table 4.1 Mechanical characteristics of metal samples for testing

Steel grade	σ_T, MPa	σ_b, MPa	ψ, %	ε_c	Source
08kp	190	270	80	1.6	Ostash et al. (1998)
U8A	266	661	49.5	0.68	Ostash et al. (1998)
65G	560	920	45	0.59	Yarema and Popovich (1985)
Kh18N10T	260	610	79	1.56	Panasyuk (1988)
30KhGSNA	1640	1780	52	0.73	Karpenko (1960)
EI-643	1690	1970	43	0.56	Karpenko (1960)

4.4 A Model for Determining the Period of Fatigue Microcrack Growth at the Hydrogenated Stress Notch Tip

The transition from the fatigue crack initiation to the stage of its growth is the passing from damages (including the formation of microcracks) scattered throughout the stressed volume of the material, to the fracture concentrated at the main macrocrack

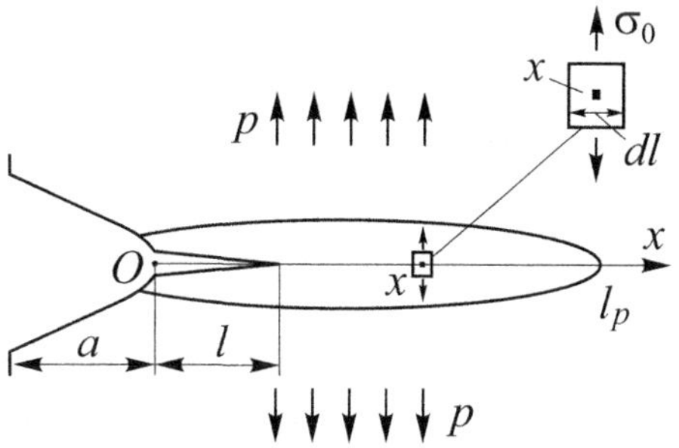

Fig. 4.14 Scheme of the plastic zone with a microcrack at the stress notch tip

front. This period corresponds to the stage of the initiated microcrack propagation to the macroscopic size. The methods of linear fracture mechanics cannot be used to calculate the kinetics of such microcrack growth, because its dimensions are usually commensurable with the dimensions of the plastic deformations zone in front of its tip. Therefore, based on the fracture mechanics energy approach, a calculation model is proposed, which allows us to estimate realistically the period of the microcrack growth to macroscopic size in the hydrogenated material at the stress notch tip.

Consider an elastic–plastic plate with a stress concentrator hydrogenated to a certain level of concentration C_{H} under action of cyclic load with force parameter p (Fig. 4.14). As a result of loading at the stress notch tip there is an initial plastic zone of length l_1, which is modeled by a rectilinear band of plasticity. The value of the initial plastic zone can be approximated, for example, by the relation proposed in Cherepanov (1967)

$$l_1 = \frac{\left(K_{\mathrm{Imax}}^{(e)}\right)^2}{2\pi\sigma_T^2} - \frac{\rho}{8}, \tag{4.35}$$

where ρ is the curvature radius at the stress notch tip and $K_{\mathrm{I\,max}}^{(e)}$ is the maximum SIF value at the equivalent crack tip ($\rho = 0$).

Let after N_1 load cycles, an elementary microcrack of length l_0 appears jump-like at the stress notch tip (Fig. 4.7). Then, the fatigue crack of length l initiates rectilinearly from the surface of the notch tip deep into the metal (Fig. 4.14). Assume that at the stress notch tip the submicrocrack will transform into a macrocrack, if during loading it have passed the entire initial static plastic zone formed in front of the stress notch tip and creates its own (typical of macrocrack) plastic zone.

The process of the macrocrack initiation can be modeled by successive fracture of elementary rectangles (calculation elements) of length dl. Based on the energy fracture criterion (4.21), the fracture condition of an arbitrary element with a center at point x (Fig. 4.14) can be written as

$$\gamma_s^{(\mathrm{H})}(x) + W_{f1}^{(\mathrm{H})}(x) + W_{f2}^{(\mathrm{H})}(x) = \alpha \cdot \gamma_c^{(\mathrm{H})}(x), \tag{4.36}$$

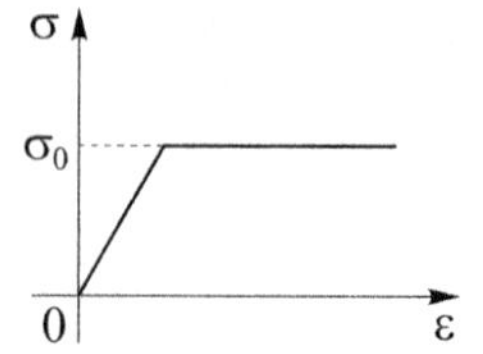

Fig. 4.15 Model representation of the tensile diagram

where $\gamma_s^{(H)}(x)$ is the density of the static component of the energy of plastic deformations dissipation in the pre-fracture zone; $W_{f1}^{(H)}(x)$, $W_{f2}^{(H)}(x)$ are cyclic components of the energy of plastic deformations scattering for N_1 load cycles before the appearance of a microcrack and for N_x load cycles during the microcrack growth to point x, respectively (Fig. 4.14); and α is the Morrow coefficient.

Let after N_x loading cycles, the microcrack, moving in the plastic zone, reaches the calculation element x and destroys it. Assuming that this element is in the zone of reversible plastic yield, the total cyclic scattering energy of the plastic deformations accumulated in this element for N_x load cycles, based on (4.27), (4.29) will be determined by the formula

$$W_{f2}^{(H)}(x) = \sigma_0(1 - R)^2 \int_o^{N_x} \varepsilon_{\max}[x - l(N),\ l(N)]\mathrm{d}N, \tag{4.37}$$

where R is the stress ratio; $\varepsilon_{\max}[x - l(N),\ l(N)]$ is the maximum plastic deformation along the axis Ox at a distance $x - l(N)$ from the tip of the microcrack of length $l(N)$ (Fig. 4.14); and σ_0 is a model value of the material yield strength, according to the δ_k-model (Panasyuk 1991) (Fig. 4.15).

Considering that for the crack initiation rate $v(l)$ the following relation is typical:

$$\mathrm{d}l = v(l)\mathrm{d}N \tag{4.38}$$

or

$$\mathrm{d}N = v^{-1}(l)\mathrm{d}l, \tag{4.39}$$

after replacing the integration variable in the integral of formula (4.37) it can be written as

$$W_{f2}^{(H)}(x) = \sigma_0(1 - R)^2 \int_0^{x} \varepsilon_{\max}[x - l,\ l]v^{-1}(l)\mathrm{d}l. \tag{4.40}$$

The value $\gamma_s^{(H)}$ can be represented as

$$\gamma_s^{(H)}(x) = \sigma_0\varepsilon_{\max}(0, x). \tag{4.41}$$

The density of the static energy of the material fracture is determined by the dependence

$$\gamma_c^{(\mathrm{H})} = \sigma_0 \varepsilon_c^{(\mathrm{H})}. \tag{4.42}$$

Taking into account (4.24) we obtain

$$\gamma_c^{(\mathrm{H})} = \sigma_0(\varepsilon_c - A_1 C_{\mathrm{H}}). \tag{4.43}$$

The value of the plastic deformation $\varepsilon_{\max}$ at the crack tip extension can be described with sufficient accuracy by the relation (Andreikiv and Lishchyns'ka 1999)

$$\varepsilon_{\max}(x - l, l) = \varepsilon_{\max}(0, l)\left(1 - \frac{x - l}{l_p}\right). \tag{4.44}$$

Here $\varepsilon_{\max}(0, l)$ is the maximum value of deformation at the tip of the microcrack of length l.

Assume [see, for example, (Cherepanov 1974)] that for sufficiently small nucleating cracks the amount of plastic deformation at its tip $\varepsilon_{\max}(0, l)$ is proportional to the crack length until it passes the entire initial plastic zone at the stress notch, i.e., becomes macroscopic. Based on this assumption, in Andreykiv and Darchuk (1992) it is shown that for the value $\varepsilon_{\max}(0, l)$ at the Mode I crack tip such formula is true

$$\varepsilon_{\max}(0, l) = \varepsilon_0(\rho) + \left[K_{\mathrm{I}\max}^2 K_c^{-2} \varepsilon_c - \varepsilon_0(\rho)\right] l \cdot l_1^{-1}, \tag{4.45}$$

where $\varepsilon_0(\rho)$ is the maximum value of deformation at the stress notch tip in the initial state (in the case of crack absence); $K_{\mathrm{I}\max}$ is the SIF at the crack tip of length l_1; and K_c is the critical SIF value.

At the moment of the nucleating microcrack appearance after N_1 load cycles in the initial plastic zone at the stress notch, the dissipation of the plastic deformations energy will have some distribution $W_{f1}(x)$. This distribution can be considered approximately as linear, exactly satisfying the start and end points of the initial plastic zone for $x = 0$ we have $W_{f1}^{(\mathrm{H})}(x) = \gamma_c^{(\mathrm{H})} - \gamma_s^{(\mathrm{H})}$,

if $x = l_1$ then $W_{f1}^{(\mathrm{H})}(x) = 0$.

That is why

$$W_{f1}(x) = \sigma_0\left[\alpha \cdot \varepsilon_c^{(\mathrm{H})} - \varepsilon_0(\rho)\right]\left(1 - \frac{x}{l_1}\right). \tag{4.46}$$

During the microcrack growth, the length of the plastic zone l_p at the stress notch will change from the initial one according to some law, which is approximately linear

$$l_p = l_1 + \beta l. \tag{4.47}$$

Here β is the coefficient of proportionality, which is found in the following way. After passing the entire initial plastic zone l_1 at the stress notch tip, the microcrack will create its own, typical of macrocracks, plastic zone, the length l_2 of which can be determined by the known formula for a self-similar crack (Panusyul 1988)

$$l_2 = \frac{K_{\mathrm{I}\max}^2}{2\pi\sigma_T^2}. \tag{4.48}$$

Then from formulas (4.47), (4.48) for β we get the expression

$$\beta = \frac{K_{\mathrm{I}\max}^2}{2\pi\sigma_T^2 l_1} - 1. \tag{4.49}$$

Having thus determined all the energies included in the energy criterion (4.36), the problem on its base is reduced to solving the following integral equation:

$$\begin{aligned}(1-R^2)\int\limits_0^x \left(\varepsilon_0 + \left(\varepsilon_{\max}^{(1)} - \varepsilon_0\right)\frac{l}{l_1}\right)\left[1 - \frac{x-l}{l_1+\beta\cdot l}\right]\upsilon^{-1}(l)\mathrm{d}l \\ = \left(\alpha\cdot\varepsilon_c^{(\mathrm{H})} - \varepsilon_{\max}^{(1)}\right)\frac{x}{l_1}, \quad 0\le x\le l_1, \quad x\ge l,\end{aligned} \tag{4.50}$$

where $\varepsilon_{\max}^{(1)}$ is the value of deformation at the minimum macrocrack tip.

Relation (4.50) is the second kind integral equation of Volterra with a convolution kernel with respect to the unknown function of the fatigue microcrack growth rate υ, as a function of its length l. Applying to its solution the analytical resolvent method (Verlan and Sizikov 1986) for the microcrack growth rate we obtain the following expression:

$$\nu(l) = (1-R)^2\frac{\varepsilon_0(l_1-l)+\varepsilon_{\max}^{(1)}\cdot l}{\alpha\cdot(\varepsilon_c - A_1C_H) - \varepsilon_{\max}^{(1)}}\left[1+\beta\frac{l}{l_1}\right]^{-1/\beta}. \tag{4.51}$$

Determining thus the fatigue microcrack growth rate (4.51), the total fatigue crack initiation period, which consists of the fatigue microcrack initiation period (4.32) and the period of its growth to macroscopic dimensions, is found by relation (4.19).

The kinetic Eq. (4.51) was compared with the experimental data obtained in Langford (1982) for short cracks in the V95 aluminum alloy (Fig. 4.16, the area is bounded by dotted lines).

Dependence of fatigue microcrack length on the number of load cycles, calculated by relations (4.51) and

$$N_2(l) = N_1 + \int\limits_{l_0}^{l} \upsilon^{-1}(y)\mathrm{d}y, \tag{4.52}$$

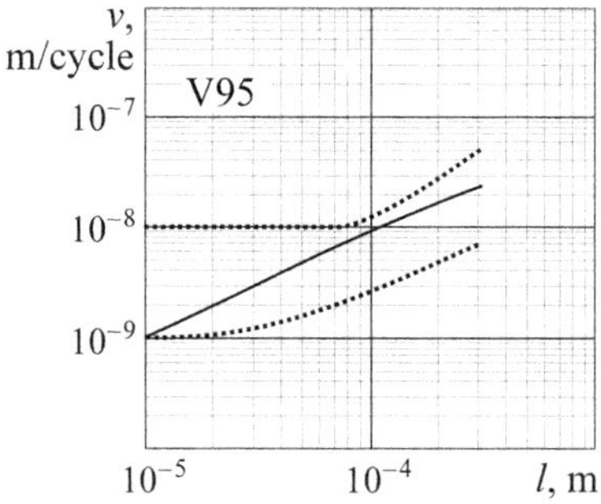

Fig. 4.16 Graphic representation of the calculation by expression (4.51) (solid line) compared to experimental data (dotted lines)

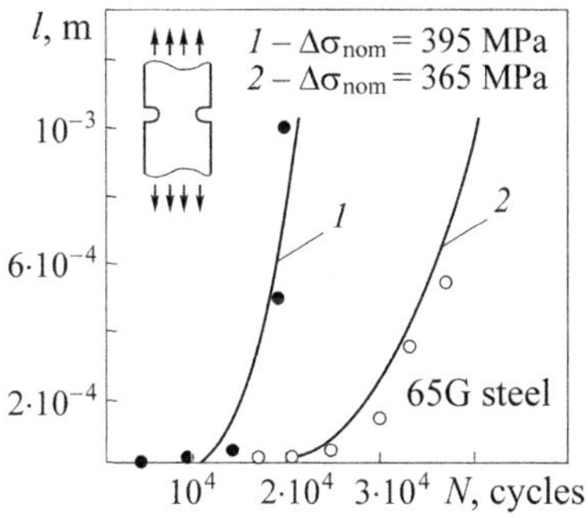

Fig. 4.17 Comparison of the calculated dependence $l = l(N)$ (lines 1, 2) with experimental data (○, ●)

is compared with experimental data obtained for 65G steel for different load levels (Yarema and Popovich 1985) (Fig. 4.17). As we can see, the results of this comparison prove the model described above.

A specific feature of the calculations of engineering metal structures from the standpoint of fracture mechanics is the ability to take into account the presence of the original defects or fatigue damage in the material that initiate during operation. Such calculations are based on the study of the kinetics of fatigue cracks propagation before they reach critical dimensions, and their main purpose is to establish the residual life of the elements of metal structures. In engineering structures, the stage of fatigue crack extension can be from 10 to 90% of the total life of the product (Pokhodnia 1998). Therefore, the calculated determination of the residual life of the structure becomes important in practice. When diagnosing with the methods of non-destructive testing, it is possible to establish the initial defect of the product, and then calculate the residual life of the structure N_d according to the known relation (Andreykiv and Darchuk 1992)

$$N_d = \int_{l_0}^{l_*} V^{-1}(l)\mathrm{d}l, \qquad (4.53)$$

where l_0, l_* are the initial and critical length of the fatigue macrocrack, respectively; $V(l)$ is the fatigue macrocrack growth rate.

In order to use formula (4.53), it is necessary to know the crack growth kinetics, which is determined from the kinetic equations of fatigue crack propagation. Known in the literature (Panasyuk 1988) fatigue fracture models are implemented mainly for cases of homogeneous materials with no account of the influence of hydrogen-containing environment, in which metal structural elements are often used. Existing calculation models are also ineffective for cases of the complex stress state, which is typical of the vast majority of elements of engineering constructions and structures. At the same time, in analytical studies, the regularities of the fatigue cracks propagation in inhomogeneous materials, in particular in welded structural elements, have been studied insufficiently (Trufiakov 1990; Lobanov 1993; Karzov et al. 1982; Schwalbe and Kocak 1997). Therefore, in this chapter we construct a calculation model for estimating the kinetics and determining the period of subcritical fatigue crack growth in hydrogenated and heterogeneous, by mechanical properties, metallic materials under the plane stress conditions. The proposed equations are obtained using the energy criterion of fatigue fracture of materials, which is based on the equation of energy balance in thermodynamics (the first law of thermodynamics) and the following hypothesis: the total energy of elastic–plastic deformations scattering due to fatigue crack propagation per unit area of the new surface is the material constant under given external conditions and temperature.

4.5 Kinetics of Fatigue Macrocrack Propagation in Hydrogenated Materials Under Plane Stress Conditions

Consider an inhomogeneous elastic–plastic plate hydrogenated to a certain level of concentration $C_{\mathrm{H}}(x,\ y)$ and weakened by a crack. Let the plate be loaded by external forces characterized by the force parameter p (Fig. 4.18a).

Let for ΔN load cycles the crack extends to the length Δl, and a cyclic plastic zone of length l_{pf} be formed at its tip. As is known (Panasyuk et al. 1994), l_{pf} is less than the length of the static plastic zone l_p (Fig. 4.18b) and depends on the stress ratio R $(R = K_{\min}/K_{\max})$

$$l_{pf} = \frac{(1-R)^2}{4} l_p. \tag{4.54}$$

To construct the kinetic equations of fatigue macrocrack growth, use the energy fracture criterion, obtained on the basis of the first law of thermodynamics (the energy conservation law) (Andreykiv et al. 2017). According to this criterion, in order to

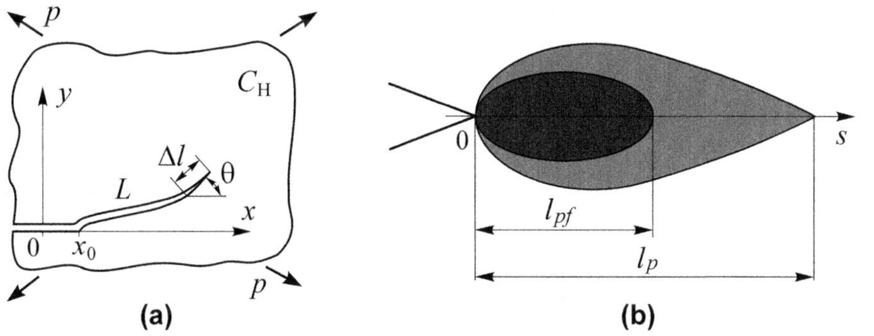

Fig. 4.18 Scheme of a plate with a crack (**a**) and a plastic zone in front of the fatigue crack tip (**b**)

extend the fatigue crack to a length Δl for ΔN load cycles, the total energy of plastic deformations scattering in the material for ΔN cycles at points $(x,\ y)$ on the crack growth path ($W = W(x,\ y)$) must reach the value of the energy of material fracture ($W_c = W_c(x,\ y)$))

$$\Delta W = \Delta W_c. \tag{4.55}$$

Expanding the total energy of plastic deformations scattering into static and cyclic components and assuming that the density of the static component and the fracture energy of the material are constant during crack propagation by a small value Δl, we write the energy criterion of fatigue fracture (4.55) as

$$\gamma_s^{(\mathrm{H})}\Delta l + \Delta N\left(w_f^{(\mathrm{H})} - w_{th}^{(\mathrm{H})}\right) = \alpha \cdot \gamma_c^{(\mathrm{H})}\Delta l, \tag{4.56}$$

where $\gamma_s^{(\mathrm{H})}$ is the density of the static component of the energy of plastic deformations scattering; $w_f^{(\mathrm{H})}$ and $w_{th}^{(\mathrm{H})}$ are the cyclic component of the energy of plastic deformations scattering for one load cycle and its threshold value; $\gamma_c^{(\mathrm{H})}$ is the density of the material fracture energy; and α is the Morrow coefficient.

The boundary transition at $\Delta N \to 0$ in relation (4.56) gives the following differential equation:

$$\frac{\mathrm{d}l}{\mathrm{d}N} = \frac{w_f^{(\mathrm{H})} - w_{th}^{(\mathrm{H})}}{\varepsilon \cdot \gamma_c^{(\mathrm{H})} - \gamma_s^{(\mathrm{H})}}. \tag{4.57}$$

Formula (4.57) is a general form of the kinetic equation for determining the fatigue crack propagation rate $V = dl/dN$ in the hydrogenated inhomogeneous material.

To determine the path of the fatigue crack propagation, we suppose that θ is the angle of the fatigue crack propagation direction. Consider that the crack propagates in the direction where its speed will be maximum (Andreikiv and Lishchyns'ka 1999). Then, equating to zero the derivative of the right-hand side of Eq. (4.57) in

Fig. 4.19 Tensile and shear model curves

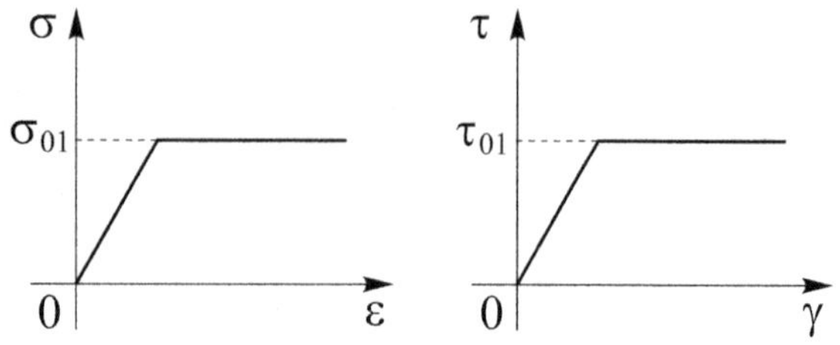

the direction of fatigue crack growth θ, we obtain the second kinetic equation for predicting the crack propagation path

$$\frac{\partial}{\partial\theta}\left(\frac{w_f^{(\mathrm{H})} - w_{th}^{(\mathrm{H})}}{\alpha\cdot\gamma_c^{(\mathrm{H})} - \gamma_s^{(\mathrm{H})}}\right) = 0 \tag{4.58}$$

under following initial and final conditions:

$$N = 0 \text{ for } l = l_0; \quad N = N_d \text{ for } l = l_*. \tag{4.59}$$

The plastic zone in front of the fatigue crack tip according to the δ_k-model is represented by a model notch to the faces of which the adhesion forces are applied. Such forces are equal to the model values of yield strength for a perfect elastic–plastic material σ_{01} and τ_{01} under tension and shear, respectively (Fig. 4.19).

Then the value of the cyclic component of the energy $w_f^{(\mathrm{H})}$ of plastic deformations scattering per one load cycle can be represented as

$$w_f^{(\mathrm{H})} = \int\limits_0^{l_{pf}}\left(\sigma_{01}\Delta\delta_{\mathrm{I}}^{(\mathrm{H})}(s) + \tau_{01}\Delta\delta_{\mathrm{II}}^{(\mathrm{H})}(s)\right)\mathrm{d}s. \tag{4.60}$$

Here

$$\Delta\delta_{\mathrm{I}}^{(\mathrm{H})}(s) = \delta_{\mathrm{I\,max}}^{(\mathrm{H})}(s) - \delta_{\mathrm{I\,min}}^{(\mathrm{H})}(s) \text{ and } \Delta\delta_{\mathrm{II}}^{(\mathrm{H})}(s) = \delta_{\mathrm{II\,max}}^{(\mathrm{H})}(s) - \delta_{\mathrm{II\,min}}^{(\mathrm{H})}(s)$$

are the ranges of the model crack opening displacement according to the δ_k-model at point s of the plastic pre-fracture zone ($0 \le s \le l_{pf}$); $\delta_{\mathrm{I\,max}}^{(\mathrm{H})}$, $\delta_{\mathrm{II\,max}}^{(\mathrm{H})}$, $\delta_{\mathrm{I\,min}}^{(\mathrm{H})}$, and $\delta_{\mathrm{II\,min}}^{(\mathrm{H})}$ are maximum and minimum values of the Mode I and Mode II fatigue crack tip opening displacement in the pre-fracture zone of the hydrogenated material.

The threshold value of the cyclic component of the energy $w_{th}^{(\mathrm{H})}$ of plastic deformations scattering can be determined using the relation

$$w_{th}^{(\mathrm{H})} = \int\limits_0^{l_{pf}}\left(\sigma_{01}\Delta\delta_{\mathrm{I}\,th}^{(\mathrm{H})}(s) + \tau_{01}\Delta\delta_{\mathrm{II}\,th}^{(\mathrm{H})}(s)\right)\mathrm{d}s, \tag{4.61}$$

where $\Delta\delta_{\mathrm{I}th}^{(\mathrm{H})}$ and $\Delta\delta_{\mathrm{II}th}^{(\mathrm{H})}$ are the threshold values of the Mode I and Mode II cracks opening displacement, respectively.

Since for one fracture event at each point of the segment Δl the crack tip opening displacement acquires the maximum value ($\delta_{\max}$), and the static component of the energy $\gamma_s^{(\mathrm{H})}$ of plastic deformations scattering is given in the form

$$\gamma_s^{(\mathrm{H})} = \sigma_{01}\delta_{\mathrm{I}\max}^{(\mathrm{H})} + \tau_{01}\delta_{\mathrm{II}\max}^{(\mathrm{H})}. \tag{4.62}$$

In (Panasyuk et al. 1994), a relation is obtained that connects the maximum and minimum values of the crack opening displacement δ per load cycle in the plastic zone in terms of the stress ratio RR

$$\delta_{\min} = \left[1 - \frac{(1-R)^2}{2}\right]\delta_{\max}. \tag{4.63}$$

The specific fracture energy of the hydrogenated material under the plane stress state, in the general case, can be written as

$$\gamma_c^{(\mathrm{H})} = \sigma_{01}\delta_{\mathrm{I}*}^{(\mathrm{H})} + \tau_{01}\delta_{\mathrm{II}*}^{(\mathrm{H})}, \tag{4.64}$$

where $\delta_{\mathrm{I}*}^{(\mathrm{H})}$ and $\delta_{\mathrm{II}*}^{(\mathrm{H})}$ are the critical values of the Mode I and Mode II opening displacement in the plastic zone, during simultaneous achievement of which the material fracture occurs.

Assuming that the material is hydrogenated to concentration C_{H} and taking into account the relationship (4.24) we rewrite formula (4.64) in the form

$$\gamma_c^{(\mathrm{H})} = \sigma_{01}(\delta_{\mathrm{I}*} - A_1 C_{\mathrm{H}}) + \tau_{01}(\delta_{\mathrm{II}*} - A_2 C_{\mathrm{H}}), \tag{4.65}$$

where A_1 and A_2 are experimental constants of the system "metal–hydrogen–containing medium".

Having thus determined the energies in kinetic Eqs. (4.57) and (4.58), we reduce those using Eqs. (4.60)–(4.65) to the form

$$\frac{\mathrm{d}l}{\mathrm{d}N} = \frac{0.5\,(1-R)^2 \int_0^{l_{pf}} \left(\sigma_{01}\delta_{\mathrm{Imax}}^{(\mathrm{H})}(s) + \tau_{01}\delta_{\mathrm{IImax}}^{(\mathrm{H})}(s)\right)\mathrm{d}s}{\alpha\cdot(\sigma_{01}\delta_{\mathrm{I}*} + \tau_{01}\delta_{\mathrm{II}*} - (A_1\sigma_{01} + A_2\tau_{01})C_{\mathrm{H}}) - \left(\sigma_{01}\delta_{\mathrm{I}\max}^{(\mathrm{H})} + \tau_{01}\delta_{\mathrm{II}\max}^{(\mathrm{H})}\right)}$$
$$- \frac{\int_0^{l_{pf}} \left(\sigma_{01}\Delta\delta_{\mathrm{I}th}^{(\mathrm{H})}(s) + \tau_{01}\Delta\delta_{\mathrm{II}th}^{(\mathrm{H})}(s)\right)\mathrm{d}s}{\alpha\cdot(\sigma_{01}\delta_{\mathrm{I}*} + \tau_{01}\delta_{\mathrm{II}*} - (A_1\sigma_{01} + A_2\tau_{01})C_{\mathrm{H}}) - \left(\sigma_{01}\delta_{\mathrm{I}\max}^{(\mathrm{H})} + \tau_{01}\delta_{\mathrm{II}\max}^{(\mathrm{H})}\right)}. \tag{4.66}$$

$$\frac{\partial}{\partial\theta}\left(\frac{dl}{dN}\right) = 0 \tag{4.67}$$

4.6 Determination of the Period of Subcritical Fatigue Crack Growth in a Hydrogenated Heterogeneous

For the convenience of application of the described model to solve practical problems, it is necessary to determine the deformation energies included in the determining Eqs. (4.66), (4.67), in terms of parameters of linear fracture mechanics—SIF for a self-similar crack. To do this, we approximate the maximum values of the crack opening displacement δ at the points of the model plastic zone of length l_{pf} in front of the fatigue crack tip by quadratic dependences (Andreikiv and Lishchyns'ka 1999)

$$\delta_{\mathrm{I\,max}}^{(\mathrm{H})}(x) = \delta_{\mathrm{I\,max}}^{(\mathrm{H})}(0)\left(1 - \frac{x}{l_{pf}}\right)^2, \quad \delta_{\mathrm{II\,max}}^{(\mathrm{H})}(x) = \delta_{\mathrm{II\,max}}^{(\mathrm{H})}(0)\left(1 - \frac{x}{l_{pf}}\right)^2. \tag{4.68}$$

Here $\delta_{\mathrm{I\,max}}^{(\mathrm{H})}(0)$ and $\delta_{\mathrm{II\,max}}^{(\mathrm{H})}(0)$ are the maximum values of the crack tip opening displacement of the Mode I and Mode II fatigue crack, which can be determined by the known relations of linear fracture mechanics (Panasyuk 1988)

$$\delta_{\mathrm{I\,max}}^{(\mathrm{H})}(0) = \frac{K_{\mathrm{I\,max}}^2}{E\sigma_{01}}, \quad \delta_{\mathrm{II\,max}}^{(\mathrm{H})}(0) = \frac{K_{\mathrm{II\,max}}^2}{E\tau_{01}}, \tag{4.69}$$

where E is the Young's modulus.

The length of the cyclic plastic zone in front of the fatigue macrocrack tip, taking into account formula (4.54), can be found from the following relations of linear fracture mechanics (Panasyuk 1991):

$$l_{pf\,\mathrm{I}} = \frac{\pi(1-R)^2 K_{\mathrm{I\,max}}^2}{32\sigma_{01}^2}, \quad l_{pf\,\mathrm{II}} = \frac{\pi(1-R)^2 K_{\mathrm{II\,max}}^2}{32\tau_{01}^2}. \tag{4.70}$$

Thus, substituting relations (4.68)–(4.70) in (4.60) and (4.61), after simple mathematical transformations for the cyclic component of the energy of plastic deformations scattering and its threshold value we obtain the expressions

$$w_f^{(\mathrm{H})} = \frac{\pi(1-R^4)(1-R)^2}{96E}\left(\frac{K_{\mathrm{I\,max}}^4}{\sigma_{01}^2} + \frac{K_{\mathrm{II\,max}}^4}{\tau_{01}^2}\right), \tag{4.71}$$

$$w_{th}^{(\mathrm{H})} = \frac{\pi(1-R)^2}{32E}\left[\frac{\Delta K_{\mathrm{I}th\mathrm{H}}^2 K_{\mathrm{I\,max}}^2}{\sigma_{01}^2} + \frac{\Delta K_{\mathrm{II}th\mathrm{H}}^2 K_{\mathrm{II\,max}}^2}{\tau_{01}^2}\right]. \tag{4.72}$$

The static component of the energy $\gamma_s^{(\mathrm{H})}$ of plastic deformations scattering will be determined by the dependence

$$\gamma_s^{(\mathrm{H})} = \frac{1}{E}\left(K_{\mathrm{I}\max}^2 + K_{\mathrm{II}\max}^2\right). \tag{4.73}$$

Expressing the values $\delta_{\mathrm{I}*}^{(\mathrm{H})}$, $\delta_{\mathrm{II}*}^{(\mathrm{H})}$, included in the relation for the fracture energy density $\gamma_c^{(\mathrm{H})}$ of the material through the corresponding SIF values, according to dependences (4.69), we obtain

$$\gamma_c^{(\mathrm{H})} = \frac{1}{E}\left(K_{\mathrm{I}*}^2 + K_{\mathrm{II}*}^2\right) - (A_1\sigma_{01} + A_2\tau_{01})\,C_{\mathrm{H}}. \tag{4.74}$$

To establish the critical SIF values $K_{\mathrm{I}*}$ and $K_{\mathrm{II}*}$, when these are reached simultaneously and the fracture occurs in the plastic zone in front of the fatigue crack tip, our considerations are as follows. The values $K_{\mathrm{I}*}$ and $K_{\mathrm{II}*}$ must satisfy the criterion of mixed-mode (I + II) fracture macromechanism (Panasyuk et al. 2000)

$$\left(\frac{K_{\mathrm{I}*}}{K_{\mathrm{I}c}}\right)^4 + \left(\frac{K_{\mathrm{II}*}}{K_{\mathrm{II}c}}\right)^4 = 1, \tag{4.75}$$

where $K_{\mathrm{I}c}$ and $K_{\mathrm{II}c}$ are critical SIF values under action of single Mode I or Mode II macromechanism of fracture, respectively.

Since the qualitative scheme of the deformation process in the pre-fracture zone at the crack tip under cyclic loading of the body remains for the case of the limit-equilibrium state in this zone, there is the equality

$$\frac{K_{\mathrm{II}*}}{K_{\mathrm{I}*}} = \frac{K_{\mathrm{II}\max}}{K_{\mathrm{I}\max}}. \tag{4.76}$$

Considering this as well as the criterion (4.75) for $K_{\mathrm{I}*}$, $K_{\mathrm{II}*}$, we obtain the expressions

$$K_{\mathrm{I}*} = \frac{K_{\mathrm{I}c}K_{\mathrm{II}c}}{\left[(K_{\mathrm{II}c})^4 + \eta_{\max}^4 (K_{\mathrm{I}c})^4\right]^{1/4}}$$
$$K_{\mathrm{II}*} = \frac{\eta_{\max} K_{\mathrm{I}c}K_{\mathrm{II}c}}{\left[(K_{\mathrm{II}c})^4 + \eta_{\max}^4 (K_{\mathrm{I}c})^4\right]^{1/4}}, \tag{4.77}$$

where $\eta_{\max} \equiv K_{\mathrm{II}\max}/K_{\mathrm{I}\max}$.

Then, according to (4.74) and (4.77), the density of the fracture energy $\gamma_c^{(\mathrm{H})}$ for the hydrogenated material under the plane stress state will be

$$\gamma_c^{(\mathrm{H})} = \frac{K_{\mathrm{I}c}K_{\mathrm{II}c}\,(1 + \eta_{\max}^2)}{E\left[(K_{\mathrm{II}c})^4 + \eta_{\max}^4\,(K_{\mathrm{I}c})^4\right]^{1/2}} - (A_1\sigma_{01} + A_2\tau_{01})\,C_{\mathrm{H}}. \tag{4.78}$$

In the case of plastic deformation the Huber–Mises plastic yield condition must be satisfied for σ_{01} and τ_{01}, which in the case of the plane stress state will look like (Katchanov 1969)

$$\sigma_{01}^2 + 3\tau_{01}^2 = \sigma_0^2, \tag{4.79}$$

where according to the δ_k-model σ_0 is the macroscopic value of the yield strength of the material under uniaxial tension.

From the condition of equality of Mode I and Mode II plastic zones (formula (4.70)) in front of the crack tip, the following relationship between the σ_{01} and τ_{01} values is obtained:

$$\tau_{01} = \eta_{\max}\sigma_{01}. \tag{4.80}$$

Then on the basis of relations (4.79), (4.80) the values σ_{01} and τ_{01} can be expressed in terms of σ_0:

$$\sigma_{01} = \frac{\sigma_0}{\sqrt{1 + 3\eta_{\max}^2}}, \quad \tau_{01} = \frac{\eta_{\max}\sigma_0}{\sqrt{1 + 3\eta_{\max}^2}}. \tag{4.81}$$

Since the characteristic σ_0 is a function of the coordinates of the point $(x, \ y)$, it will vary along the entire length of the cyclic plastic zone l_{pf}. However, due to the infinitesimality of this zone we will assume as the yield strength σ_0 in the direction θ of the crack propagation its average value in length l_{pf} (Fig. 4.20)

$$\tilde{\sigma}_0(x, y, \theta) = \frac{1}{2}\big(\sigma_0(x, y) + \sigma_0(x + l_{pf}\cos\theta, y + l_{pf}\sin\theta)\big). \tag{4.82}$$

Expanding the expression (4.82) into the Taylor series and rejecting the summands of the higher order of infinitesimality, we obtain

$$\tilde{\sigma}_0(x, y, \theta) = \sigma_0(x, y) + \frac{l_{pf}}{2}\left(\frac{\partial\sigma_0(x, y)}{\partial x}\cos\theta + \frac{\partial\sigma_0(x, y)}{\partial y}\sin\theta\right). \tag{4.83}$$

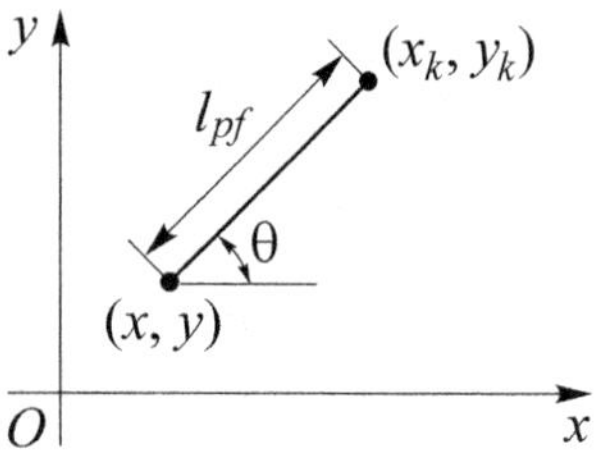

Fig. 4.20 Scheme of the plastic zone

Having written down the energies included in the determining relations (4.66) and (4.67) through the parameters of the linear fracture mechanics and taking into account relations (4.81), (4.83), we obtain

$$V(x, y, \theta) = \frac{\frac{\pi(1-R)^2}{32}\frac{(1+3\eta_{\max}^2)}{\tilde{\sigma}_0^{-2}}\left(\frac{1-R^4}{3}\left(K_{\mathrm{I}\max}^4 + \frac{K_{\mathrm{II}\max}^4}{\eta_{\max}^2}\right) - \Delta K_{\mathrm{I}th\mathrm{H}}^2 K_{\mathrm{I}\max}^2 - \frac{\Delta K_{\mathrm{II}th\mathrm{H}}^2 K_{\mathrm{II}\max}^2}{\eta_{\max}^2}\right)}{\alpha\cdot\left(\frac{K_{\mathrm{I}c}^2 K_{\mathrm{II}c}^2(1+\eta_{\max}^2)}{\sqrt{(K_{\mathrm{II}c})^4+\eta_{\max}^4(K_{\mathrm{I}c})^4}} - E(A_1\sigma_{01} + A_2\tau_{01})C_{\mathrm{H}}\right) - K_{\mathrm{I}\max}^2 - K_{\mathrm{II}\max}^2}, \tag{4.84}$$

$$\frac{\partial}{\partial\theta}V(x, y, \theta) = 0, \tag{4.85}$$

where the Morrow coefficient α, according to (4.1) and (4.2), was calculated by the formula

$$\alpha = \frac{\sigma_b^4}{(1-\psi^{(\mathrm{H})})^4\left(\sigma_T + B\left[K_{\mathrm{I}\max}/K_{\mathrm{I}c}\right]^{2n}\varepsilon_c^n\right)^4}. \tag{4.86}$$

The angle of the propagation direction of the fatigue crack θ, as can be seen from Fig. 4.18, is the geometric meaning of the derivative $y'(x)$ value. Then, substituting after differentiation by θ the value of $y'(x)$ for θ in the relation (4.85), we obtain a nonlinear first-order differential equation to determine the fatigue crack propagation path. Integrating this differential equation with the initial condition

$$y(x_0) = y_0, \tag{4.87}$$

we obtain the path equation in the form

$$F(x, y) = 0. \tag{4.88}$$

Evaluating the fatigue crack propagation rate based on relations (4.84)–(4.88), the subcritical period of its growth is determined from formula (4.81).

The calculated kinetic Eq. (4.84) was compared with the data (Panasyuk et al. 2000) obtained from the fatigue crack growth testing in air under mixed-mode (I + II) fracture macromechanism (Fig. 4.21). The essence of this experiment was as follows. A long cylindrical thin-walled specimen made of 20 steel (see Appendix), with two symmetrical through cracks (Fig. 4.21), was cyclically loaded simultaneously with torque M and tensile force T, thus providing the plane stress state at the crack tip for a constant value of $\eta_{\max} = 0.8$.

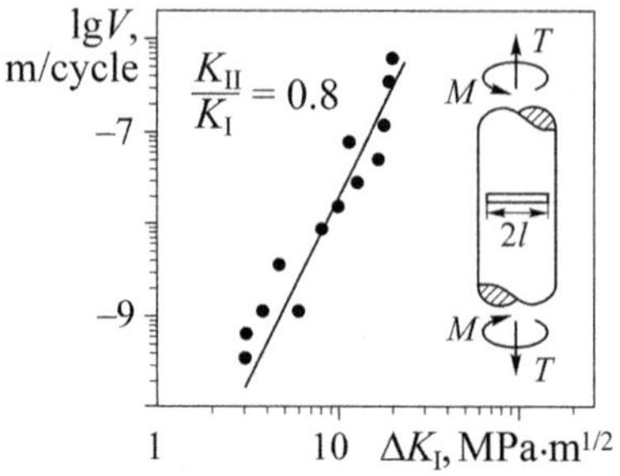

Fig. 4.21 Graphical representation of the calculated dependence (4.84) in comparison with experimental data (circles) of fatigue macrocrack propagation under mixed-mode (I + II) fracture macromechanism of a 20 steel cylindrical sample

Kinetic Eq. (4.84) for the case of uniaxial tension ($K_{II} = 0$) was compared with the experimental data (Dauskardt and Ritchi 1986) during cycling of a compact sample of SA387-2-22 steel in hydrogen (Fig. 4.22a), as well as for samples made of different zones of welded joints of this steel (Fig. 4.22b).

As can be seen at (Figs. 4.21 and 4.22), the calculation and experimental data agree well.

Mechanical characteristics of the mentioned steels are presented in Table 4.2.

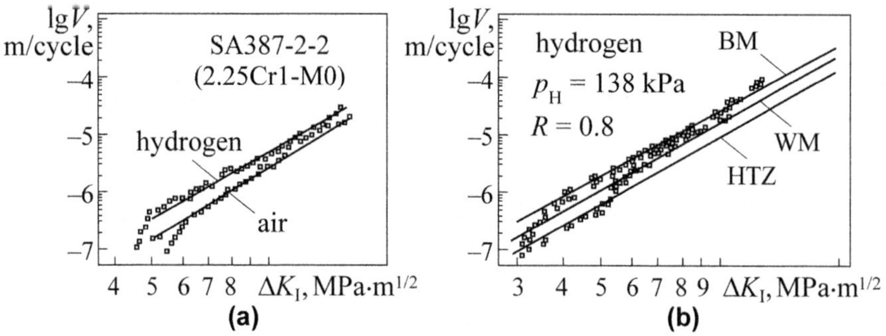

Fig. 4.22 Comparison of experimental (points) and calculated by formula (4.84) (solid lines) kinetic diagrams of fatigue fracture of SA387-2–2 steel (**a**) and welded joints of this steel in hydrogen (**b**)

Table 4.2 Mechanical characteristics of metal samples for testing (Dauskardt and Ritchi 1986)

Steel grade	σ_T, MPa	σ_b, MPa	ψ, %	K_{Ic}, MPa · m$^{1/2}$	K_{IIc}, MPa · m$^{1/2}$
SA387-2-2 steel	290	500	76	286	–
20 steel	270	460	50	101	210

Thus, relations (4.81), (4.84)–(4.87) form a computational model for determining the path and estimating the period of subcritical fatigue macrocrack growth in hydrogenated inhomogeneous, by mechanical characteristics, elements of metal structures under the plane stress conditions.

References

Achenbach JD, Harris JG (1979) Acoustic emission from a brief crack propagation event. ASME J Appl Mech 46(1):107–112

Aki K, Richards PK (1983) Kolichestvennaia seismologiia: teoriia i metody (Quantitative seismology: theory and methods). Mir Publishing House

Andreikiv OÉ, Sergiénko OM, Lisak MV, Skal's'kii VR (1993) The contribution of Rayleigh waves to the acoustic field arising from internal-defect growth. Mater Sci 29(2):115–120

Andreikiv OE, Nazarchuk ZT, Skal's'kyi VR, Rudavs'kyi DV, Serhienko OM (2008) Stress intensity factors caused by magnetic fields in ferromagnets. Mater Sci 44(3):456–460

Andreikiv OE, Lishchyns'ka MV (1999) Equations of growth of fatigue cracks in inhomogeneous plates. Mater Sci 35(3):355–362

Andreykiv OYe, Darchuk AI (1992) Ustalostnoie razrusheniie i dolgovechnost konstruktsii (Fatigue fracture and life time of structures). Naukova dumka Publishing House

Andreykiv OYe, Skalskyi VR, Dolinska IYa (2017) Zapovilnene ruinuvannia materialiv za localnoi povzuchosti (Delayed fracture of materials due to local creep). Publishing House of I. Franko National University of Lviv

Cagniard I (1962) Reflection of progressive seismic waves. McGraw-Hill

Cherepanov GP (1967) O razprostranenii treshchin v sploshnoi srede (On crack propagation in continuous medium). Appl Math Mech 31(3):476–488

Cherepanov GP (1974) Mekhanika khrupkogo razrusheniia (Brittle fracture mechanics). Nauka Publishing House

Dauskardt RH, Ritchi RO (1986) Fatigue crack propagation behavior in pressure vessel steels for high pressure hydrogen service. In Nisbett EG (ed) Properties of high-strength steels for high-pressure containments. The American Society of Mechanical Engineers

Gerberich WW, Atteridge DG, Lessar JF (1975) Acoustic emission investigation of microscopic ductile fracture. Metall Trans A 6A(2):797–801

Harris JG, Pott J (1984) Surface motion excited by acoustic emission from a buried crack. Trans ASME J Appl Mech 51(1):77–83

Karpenko HV (1960) Vplyv vodniu na mechanichni vlastyvosti stali (Hydrogen effect on mechanical properties of steels). Vydavnytstvo Akademii Nauk USSR (Publ House of Ukr SSR Academy of Sciences), Kyiv

Karpenko HV (1985) Rabotosposobnost konstruktsionnykh materialov v aggresivnykh sredakh (Serviceability of structural meterials in aggresive media). Naukova dumka Publishing House

Karzov GP, Leonov VP, Timofeyev BT (1982) Svarnyie sosudy vysokogo davleniia: Prochnost i dolgovechnost (Welded high-pressure vessels: Strength and durability). Machinebuilding Publishing House

Katchanov LM (1969) Osnovy teorii plastychnosti (Bases of plasticity theory). Nauka Publishing House

Langford J (1982) The growth of small fatigue cracks in 7075–T6 aluminium. Fatigue Eng Mater Struct 5(2):233–248

Lobanov LM (ed) (1993) Svarnyie stroitelnyie konstruktsii. T. 1: Osnovy proektirovaniia konstrukysii (Welded building constructions. Vol. I.: Bases of structures design). Naukova dumka Publishing House

McMeeking RM (1977) Finite deformation analysis of crack tip opening in elastic-plastic materials and implication for fracture. J Mech Phys Solids 25(5):357–381
Nazarchuk ZT, Skalskyi VR (2009) Akustyko-emisiine diagnostuvannia elementiv konstruktsii. T. 1: Teoretychni osnovy metodu akustychnoi emissii (Acoustic-emisical diagnostics of structural elements. Vol. 1: theoretical foundations of the acoustic emission method). Naukova dumka Publishing House
Nazarchuk Z, Skalskyi V, Serhiyenko O (2017a) Analysis of acoustic emission caused by internal cracks. In: Acoustic emission. methodology and application (Foundations in engineering mechanics, pp 75–105). Springer, Cham
Nazarchuk Z, Skalskyi V, Serhiyenko O (2017b) Some methodological foundations for selecting and processing AE signals. In: Acoustic emission. Methodology and application (Foundations in engineering mechanics, pp 107–159). Springer, Cham
Ostash OP, Panasyuk VV, Kostyk EM (1998) Unified model of nucleation and growth of fatigue macrocracks. Part 2. Application of deformational parameters of the fracture mechanics of materials in the stage of crack initiation. Mater Sci 34(3):351–364
Panasyuk VV (ed) (1988) Mekhanika razrusheniia i prochnost materialov. T. 1–4 (Fracture mechanics and strength of materials, vol 1–4). Naukova dumka Publishing House
Panasyuk VV, Andreykiv OY, Obukhivskyi OI (1982) Raschetnaia model rosta treshchiny v metallakh pri vozdeistvii vodoroda (Calculation model of crack propagation under hydrogen effect). Fiziko-Khimicheskaya Mekhanika Materialov (Physicochemical Mechanics of Materials) 3:113–115
Panasyuk VV (1991) Mekhanika kvazikhrupkogo razrusheniia materialov (Mechanics of quasibrittle fracture of materials). Naukiva dumka Publishing House
Panasyuk VV (ed) (1988) Mekhanika razrusheniia i prochnost materialov. T. 4: Romaniv OM, Yarema SYa, Nykyforchyn HM. Ustalost i tsyklicheskaia treshchinostoikost konstruktsionnykh materialov (Fracture Mechanics and Strength of Materials. Vol. 4: Romaniv OM, Yarema SYa, Nykyforchyn HM. Fatigue and cyclic crack growth resistance of structural materials. Naukova dumka Publishing House, Kyiv
Panasyuk VV, Andreykiv OYe, Darchuk OI, Kun PS (1994) Analysis of short and long fatigue cracks growth kinetics under non-regular loading. Structural integrity: experiments, models, applications. In: Schwalbe KH, Berger C (eds) Proceedings of 10 European Conference on Fracture (ECF-10). EMAS, vol 2, pp 1271–1276
Panasyuk VV, Ivanytskyi YaL, Andreykiv OYe (2000) Metody otsinky tsyklichnoi trishchynostiikosti materialiv pry realizatsii zmishanykh mekhanizmiv ruinuvannia. Otsinka i obhruntuvannia prodovzhennia resursu elementiv konstruktsii (Methods of assessing the cyclic crack growth resistance of materials when implementing mixed macro-mechanisms of fracture. Evaluation and substantiation of the extension of the life time of structural elements). Materialy Mazhnarodnoi konferencii (Proceedings of International Conference), Kyiv, 6–9 June, vol 1, pp 43–52
Pokhodnia IK (1998) Problemy svarki vysokoprochnykh nizkolegirovannykh stalei. Suchasne materialoznavstvo XXI storichcha (Problems of welding of high-strength low-alloy steels. Modern materials science of XXI century). Naukova dumaka Publishing House
Rice JR (1967) Mechanics of crack tip deformation and extension by fatigue. Fatigue crack propagation. Philadelphia (Pa) Am Soc Testing Mater STP 415:247–309
Schwalbe K-H, Kocak M (eds) (1997) Mis-matching of Interfaces and Welds. GKSS, Geesthacht
Scruby CB, Wadley HNG, Rusbridge KL (1983) Origin in acoustic emission in Al-Zn-Mg alloys. Mater Sci Eng 59(2):169–183
Seismov VM (1976) Dinamicheskiie kontaktnyie zadachi (Dynamic contact problems). Naukova dumka Publishing House
Shibata M (1984) A theoretical evaluation of acoustic emission signals. The rise-time effect of dynamic force. Mater Eval 42(1):107–120
Sinclair JE (1979) Epicentre solution for point multipole sources in an elastic half-space. J Phys D Appl Phys 12(8):1309–1315

Skalskyi VR, Andreikiv OE, Serhiyenko OM (1999) Otsinka vodnevoi poshkodzhenosti za amplitudamy syhnaliv akustychnoi emisii (Estimation of hydrogen resistance of materials by signal amplitudes and acoustic emission), Techniczeskaia diagnostika i nerazrushaiushczij control (Technical Diagnostics and Non-destructive Testing) 1:17–27

Takahashi H, Khan MA, Kikuchi M, Suzuki M (1981) Acoustic emission crack monitoring in fracture toughness for AISI 4340 and SA 533B steels. Experimetal Mech 21(3):89–99

Troshchenko VT (1981) Deformirovaniie i razrusheniie metallov pri mnogotsyklovom nagruzhenii (Deformation and fracture of metals under low-cycle loading). Naukova dumka Publishing House

Trufiakov VI (ed) (1990) Prochnost svarnykh soiedinenii pri peremennykh nagruzkakh (Strength of welded joints under variabe loads). Naukova dumka Publishing House

Verlan AF, Sizikov VS (1986) Integralnyie uravneniia: metody algoritmy pogrammy (Integral equations: methods of algorythms of program). Naukova dumka Publishing House

Yarema SY, Popovich VV (1985) Influence of the structure and stress concentration on the period of fatigue crack origin in 65G steel. Mater Sci 21(2):133–138

Chapter 5
Estimation of Hydrogen Effect on Metals Fracture

The problem of hydrogen effect on the physical and chemical properties of metals belongs to the actual directions of modern fracture mechanics of materials. The importance of this problem is due to the growing needs of the development of hydrogen energy and hydrogen technologies. At the same time, hydrogen has a significant negative property—it vitally reduces the resistance of metals to fracture. In this case there is a typical situation when machines and equipment with a sufficiently high strength in normal operating conditions undergo fracture under the influence of hydrogen, with premature and, at first glance, unpredictable failure. Hydrogen degradation of metallic materials and structures often results from such harmful effects of hydrogen on their behavior and integrity.

5.1 Interaction of Hydrogen with Steels and Emission of Elastic Waves

Local stress notches are always formed in structural materials due to various treatment modes and in the presence of defects. In steels, these include, for example, non-metallic inclusions that are formed at high temperatures and cooled together with the metal. Then, due to the difference between the coefficients of linear expansion of steel and inclusions, thermal stresses occur in the metal. It is shown that different inclusions, having their own specific shape and size, are characterized by a high level of microstresses, even without the application of external forces (Volchok 1993). This is explained not only by different coefficients of linear expansion but also by different moduli of elasticity of the metal matrix and inclusions. Thus, the latter are stress notches in alloys, and the value of these stresses depends on the chemical composition, the size of the inclusions, as well as the value of the external load applied to the material.

V. Skalskyi and Z. Nazarchuk, *Magnetoelastic Acoustic Emission*, Springer-AAS Acoustics Series, https://doi.org/10.1007/978-981-99-4032-5_5

Real products or some of their elements in the operation conditions are exposed to various types of loading and the influence of aggressive working environments. Therefore, these defects contribute to the appearance and propagation of cracks in the material of structures. In the presence of the latter, there is a corresponding redistribution of stresses (McMeeking 1977; Dal and Anton 1986; Neimitz 1998). As a result, a special stress–strain state occurs at the crack tip, which under the influence of corrosion and mechanical factors as well as the material's own microstructure contributes to further local fracture. The latter occurs inside the material and excludes the possibility of visual detection of such processes. That is why, considering the fact that the processes of local fracture are accompanied by elastic wave emission, the phenomenon of AE is widely used for their diagnosis.

It is known that at the boundary of condensed matter and gas due to the asymmetry of the force field and the specific interaction between the particles, there is a two-dimensional storage of gas molecules or products of their dissociation, i.e., the phenomenon of adsorption. Its character depends on the nature of the reagents interaction as well as pressure and temperature. Hydrogen interacts with almost all metals. When metal comes into contact with gaseous hydrogen or a mixture of gases containing hydrogen, its molecules (as well as other gases), interacting with the metal surface, are partially adsorbed. Some of the molecules that were previously adsorbed may desorb after some time, and some may dissociate into atoms and penetrate the crystal lattice of the metal (absorption).

Since the activation energy of the dissociation process is close to the energy of dissociation itself, it is obvious that the rate of hydrogen molecules decay into atoms on the surface will be much higher than in the volume, so the concentration of gas on the body surface is always higher than in the gas phase (Archakov 1985).

Therefore, the process of interaction of hydrogen with metals begins from the surface, where a thin layer of adsorbed hydrogen is formed according to the following scheme:

$$H_{2(\text{gas})} \leftrightarrow H^{+}_{2(\text{gas})} + e^{-};\ H^{+}_{2(\text{gas})} + 3e^{-} \leftrightarrow 2H^{-}_{(\text{ads})}\,. \qquad (5.1)$$

According to Gibbs, the adsorption of gases on the metal surface causes a decrease in the surface energy of the metal and promotes the diffusion of hydrogen (protons) into the metal (Zakharov 1987).

In (Druchenko et al. 1977), a qualitative description of AE generated by the adsorption and desorption of hydrogen during stress corrosion cracking (SCC) is given. At present, it is difficult to determine the absolute value of the AE parameters caused by the attenuation or reflection from interference or free surfaces during the propagation of the elastic waves and due to modulation of their shape and frequency spectrum which appears as a result of the dependence of AE signals (AES) on primary AE converter (AEC). In (Yuyama 1986) it is shown that the adsorption and desorption of hydrogen still provide a sufficiently high AE energy (in some cases, AES reach several hundred μV) to record it in the laboratory.

The problem of metal hydrogen charging (e.g., steel) according to Karpenko and Krypyakevych (1962) can be considered in three aspects: the diffusion of hydrogen into metal, the solubility of hydrogen in steel, and the effect of hydrogen on the physical properties of steel. Penetration of hydrogen into steel in a molecular form is impossible. It occurs in the presence of ionized hydrogen (protons) on the metal surface, which is formed during dissociation. Defects in the steel structure, which include defects of the polycrystalline body: microcavities, cracks, flaws, non-metallic inclusions, intergranular boundaries, and ultramicroscopic defects of the steel crystal lattice—vacancies, their accumulation (or coagulations) and dislocations play an important role in the distribution of hydrogen absorbed by steel. All of them play the role of collectors (traps), where hydrogen absorbed by steel collects, which can be in two states: ionized (protons dissolved in the crystal lattice) and molyzed (accumulates in traps). Reaching the surface (of the cracks, pores, etc.) and capturing electrons from the conduction band of the metal, the protons are converted into atomic hydrogen, which is again adsorbed on the surfaces. Then it recombines, forming the molecular hydrogen. The effective radius of hydrogen molecules exceeds the size of the crystal lattice, and molecular hydrogen cannot penetrate the metal, accumulating in traps.

At normal temperature, almost all the hydrogen in the crack is in the molecular state. Thus, even prior to the equilibrium between the hydrogen in the crack and in the lattice (equilibrium can occur at a molecular hydrogen pressure in the crack of approximately 103 MPa), the crack surface acts as a semipermeable membrane, and the crack itself is a perfect drain for hydrogen dissolved in the metal. Even before the beginning of equilibrium conditions, the accumulation of hydrogen in the crack can lead to its propagation (Goldshtein et al. 1977).

Hydrogen capture combines the phenomena of dissolution [often referred to in the literature as absorption or occlusion (Paisl 1981)] and the filling of defects within the metal with molecular hydrogen. After penetration into the metal, for example, steel, hydrogen reacts with carbon in cementite, which leads to the reaction of hydrogen charging of cementite (decarbitization of steel):

$$Fe_3C + 4H^+ + 4e \rightarrow 3Fe + CH_4, \tag{5.2}$$

or for molecular hydrogen segregated in voids

$$Fe_3C + 2H_2 = 3Fe + CH_4. \tag{5.3}$$

The more methane formed, the higher the gas pressure and stress state along the grain boundaries up to the formation of a plastically deformed volume and thus microcracks. The latter, in turn, combining, form a macrocrack nuclei, the development of which causes failure. During these processes, the elastic relaxation occurs [Snook effect (Paisl 1981)], associated with the reorientation of the elastic dipole, which appears due to the lattice deformation by the hydrogen atom. Hydrogen-created lattice defects are able to move in the presence of a deformation gradient. The diffusion relaxation takes place [Gorsky effect (Paisl 1981)]. All this together also leads to the generation of the AE elastic waves with amplitudes greater than the signals generated by the mechanisms of adsorption phenomena (Schnitt-Thomas and Stengel 1983; Andreikiv et al. 1991).

In the case of hydrogen embrittlement, the local formation of microcracks in an isolated volume of a solid or at the macrocrack tip is particularly typical (Kott 1978). As mentioned above, this is due to the increase in local hydrogen concentration, at which the development of microdefects caused by reduced lattice cohesion, surface energy and high molecular hydrogen pressure in microcavities, is possible in metal microvolumes.

Local regions of the metal are enriched in hydrogen in various ways, such as directed diffusion into the region of maximum triaxial tensile stresses or hydrogen transfer by dislocations when in the head of the dislocation cluster formed on different obstacles, the local increase in hydrogen is much higher than its average concentration in the metal volume. However, even in the case of a perfect lattice, the uniform distribution of hydrogen atoms in the volume is energetically unfavorable. Therefore, they can cluster, reducing the energy of the "metal–hydrogen" system. The possibility of the existence of hydrogen clusters also follows from direct quantum-mechanical calculations of the configurational part of the average energy of interaction of the ion-proton system (Vavrukh and Solov'yan 1986). Although because of the rather large potential barrier in the metal, the formation of clusters is difficult, as a result of the application of external tensile forces, the value of this barrier is significantly reduced.

If there is a crack in the metal, its growth depends not only on the stress–strain state around it and the properties of the metal but also on the specific features of the hydrogen transfer process in local areas. Analytical description of the diffusion distribution of hydrogen in front of the crack tip shows (Fig. 5.1) that the hydrogen concentration depending on the metal-hydrogen system reaches a maximum at a distance from the crack edge, which is approximately twice the value of its opening displacement (Panasyuk et al. 1982a). Thus, both under the action of structural and mechanical factors and under the influence of hydrogen, the microcrack propagation occurs by the formation of isolated microcracks—both in local regions of the solid and around its tip (front) with the AE elastic waves generation (Skalskyi and Koval 2007).

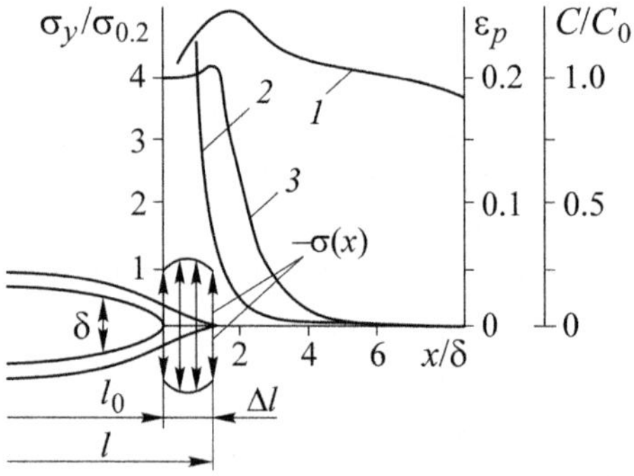

Fig. 5.1 Distribution of stresses and strains [curves *1* and *2*, respectively, (McMeeking 1977)], as well as hydrogen concentration [curve *3*, (Panasyuk et al. 1982a)] in front of the macrocrack tip

5.2 Types of Hydrogen Degradation of Metals

Since hydrogen degradation is exhibited in different ways, several options for classifying its types have been proposed. According to Kolachev's classification (Kolachev 1985), there are two kinds and seven types of hydrogen brittleness, according to the sources that cause it and the conditions of development. However, the term "hydrogen brittleness" is not entirely successful, because quite often hydrogen does not lead to purely brittle fracture. Therefore, it is better to use the term "hydrogen degradation", which includes the whole set of negative phenomena caused by high hydrogen content in the metal. According to Archakov (1985), Tkachov (1999), Andreykiv and Hembara (2008), the phenomenon of hydrogen degradation by its external signs (manifestations) can be divided into three groups of effects: hydrogen embrittlement, hydrogen degradation of transformation, and hydrogen destruction.

Hydrogen embrittlement is a phenomenon caused by the physical influence of hydrogen on the deformation microprocesses in metals under load. They occur both in the presence of dissolved hydrogen in metals and under the effect of hydrogen-containing media on them. Most steels are inclined to this type of degradation in the temperature range close to room temperature, i.e., in the range of 370 K.

Hydrogen degradation of transformation is a phenomenon of degradation of chemical and physicochemical nature. This is a consequence of either the reactions of the formation of new phases of hydrogen-containing compounds in alloys or the implementation of phase and structural transformations, which are possible only in the presence of hydrogen in metals (i.e., in which hydrogen acts as a transformation catalyst). These include, in particular, such phenomena as hydrogen corrosion of carbon steels at elevated temperatures and hydrogen activity, "hydrogen disease" of copper, hydride embrittlement of alloys of titanium, zirconium, and other metals. Effects of this kind can be activated by thermomechanical processes. Some of the phenomena in this group are sometimes referred to as "reactive hydrogen embrittlement" to emphasize their association with chemical reactions.

Hydrogen degradation is a phenomenon of the formation of discontinuities in metals due to their supersaturation (possibly local) with hydrogen, which is not caused by stresses or transformations and chemical reactions. These include such effects as clustering, the occurrence of delaminations in rolled products, delaminations in bimetals and in metals with surfacing and coatings, etc.

Each of the effects of these groups is revealed under appropriate conditions of hydrogen action on metals. The analysis of the conditions of hydrogen interaction with metals in the presence of external fields (force, temperature), taking into account their real chronology, showed the possibility of implicit (hidden, unobvious) formation of favorable conditions for hydrogen degradation in (micro) local areas, when global parameters of the "metal–environment" system show no concern about the possible risk of hydrogen degradation (Kronshtal' and Kharin 1992; Panasyuk et al. 1989; Andreikiv et al. 1978). Typical examples of this are the possibility of hydrogen degradation in cryogenic systems, as well as the creation of preconditions for degradation due to non-stationary temperature fields and heterogeneity of materials. The same material under different external conditions (temperature, pressure, stress, etc.) can be sensitive to different types of hydrogen degradation.

As mentioned above, the forms of realization of hydrogen degradation of metals in specific situations are quite diverse and may come as a surprise. This can occur during operating cycles in structural elements under load in the form of initiation and propagation of cracks of hydrogen embrittlement and can manifest itself in the form of hydrogen destructions—microdiscontinuities and delaminations—during the rest of the units, after leaving the operating mode and without further contact with hydrogen. Such local destructions during the re-entry into the technological mode can be the initiators of complete failure.

In order to assess the serviceability and justify the service lifetime of the structure, it is not enough to make calculations for a particular mode of the unit operation, which is considered the most dangerous by the stress level or external activity of hydrogen (in the environment). For the reliability of the results and conclusions, it is necessary to analyze the history of interaction with hydrogen and changes in temperature and the stress–strain state of structures for all components of the cycles "start—operating mode—required stop—period between starts", as well as for possible emergencies. This is due to the fact that at any of these stages, favorable conditions can be created for the implementation of a catastrophic combination of the external effects and factors of hydrogen degradation of the material in the structure, which can exhibit themselves now, or be a cause of further fracture.

It should be borne in mind that hydrogen degradation of metals is associated with two types of processes: kinetic and destructive (Tkachov 1999). The kinetic ones include the factors and processes that control the hydrogen penetration into the potential degradation zones. Analysis of kinetic aspects makes it possible not only to assess (predict) the rate of degradation development (i.e., the kinetics of damage accumulation, crack rate, etc.) and tendency of its change, but also to predict the formation of conditions promoting the hydrogen degradation. However, the kinetic analysis does not allow us to assess the extent to which the material is subject to a particular type of hydrogen degradation, to judge the true mechanisms of hydrogen damage and the real hydrogen resistance of materials and structures made of them. Destructive processes and mechanisms of formation and development of damages in metals include defects, unfavorable phases, and discontinuities caused by hydrogen.

Whichever type of hydrogen degradation is involved, the diffusion of hydrogen in the metal can almost always be considered a necessary kinetic factor. The influence of the inhomogeneous stress field on the diffusion of hydrogen in metals is considered as an effect of primary importance for the growth of hydrogen-induced cracks. Meanwhile, this is just one of the manifestations of a more universal effect, which is based on the heterogeneity of arbitrary fields in the metal volume, which affect the local effective solubility of hydrogen in it. These are, first of all, inhomogeneous chemical and phase compositions of the material, its microstructure, temperature distribution, plastic deformations and, probably, others. Diffusion of hydrogen in metals in some cases may play a role of the main factor in the hydrogen degradation.

5.3 The Main Mechanisms of Hydrogen Fracture of Metals

The mechanisms of hydrogen interaction with metals are analyzed in detail in a number of reviews and monographs (Kronshtal' and Kharin 1992; Panasyuk et al. 1989, 2000, 2001, 1994, 1984, 1982b; Andreikiv et al. 1978; Pokhodnia et al. 2004; Andreykiv 2003; Andreykiv et al. 2001, 1980, 1987; Pokhmurskii and Fedorov 1998). The embrittlement action of hydrogen, according to many authors, is due to the following factors: the specific feature of the hydrogen solubility in metal; diffusion anomalies; interaction of dissolved hydrogen with defects of the crystal lattice; its chemical interaction with steel components; adsorption phenomenon; the pressure of hydrogen and other gases in microcavities; the effect of hydrogen on the bonding forces of iron atoms in the crystal lattice; hydrogen chemisorption on internal surfaces (Andreykiv et al. 1980). Existing hypotheses of hydrogen degradation, which use these factors in different combinations, can be divided into several groups (Tkachov 1999).

The first group includes the hypothesis of "high pressure" of molecular hydrogen in the internal microcavities. It is based on the phenomenon that temporary local supersaturation zones can occur in the metal during hydrogen saturation. The latter can become critical to the integrity of bodies due to the high risk of formation and development of hydrogen damage (hydrogen degradation) hidden inside the material. An idea about the extent of this risk can be obtained from the following assessment.

Hydrogen degradation can develop if, in the above situation, there is a microcavity in the zone of maximum hydrogen supersaturation. Hydrogen will leak into the microcavity from the metal, creating pressure P_D. When the cavity is small enough, the equilibrium between the hydrogen in it and in the surrounding metal is established fairly quickly. It can be considered in accordance with the Sieverts' law that $P_D = C_{\max}^2/K_s^2(T_1)$. As a result, the pressure P of the medium during the metal hydrogen charging and the pressure in the defect after supersaturation can be related by the formula (Kronshtal' and Kharin 1992):

$$\frac{P_D}{P} = \left(\frac{C_{\max}}{C_e}\frac{K_S(T_0)}{K_S(T_1)}\right)^2 = \left(\frac{C_{\max}}{C_e}\right)^2 \exp\left[\frac{2\Delta \mathrm{H}_L}{R}\left(\frac{1}{T_1} - \frac{1}{T_0}\right)\right]. \tag{5.4}$$

In this case, for the steel sample, we have $\Delta \mathrm{H}_L = 20.1$ kJ/mol (Nykyforchyn et al. 1998). If hydrogen charging occurs with a fairly conservative estimate $C_{\max}/C_e = 2\ldots4$, then from relation (5.4) we obtain $P_D/P \approx 400-1700$. That is, if $P = 5$ MPa, the pressure in the defect is estimated by the value of 2–85 GPa. This pressure can exceed even the theoretically calculated strength value. It is clear that this is an overestimation because, at the pressures of this order, it is necessary to take into account the imperfection of the hydrogen gas (Leeuwen 1985). However, such pressure can cause hydrogen destruction, i.e., material rupture.

The implementation of the described heat and mass transfer in carbon steels can lead to the development of hydrogen degradation in the form of hydrogen corrosion inside the metal, and not from the surface, as usually occur (Nykyforchyn et al. 1998).

Note (Kronshtal' and Kharin 1992) that during the interaction of hydrogen with the deformed metal, none of the stationary factors can lead to the formation of a pressure in the micropores that would exceed the equilibrium one, i.e., greater than the ambient pressure of hydrogen or equilibrial enforcement for a given concentration of hydrogen in the metal in a stationary state. Stationary factors are those that can cause a local increase in the concentration of hydrogen in the metal (stress concentration, stationary temperature field, etc.). The lack of increased pressure in the cavities located in areas with high hydrogen concentration is due to the fact that the increase in concentration here is caused by a local increase in its solubility in the metal, while pressure in the cavity in other areas of metal can reach maximally the state of the "metal-environment" system.

Thus, the creation of high hydrogen pressures in microcavities in the metal can be ensured only by dynamic factors. These values themselves can be caused by nonequilibrium processes, which require the continuation of the corresponding kinetic mechanisms. These can be either transported by dislocations or non-stationary inhomogeneous, in particular, temperature fields.

Therefore, the hypothesis of molecular pressure cannot represent the real mechanism of hydrogen embrittlement. The situation is not improved even by the modernization of its classical version with an account of hydrogen transfer to the crack volume by dislocations (Mnushkin and Kopelman 1980), since the increase in the pressure which is possible during such a transfer intensifies the reverse process—the dissolution of hydrogen in the crystal lattice of metal (Shapovalov and Trofimenko 1987). This does not mean that the hydrogen pressure in the microcavities does not play any role in hydrogen embrittlement, but it cannot be decisive, especially during crack initiation, with which most authors agree. Therefore, attempts to explain the mechanism of crack initiation based on the molecular pressure hypothesis proved to be unsuccessful.

The second group of hypotheses is based on the conception of the reduction of interatomic bonds in metal under the action of dissolved hydrogen, i.e., on the idea of the drop of so-called cohesive strength. It is also assumed that decohesion takes place only in special tensioned volumes of the crystal lattice, where the amount of hydrogen can be several times higher than its average concentration. It is established that a noticeable decrease in cohesion forces occurs at approximately equal ratios of the number of iron and hydrogen atoms (Kolachev 1985). Thermodynamic analysis shows that hydrogen does have to diffuse toward the deformed metal volumes. However, there are no experimental data that could clearly prove the validity of the decohesion hypothesis. Such experiments are complicated because the decohesion effect of hydrogen in the metal lattice must be detected both against the background of its interaction with various defects of the crystal structure, and in the presence of the internal stresses. At the same time, simple estimates have shown that with a total hydrogen capacity of up to $10\,\text{cm}^3/100\,\text{g}$ of metal and its uniform distribution in the lattice (that is unrealistic due to the fact of hydrogen accumulation in defects), one hydrogen atom is per thousand metal atoms (Tkachov 1999). This relationship is considered the most significant objection to the decohesion hypothesis and indicates its imperfection (Tkachov 1999). However, the decohesion hypothesis, supplemented by the notion of directed bulk diffusion in local regions of triaxial stresses that form macrocracks or other concentrators, is often used to explain the peculiarities of hydrogen embrittlement (Kasatkin et al. 1986).

The hypotheses of the third group are based on the assumption of a decrease in surface energy inside the crack during hydrogen adsorption, which should lead to a drop in destructive stress. Thermodynamic estimates using the Gibbs equation indicate a very strong effect of the surface energy reduction due to adsorption. In the classical study (Peth 1956), the mechanical characteristics calculated on the basis of the Griffith theory, taking into account this effect, were in good agreement with the experimental data, which confirmed the correctness of the explanation that hydrogen embrittlement was a result of hydrogen adsorption. However, such ideas have not been developed due to the shortcomings of the Griffith theory, which assumes the initiation of cracks in the initial material without explaining their origin. A number of arguments against the adsorption hypothesis are as follows (Hirth 1980): the idea of adsorption reduction of the surface energy leads to a significant underestimation of the work of fracture; the adsorption hypothesis cannot explain the jump-like crack propagation detected by the methods of acoustic emission, the reversibility of hydrogen embrittlement, and why hydrogen causes embrittlement, while oxygen, which has a higher adsorption heat, not only decelerates but even stops cracking (if, for example, it is added to the pure molecular hydrogen). However, all these objections are removed if we take into account the hydrogen atoms transport to the sources of fracture initiation (Kolachev 1985).

The fourth group includes hypotheses that consider the interaction of hydrogen with dislocations (Kolachev 1985). It is assumed that the properties of the metal in the presence of dissolved hydrogen are determined by the transport of hydrogen atoms by mobile dislocations during plastic deformation. As a result, at the grain boundaries, interfacial boundaries, and other barriers where dislocations accumulate, the hydrogen concentration becomes sufficient to sharply accelerate the fracture by one or another mechanism. The possibility of hydrogen condensation at dislocations is confirmed by thermodynamic calculations and the results of direct observation of its distribution by electron microscopic autoradiography (Bokstein and Ginzburg 1978). The main argument in favor of the dislocation hypothesis is that it, in contrast to other hypotheses, easily explains many features of the hydrogen embrittlement, including such anomalous ones as the dependence of its degree on the temperature and strain rate (Kolachev 1985).

For all the attractiveness of the notion of hydrogen transfer by dislocations, this process is only a separate link in the general mechanism of hydrogen embrittlement, if we consider its beginning the solubility of hydrogen in the metal, and the end—the formation of macrocracks. It turns out that the dislocation hypothesis describes in detail only the preparatory stage of hydrogen embrittlement, without explaining the most important point—the formation of a crack: everything that happens later after hydrogen delivery to the crack origin is considered obvious. Such fragmentarity, however, inherent in other hypotheses of hydrogen embrittlement, is considered to be its main drawback.

The fifth group consists of hypotheses that use elements of both existing and new ideas. These include the dislocation-decohesion hypothesis of hydrogen embrittlement, proposed by the authors (Shved 1985). It is based on the idea that at the temperatures of this process manifestation, according to calculations, most of the hydrogen in metals is concentrated in dislocations, and in the area of the dislocation nucleus, its atomic concentration can reach the values of the order of unity. It is assumed that this ensures the possibility of a local decohesion effect, which is manifested in the reduction of interplanar adhesion. In the presence of a barrier to dislocation movement, this leads to an increase in the compressive strength of a number of similar dislocations and increases the interaction between their nuclei. As a result, the equilibrium of accumulation is disturbed, and it moves to the delayed dislocation—the elementary event of plastic deformation is realized.

For a certain relationship of the accumulation force, external stresses and hydrogen concentration in the dislocation nuclei, this process causes the formation of a macrocrack and its further propagation by absorbing the accumulation dislocations. Another of hydrogen embrittlement factor in metals is also assumed to be the nonequilibrium pressure of hydrogen molized in the macrocrack after its release from dislocations. However, the evaluation of its values (according to the known total number of atoms associated with hydrogen dislocation) confirmed that the role of pressure is insignificant at the stage of the initiated macrocrack growth. This concept proposes two new ideas: decohesion action of hydrogen localized in the dislocation nuclei; the result is a loss of stability in dislocated families before fracture. This eliminates the main objection to the classical version of the decohesion hypothesis, which is associated with a small value of the average macrovolume atomic concentration of hydrogen, and also the micromechanism of the hydrogen-induced fracture is explained. This forms a broader basis for explaining the mechanism of hydrogen embrittlement, which favorably distinguishes the dislocation-decohesion hypothesis from those proposed earlier. Note that in this hypothesis the mechanism of hydrogen transport to the place of the macrocrack initiation, contrary to the dislocational hypothesis, plays a secondary role, because the effect of hydrogen on the existing accumulation formed in the metal as a result of microplastic deformations and which before hydrogen arrival was in the stable equilibrium state, is considered.

There is also no experimental confirmation of both the decohesion action of hydrogen itself and the imaginary activation of microplastic deformations by hydrogen. The latter assumption in itself is interesting because in this case the elementary event of plastic deformation is used to explain the opposite effect—the embrittlement. Since the dislocation movement causes not only micro but also macroscopic deformation, the increase in their mobility under the influence of hydrogen should lead not to embrittlement, but to the increase in macroplasticity. As a last resort, hydrogen charging should reduce the yield strength, which is determined by mechanical investigations. However, the experimental data contradict this, and in some cases, there was even an increase in the yield strength in the presence of hydrogen (Shved 1985). In addition, it is considered that no matter how perfect the knowledge about the regularities of hydrogen effect on microscopic deformation processes, it is currently impossible to establish a reliable relation between micromechanisms and the development of macrocracks that lead to fracture (Pokhmurskii and Fedorov 1998). This assessment is valid for all existing hypotheses of the hydrogen embrittlement.

Based on literature data and the results of experimental research performed recently in the Ye.O. Paton Electric Welding Institute on this problem for structural steels and welds, a new model of hydrogen embrittlement was proposed (Pokhodnia et al. 1978, 2004; Pokhodnia 1972, 1998; Pokhodnya and Shvachko 2001; Pokhodnya et al. 2002). According to this model, the effect of hydrogen on the fracture is as follows. At typical for steel concentrations of hydrogen and in the absence of irreversible traps (pores, cracks, interfacial boundaries), hydrogen dissolved in the metal is most likely to be in reversible traps, which are dislocations. With the beginning of plastic deformation, hydrogen, which has an abnormally high diffusion coefficient in iron, will be easily transported by mobile dislocations to the site of a submicrocrack initiation. Well-developed dislocation theory proposes many models of dislocation rearrangements that can lead to the formation of extremely sharp, atomic dimensions, initiated crack (submicrocrack). However, the experimental data currently used by the physics of fracture do not allow to prefer any of these models. Since the main stages of the evolution of submicrocracks are invariant with respect to different types of dislocation rearrangements, we can imagine the appearance of submicrocracks according to the classical Ziner–Straw model. According to it, a submicrocrack is formed at the tip of the accumulation of dislocations, determined by the grain boundary or other obstacles. The further behavior of the submicrocrack in the field of external stresses is determined by the energy of the "dislocation accumulation – submicrocrack" system: it can close, remain in the state of elastic equilibrium or grow infinitely.

Hydrogen, which is released from dislocations during crack initiation and enters its volume, will primarily be chemisorbed on the newly formed juvenile surfaces. Direct experiments performed by the method of secondary ion mass spectrometry have shown that hydrogen atoms on the iron surface are negatively charged (Kikuta 1976). This layer of negative ions decreases the density of free electrons near the surface. This, in turn, changes the relations that form the energy of the "dislocation accumulation—submicrocrack" system and makes it easier for a crack to overcome the potential barrier, which is an obstacle to its propagation. The particularly strong influence of the hydrogen should be expected at the stage of the submicrocracks growth in the stress field created by dislocation accumulation. In this case, even a small number of hydrogen atoms required to fill the monoatomic surface of the resulting crack will cause the loss of its elastic equilibrium as a result of reducing the surface energy. Therefore, the transition of the submicrocrack to the unstable autocatalytic propagation will be possible with less normal external stress than in the absence of hydrogen. At the macrolevel, this will be the embrittlement effect of hydrogen. If the amount of hydrogen released from the dislocations is sufficient not only for the chemisorption layer but also to create pressure in the submicrocrack volume, there will be additional stress, which reduces fracture resistance.

5.4 Fracture of Metals Under the Action of Hydrogen-Mechanical Factor

Despite the fact that various types of hydrogen degradation have been identified so far, many sources of metal saturation with hydrogen have been identified, various hypotheses have been proposed to explain the studied phenomena, there is still no accepted theory of hydrogen embrittlement. None of the existing theoretical models is able to take into account a significant part of the experimentally established effects. The reason for this situation is the lack of a complete and consistent theory of the interaction of hydrogen with metals, starting from the surface phenomena to the interaction of absorbed hydrogen with the atoms of crystal lattice and defects in the structure of metals, as well as the lack of a complete physical model of the fracture process, which allows us to consider such parameters of the material that control the process and are affected by hydrogen interacting with the metal.

The formation of the quantitative theory of hydrogen embrittlement of metals meets further complications when reliably determining the actual parameters of the interaction of hydrogen with a deformed metal (local hydrogen concentration in the region of fracture initiation, hydrogen pressure in microcavities, etc.). However, the creation of the foundations of the quantitative theory of hydrogen embrittlement is an important part of the development of this complex problem, which has recently developed intensively.

It is established that hydrogen embrittlement is represented by not only changes in mechanical characteristics of the material—plasticity and strength (ψ, σ_b) and other values (Moroz and Chechulin 1967), but also by the fact that in the presence of hydrogen either in the environment or dissolved in the metal lattice, so-called delayed fracture under load is observed, which in the absence of hydrogen is not dangerous (Pokhodnia et al. 2004; Andreykiv 2003; Andreykiv et al. 2001; Barth and Steigerwald 1970; Steklov 1992; Bernstein 1970; Johnson et al. 1958; Fidelle et al. 1974; Steigerwald et al. 1959; Williams 1970). The presence of an intermediate stage of subcritical propagation of the defect in the process of fracture is due to the fact that embrittlement (and fracture) requires transportation of hydrogen from its initial location to some local area, where, when reaching the critical ratio of hydrogen and mechanical stress, there is an elementary event of a fracture. Further development of the process is controlled by the delivery of additional portions of hydrogen to the newly created critical areas. The crack growth takes place in successive jumps until its critical length is reached and there is a rapid and complete fracture (Andreikiv et al. 1978). The effect on embrittlement of such factors as strain rate, temperature, type of load, and others is often explained by participation in the process of hydrogen transfer reactions (Andreykiv et al. 1987; Moroz and Chechulin 1967).

Since the change in mechanical characteristics (stress and strain fields) at the crack tip under certain restrictions (Panasyuk et al. 1989) is characterized by a change in a single parameter—stress intensity factor K_{I}, which contains information about the geometry of the body and applied load, the use of the fracture mechanics methods to study hydrogen embrittlement is a very convenient and effective way to quantify the process regularities. In this case the key points of the analytical study of the problem are as follows:

– establishing the criterion of local fracture at the crack tip, i.e., determining the critical combination of the K_{I} value and the amount of hydrogen (e.g., concentration C), which corresponds to the elementary event of fracture:

$$K_{\mathrm{I}cr} = f_1(C) \text{ or } C_{cr} = f_2(K_{\mathrm{I}}) \tag{5.5}$$

where $K_{\mathrm{I}cr}$ (or C_{cr}) is the critical value of SIF (or concentration) for the given value of C (or K_{I});

– determination of the crack growth kinetics, i.e., estimation of crack growth rate $V = \mathrm{d}l / \mathrm{d}t$ (here l is crack length, t is time) for different values of SIF:

$$V = g(K_{\mathrm{I}}), \tag{5.6}$$

depending on the main physicochemical characteristics of the interaction in the "metal–hydrogen" system.

The creation of a complete theoretical model of hydrogen embrittlement should include the solution to these two problems. The form of dependence (5.5) has not yet been reliably established experimentally, but based on the available data (Steklov 1992; Gerberich et al. 1975) it can be assumed that C_{cr} should be a decreasing function of K_{I}, and $C_{cr} \to 0$, if $K_{\mathrm{I}} \to K_{\mathrm{I}c}$ are the fracture toughness (crack growth resistance) of metal in the absence of hydrogen. Typical experimentally obtained kinetic curves have a number of typical features (see Fig. 5.2).

As mentioned above, the kinetics of hydrogen embrittlement is controlled by the kinetics of hydrogen transport. It is difficult to determine the exact type of transfer reactions in real systems, as the interaction of hydrogen with metal is significantly

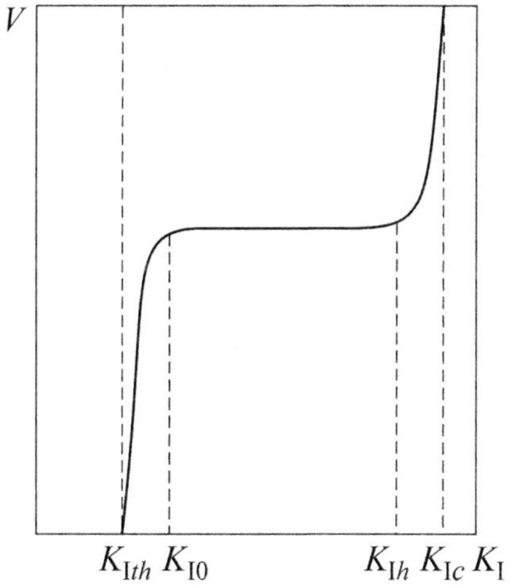

Fig. 5.2 General view of the kinetic curve of crack growth rate V versus stress intensity factor K_{I}

influenced by the composition and structure of the metal, the composition of the hydrogen-containing medium and other factors (Andreikiv et al. 1978).

Currently, the process of hydrogen interaction with metals is divided into the following main stages (Panasyuk et al. 1989, 1984; Andreikiv et al. 1978; Andreykiv et al. 1987; Pokhmurskii and Fedorov 1998):

1. Diffusion of molecular hydrogen in the gas phase to the metal surface.
2. Condensation of gaseous hydrogen on the metal surface, dissociation and adsorption of atomic hydrogen.
3. Transfer of hydrogen atoms to the metal surface and chemisorption.
4. Dissolution of hydrogen in metal, absorption.
5. Transfer (transport) of hydrogen atoms inside the metal.

The overall rate of hydrogen entry into local critical areas, and hence the rate of embrittlement, is obviously determined by the slowest of those elementary transport processes, which are a necessary precondition for embrittlement in a particular metal–hydrogen system under certain circumstances.

In many studies investigating the phenomenon of hydrogen embrittlement of metals, the process of its rate control is often referred to as the diffusion of hydrogen within the metal (Barth and Steigerwald 1970; Johnson et al. 1958; Steigerwald et al. 1959), although a number of experiments in the gaseous hydrogen cause doubts on the diffusion universality as a type of hydrogen transfer, which determines the crack growth kinetics (Williams 1970). However, in the development of theoretical models and obtaining analytical expressions of dependence (5.6) in many studies hydrogen diffusion is considered a process that controls of crack growth rate (Panasyuk et al. 1989; Andreikiv et al. 1978; Pokhodnia et al. 2004; Andreykiv et al. 1987; Pokhmurskii and Fedorov 1998; Cherepanov and Nelson 1973; Alymov 1976; Gerberich and Chen 1975; Gerberich et al. 1975; Leeuwen 1974a, 1974b, 1975; Ochiani et al. 1975).

One of the first publications in this direction is Cherepanov's paper (Cherepanov and Nelson 1973), in which it is considered that in a cracked body, the stress–strain state of which is described by SIF K_{I}, a point source of hydrogen of constant power Q is attached at the notch tip. In this case, the hydrogen diffusion occurs, which is considered to depend only on the gradient of its concentration and is described by the equation of the form:

$$\partial C/\partial t = D\nabla^2 C, \tag{5.7}$$

where ∇^2 is the Laplace operator for spatial coordinates.

Diffusing hydrogen is accumulated inside the pre-fracture zone at the crack tip, where a metal embrittlement zone is formed—an "elastic core"—an area where the hydrogen concentration C exceeds some value C_c, required for the transition from the plastic to elastic–plastic state. The theoretical relationships obtained in this approach were compared with the results of experimental studies of the kinetics of stress corrosion cracking of metals, which is often associated with the influence of hydrogen formed during corrosion. Although such a comparison gives a good agreement between the calculation and the experiment, not all the characteristic features of the typical kinetic diagram of delayed fracture of metals are described by theoretical dependences. In (Panasyuk et al. 1989), this is explained by the action of the factors that are not directly related to the effect of hydrogen on the metal and are therefore not included to this model.

One of the possible idealizations of the hydrogen charging process is to represent the metal hydrogen charging conditions through the introduction of a source of hydrogen at the crack tip. Another one, physically more obvious way of representing the boundary conditions in the problem on hydrogen diffusion into the metal, is to set on the juvenile surface of the metal at the crack tip some equilibrium concentration C_0 (which when hydrogenating in a gaseous medium) can be related to hydrogen pressure P using known Sieverts relation:

$$C_0 \sim \sqrt{P}. \tag{5.8}$$

This consideration, together with the postulation of a local crack instability [i.e., the representation of a specific type of dependence (5.5)] forms the basis of the Alimov's theory (Alymov 1976). However, an obvious drawback of this model is its inability to describe the delayed fracture of pre-hydrogen-charged samples, in which hydrogen is preliminary uniformly distributed throughout the volume and has concentration C_B. This is due to the fact that when using the diffusion Eq. (5.7) in the model in the absence of the initial gradient of hydrogen concentration and exchange with the environment, time as one of the process parameters is excluded from the mathematical description of the phenomenon.

To eliminate this shortcoming, the following ways are proposed: firstly, to introduce, as a driving force of diffusion displacement instead of the concentration gradient, the gradient of the chemical potential of hydrogen, which is in some way related to the gradient of mechanical stresses (Shewmon 1974); secondly, to consider the redistribution of hydrogen atoms in metal due to the interaction of its atoms with defects in the structure of metals (Moroz and Chechulin 1967; Shewmon 1974), thirdly, to consider specific transport mechanisms in deformed metal—transfer of hydrogen atoms by dislocations (Leeuwen 1975). However, it is possible that to model the delayed fracture in a hydrogen-containing medium, when the area of the process is limited by a small area of the crack tip, due to a slight change in stress in the plastic zone at the crack tip, we can consider the influence of the concentration gradient only on the process of hydrogen atoms transfer. The mathematical formulation of the model will remain the same, although the values of the constants C_0 and D for the plastically deformed region at the crack tip will be different than that ones for the undeformed metal.

Thus, the model of crack growth in a hydrogen-containing environment proposed in Alymov (1976) may be acceptable, but to justify the simplifications, it is necessary to obtain a solution to the problem in a more accurate formulation, taking into account the action of various physical factors.

As mentioned above, one of these factors is the effect of mechanical stresses on hydrogen diffusion. For the first time, the assumption that hydrogen tries to diffuse in the region of the greatest tensile stresses was proposed in Johnson et al. (1958); Steigerwald et al. 1959). Considering the effect of the gradient of the chemical potential μ of hydrogen in solution on the diffusion displacement of hydrogen atoms, the following expression for the flow (Kronshtal' and Kharin 1992) was obtained:

$$\vec{J} = -D\nabla C - (D/RT)C \cdot \nabla\mu, \tag{5.9}$$

where $\vec{J}$ is hydrogen flow; ∇ is gradient operator; R is gas constant; T is absolute temperature.

It is shown that the chemical potential μ is related to the mechanical stress by the relation (Steklov 1992):

$$\mu = \mu_0 - (V_H/3)\sum_{i=1}^{3}\sigma_i + (V_H/2)\sum_{i=1}^{3}\left(\sigma_i^2/E\right), \tag{5.10}$$

where μ_0 is the chemical potential of hydrogen in the absence of mechanical stresses; σ_i are the main stresses; V_H is partial molar volume of hydrogen in the metal.

In practice, in most cases it is possible to neglect the terms containing σ_i/E in expression (5.10), and further to use a simpler relation:

$$\mu = \mu_0 - V_H\sigma, \tag{5.11}$$

where $\sigma = (\sigma_1 + \sigma_2 + \sigma_3)/3$ is hydrostatic stress.

Using expressions (5.9) and (5.11), it is easy to obtain an equation that describes the process of hydrogen diffusion in the stress field (Cotterill 1963; Shewmon 1976; Shober and Vepul 1981; Geld and Riabov 1974; Cermak and Kufudakis 1966):

$$\partial C/\partial t = D\nabla^2 C - (DV_H/RT)(\nabla C \cdot \nabla\sigma). \tag{5.12}$$

When deriving this equation, it is considered that for the elastic stresses filed in the absence of bulk forces we have $\nabla^2\sigma = 0$. The stress distribution in the body required to solve Eq. (5.12) is found from the solution of the standard equations of solid mechanics. In other words, the problem of diffusion in a deformable body is not considered as the interrelated, because the effect of dissolved hydrogen on the metal deformation is ignored. This is considered to be acceptable due to the generally low amount of hydrogen, which causes embrittlement. Note also that the solution of the interrelated problem is much more mathematical complicated than the solution of Eq. (5.12).

Liu (1972) obtained the solution of the diffusion Eq. (5.12) for the steady state (i.e., when $\partial C/\partial t = 0$) for a body with a crack on the surface of which a constant hydrogen concentration C_0 is given. The field of elastic hydrostatic stresses required to solve this problem is determined from the known solution of the canonical singular problem of the theory of elasticity in polar coordinates r, θ (introduced at the crack tip) by the relation (Panasyuk et al. 1989):

$$\sigma = \frac{2(1+\nu)K_{\mathrm{I}}}{3\sqrt{2\pi r}}\cos\frac{\theta}{2}. \tag{5.13}$$

Here ν is the Poisson's ratio.

The final expression for the stationary concentration distribution is written as:

$$C_{\mathrm{eq}} = C_0 \exp\left(\frac{2(1+\nu)V_{\mathrm{H}}K_{\mathrm{I}}}{3RT\sqrt{2\pi r}}\cos\frac{\theta}{2}\right), \tag{5.14}$$

or

$$C_{\mathrm{eq}} = C_0 e^{2B\sigma}, \tag{5.15}$$

where $B = V_{\mathrm{H}}/2RT$.

Direct verification shows that the relationship of type (5.15) satisfies Eq. (5.12) regardless of the type of function $\sigma(r, \theta)$, and that relations (5.14) and (5.15) determine the stationary distribution of hydrogen in pre-hydrogen-charged samples, if only the surface concentration C_0 is replaced by concentration C_B of hydrogen, uniformly distributed throughout the volume of the sample. In future, in cases where the mathematical formulation does not depend on whether hydrogen environmental embrittlement or prematurely introduced into the metal is considered, we will use one notation C_0, meaning C_0 and C_B (discussing special cases where this is not possible).

To obtain the dependence of the subcritical crack growth rate on SIF, the author (Liu 1972) believes that embrittlement is caused by some chemical reaction involving hydrogen (e.g., the formation of a hybrid), the rate S of which is proportional to the concentration. Assuming that the rate of this "embrittlement reaction" is much lower than the rate of hydrogen delivery, it can be concluded that S is determined by the steady-state concentration value given by expression (5.14) and the crack growth rate is determined by the reaction rate and is proportional to it.

Liu then suggests two ways to use these provisions to interpret data on the crack growth kinetics. The first one is based on the fact that the concentration stress and the reaction rate S in a solid at distances from the crack tip, proportional to K_{I}^2, are the same for different values of SIF. Then, assuming that the amount of "hydride" required for fracture does not depend on the stress level at the crack tip (i.e., does not depend on K_{I}), we can conclude that time Δt of crack propagation over a distance $\Delta l \sim K_{\mathrm{I}}^2$ does not depend on the SIF. In this case we will have (Liu 1972):

$$V = \Delta l/\Delta t = A_1 K_{\mathrm{I}}^2. \tag{5.16}$$

Since the reaction rate is $S \sim C_0$, we obtain from Eq. (5.16)

$$V = A_1' C_0 K_I^2, \tag{5.17}$$

where A_1 and A_1' are coefficients of proportionality.

Note that the Liu formula (5.17) can be easily replaced by a more general relation:

$$V = A_1' C_0 K_I^{a_1}, \quad a_1 \geq 2, \tag{5.18}$$

if we assume that the amount of reaction product of the metal (or some component of the alloy) with hydrogen required for fracture is not constant, but decreases with increasing SIF, which is true. The reaction time (embrittlement time) Δt also decreases, and in the first approximation we can assume that:

$$\Delta t = \text{const}/K_I^{a_2}, \quad a_2 \geq 0, \tag{5.19}$$

whence by analogy with formula (5.16) we can obtain the expression (5.18).

According to the second method, the crack growth rate is considered to depend on the reaction rate in some limited region of size d (structural parameter) and proportional to the reaction rate at the boundary of this region. Then, taking into account expression (5.14) we obtain:

$$V = A_2 C_0 \exp\left(\frac{2(1+\nu) V_H K_I}{3RT\sqrt{2\pi d}}\right), \tag{5.20}$$

where A_2 is some constant.

In this case we can get a more general relation, as it is done for Eq. (5.16). The rejection of the Liu's assumptions about the constancy of the amount of reaction product, required for the fracture, gives wide opportunity to obtain relations that are more consistent with the experimental data.

However, the most serious remarks concern Liu's physical models of the predominant effect of some slow "embrittlement reaction" on crack growth kinetics.

First, for many metallic materials (iron-, nickel-based alloys) the assumption of "embrittlement reactions" as chemical interactions that lead to new compounds, for a wide range of operating conditions, is not true (Kolachev 1985). The embrittlement of such metals is considered to be the result of relatively rapid physical interaction, the mechanism of which is not fully understood. Note that according to Cherepanov and Nelson (1973), the kinetics of steel embrittlement can be determined by the rate of hydrogen and carbides reaction, which may be true for some metals under known conditions, but is not a general rule.

Second, even for hydride-forming metals (titanium, zirconium, etc.), the hydride formation time can hardly be much longer than the time of establishment of the steady hydrogen concentration, which is the main assumption of the model. This provision may be correct under certain partial conditions.

Thus, within the framework of this model it is possible to construct relations capable of formally displaying experimental diagrams $V - K_I$ with a certain choice of constants included to them. However, further research is needed to establish the conditions under which these relationships will represent real processes.

The opposite approach to the description of the crack growth kinetics was proposed in the papers by Gerberich (Gerberich and Chen 1975; Gerberich et al. 1975) and Van Leuven (1974a, 1974b, 1975). It is considered that embrittlement occurs almost instantaneously when hydrogen enters critical areas. The degree of such embrittlement depends on the amount of hydrogen available there. In this case, the rate of the process depends on the rate of hydrogen supply, i.e., diffusion.

Within this approach, the critical stress intensity K_{Ith} is determined as the SIF value at which at some point $x = r_0$ in the crack path the steady-state value of the hydrogen concentration reaches the critical value defined by Eq. (5.5), i.e., for K_{Ith} we obtain equation (Gerberich and Chen 1975):

$$C_{cr}(K_{Ith}) = C_{eq}(K_{Ith}, C_0, r_0), \tag{5.21}$$

which, taking into account relation (5.14) gives:

$$K_{Ith} = \ln[C_{cr}(K_{Ith})/C_0]\frac{3RT\sqrt{2\pi r_0}}{2(1+\nu)V_H}. \tag{5.22}$$

Note that in the papers by Gerberich and Chen (1975) C_{cr} is a certain value of concentration, and not the SIF function (5.5). In this case, a similar formula from Roldugin and Martynov (1988) is an expression for determining K_{Ith}, not an equation, which it really is.

For theoretical determination of K_{Ith} by solving Eq. (5.22), it is necessary, in addition to dependence (5.5), to specify the position (coordinate) of the critical point. It is believed (Troiano 1959; Snape 1969; Loginow and Pholps 1975) that this point is located near the boundary between the elastic and plastic regions at the crack tip, where the tensile hydrostatic stress and concentration are maximum. If to use any of the known approximate relations that relate the size of the plasticity zone r_y to the SIF (Panasyuk et al. 1989):

$$r_y = K_I^2/(2\pi\sigma_{Tc}^2), \tag{5.23}$$

where σ_{Tc} is the stress of the plastic yield limited by the triaxial stress state, then from the expression (5.14) for the equilibrium hydrogen concentration C_{eq} when $r_0 = r_y$ we obtain

$$C_{eq} = C_0 \exp[2(1+\nu)V_H\sigma_{Tc}/(3RT)]. \tag{5.24}$$

That is, hydrogen concentration at the boundary of the region of plasticity depends only on the metal properties, and not on the stress level (SIF), which is not true. To remove this shortcoming, one can either find an explicit dependence of the yield stress σ_{Tc} on the SIF, which is defined as.

$$\sigma_{Tc} = p\sigma_T, \tag{5.25}$$

where p is the coefficient of plasticity limitation, which is equal to the ratio of the maximum principal stress to the yield stress, or directly determine the hydrostatic stress at the boundary of the plasticity region as a function of SIF.

The last requires the solution of the corresponding problem of a notched body deformation taking into account not only physical but, as we can see, geometric nonlinearity. Unfortunately, the complete solutions of such problems for elastic–plastic bodies are unknown.

In (Gerberich and Chen 1975), a well-known solution to the problem of deformation of a rigidly plastic body with a notch was used to determine the stress field in the plastic region at the crack tip along the x axis:

$$\sigma = \sigma_T\left[ln\left(1+\frac{x}{\rho}\right)+\frac{1}{2}\right] = \sigma_T\left(\rho-\frac{1}{2}\right), \tag{5.26}$$

where ρ is a radius at the notch tip.

In the elastic region $x > r_y$, the stresses are determined by the known linear-elastic asymptotics (Fig. 5.3) (Gerberich and Chen 1975; Gerberich et al. 1975). Note that formula (5.26) can be considered as an approximate solution of the problem for a notched body taking into account geometric nonlinearity, when the boundary conditions are satisfied on a deformed notch surface with a nonzero curvature radius at the tip. The change in stresses along the x-axis qualitatively represents the typical features of the elasto-plastic problem solution, taking into account the considerable deformations obtained by the finite element method in Yousscj and Jaeger (1974). To express the dependence of the coefficient of plasticity limitation on SIF in Gerberich and Chen (1975), the empirical relationship of Hahn and Rosenfield (1966) was used:

$$p \approx 1 + a(K_{\text{I}}/\sigma_T), \tag{5.27}$$

where a is a constant of appropriate dimension, σ_T is a yield limit.

Based on relations (5.15), (5.21), (5.26), and (5.27) the following formula was obtained for the critical stress intensity (Gerberich and Chen 1975):

$$K_{\text{I}th} = \frac{RT}{aV_{\text{H}}}\ln\frac{C_{cr}(K_{\text{I}th})}{C_0} - \frac{\sigma_T}{2a}. \tag{5.28}$$

Expression (5.28), like (5.22), is an equation for determining $K_{\text{I}th}$, and not for its calculating. A similar result can be obtained using relations (5.24), (5.25), (5.27).

Comparison in Gerberich and Chen (1975) of the theoretical and experimental data is incorrect, because it is currently impossible to determine the C_{cr}/C_0 value experimentally. The choice of its value is done by the author unreasonable due to which a good correspondence of the theory and experiment is achieved. Therefore, it is not yet possible to reliably estimate the relation (5.28).

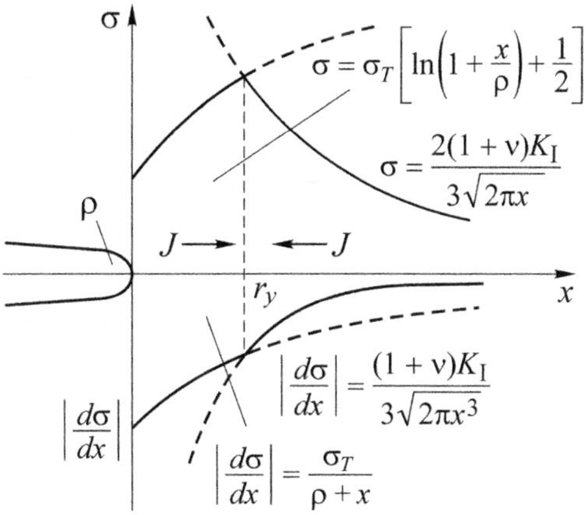

Fig. 5.3 Character of the change in the hydrostatic stress σ and its gradient $d\sigma / dx$ along the crack line (arrows indicate the direction of hydrogen flow)

The paper by Herberich and others (Gerberich et al. 1975), in which an attempt is made to construct a theoretical diagram $V - K_I$ is a direct continuation of the research presented in Gerberich and Chen (1975). Neglecting the effect of the concentration gradient on diffusion, i.e., assuming, according to relations (5.9) and (5.11), that the hydrostatic stress gradient to be the only driving force of the hydrogen transport process and comparing the absolute values of the diffusion driving force along the crack line $d\sigma / dx$ in elastic and plastic zones (Fig. 5.3) for $x = r_y$, the authors come to the conclusion that near the elastic–plastic boundary the flow of hydrogen is determined by the gradient of the elastic stress field for small r_y and of the plastic field—for large r_y.

Due to the fact that at small SIF (small r_y) the flow J of hydrogen in accordance with the sign $d\sigma / dx$ is directed from the middle of the metal to the crack tip (see Fig. 5.3), and hydrogen transfer from the crack surface is excluded, this theory is unacceptable to describe embrittlement in a hydrogen-containing medium.

The authors further assume that fracture initiates in some elementary volume of characteristic size d (it is identified with grain size). Estimating the size of the region from which hydrogen can diffuse into the "dangerous grain" for the time between crack jumps, determining the increase in concentration ΔC in the dangerous zone for time Δt, depending on the value of hydrogen flow and identifying Δt with time between crack jumps and the concentration C with its critical value C_{er}, they received for the crack growth rate:

$$V = \frac{\Delta l}{\Delta t} = \frac{3C_B D V_H \Delta l}{dRT(C_{cr} - C_B)} \cdot \frac{d\sigma}{dx}. \tag{5.29}$$

Assuming that in section I of the $V - K_I$ curve (see Fig. 5.2) $\Delta l = d$, and determining the average value of the gradient of elastic hydrostatic stress in the interval $d/2 \leq x \leq 3d/2$, it is easy to obtain the expression for the crack growth rate in section I from relation (5.29):

$$V_{\mathrm{I}} = \frac{2(1+\nu)C_B V_{\mathrm{H}} K_{\mathrm{I}}}{3d^{3/2}RT(C_{cr} - C_{B)}}. \tag{5.30}$$

Assuming that in section II of the $V - K_{\mathrm{I}}$ curve the development of cracks is controlled by a plastic stress field, and determining the corresponding average value of the gradient $\mathrm{d}\sigma / \mathrm{d}x$ at constant other assumptions, we can write:

$$V_{\mathrm{II}} = \frac{9C_B V_{\mathrm{H}} \sigma_T}{2dRT(C_{cr} - C_B)}. \tag{5.31}$$

Finally, in section III, it is assumed that (Zakkay et al. 1973):

$$\Delta l \approx K_{\mathrm{I}}^2/(E\sigma_T). \tag{5.32}$$

This allows from expression (5.29) similarly to Eq. (5.31) to obtain:

$$V_{\mathrm{III}} = \frac{9C_B D V_{\mathrm{H}} K_{\mathrm{I}}^2}{2d^2 ERT(C_{cr} - C_B)}. \tag{5.33}$$

The introduced theoretical relations in general reproduce the typical features of the typical $V - K_{\mathrm{I}}$ curve. However, the construction of the kinetic fracture curve in Gerberich et al. (1975) is not based on the correct solution of the diffusion equation. Only a rather rough approximation of this solution is given on the basis of a large number of assumptions, averages, simplifications, the error of which is not subject to preliminary assessment. Therefore, the corresponding theoretical dependence $V - K_{\mathrm{I}}$ (5.30), (5.31), (5.33) cannot serve as a reliable part of the calculation model of hydrogen-induced delayed fracture of metals.

To create a more correct theory, it is necessary to obtain a solution of Eq. (5.12), which determines the hydrogen concentration, using a minimum number of additional assumptions. One of the ways to solve this problem, when only the stress distribution around the crack tip is approximate (since the exact solution of the corresponding elastic–plastic problem is unknown), was proposed by Van Leeuwen (1974a, 1974b, 1975).

The following linear model of hydrogen embrittlement is proposed in Geld et al. (1979). It is believed that along the hydrogen crack propagation with period d there are elements of the structure that are sensitive to hydrogen. When a critical hydrogen concentration is reached on such an element, it fails under the action of stresses, and the microcrack that has arisen there merges with the main crack. The authors take into account the possibility of three variants of interaction of the stress field at the crack tip with the hydrogen maximum left behind the crack.

In one of the cases the incubation period is absent, and the crack growth rate is determined by the relation:

$$V = L / t_n, \tag{5.34}$$

where $L = nd$ is the length of the jump; n is the ordinal number of the element, sensitive to hydrogen; t_n is the diffusion time of hydrogen to the region of the hydrogen maximum, the analytical expression is proposed to determine such a maximum.

In another case, the crack growth rate is equal to the rate of hydrogen movement under the action of stress fields. It is determined by the value of the stress gradient between the points $x = 0$ and $x = d$ from the crack tip:

$$V = D\frac{V_{\mathrm{H}}}{3RT}\frac{\varepsilon(d)}{d};\quad d > 2\delta_p$$
$$V = D\frac{V_{\mathrm{H}}}{3RT}\frac{\varepsilon(2\delta)}{K_{\mathrm{I}}^2}E\sigma_{0.2};\quad d < 2\delta_p \tag{5.35}$$

Here δ_p is the crack opening displacement under the action of the intensity load p.

In all these cases the theoretical construction of the kinetic curve of hydrogen cracking requires data from microstructural studies on determination of the critical hydrogen concentration and distance d.

The authors of (Andreikiv et al. 1978; Andreykiv et al. 1987; Pokhmurskii and Fedorov 1998; Panasyuk et al. 1984) form the theory of crack growth in two stages. It is believed that there is an alternation of hydrogen accumulation by diffusion into the pre-fracture zone and the subsequent loss of hydrogen at this point during the jump.

The first step in the formation of the theory of hydrogen-induced crack growth is to establish the dependence:

$$C_* = C(t, a_i), \tag{5.36}$$

where C is the hydrogenconcentration in the fracture region; a_i are parameters of material, environment, load.

The authors obtained analytical dependences of the hydrogen concentration in the pre-fracture zone taking into account the stress–strain state at the crack tip.

The second stage is the construction of the criterion of local instability of the metal at the tip. This criterion is written as:

$$C(x_m, t) = C_*(x_m, K_{\mathrm{I}}), \tag{5.37}$$

where C_* is the critical concentration of hydrogen, at which there is the local fracture event for a given value of SIF at a distance $x = x_m$.

Assuming that the local fracture occurs at the boundary of the region of severe plastic deformation, as well as using the obtained analytical data on the distribution of hydrogen in the pre-fracture zone, the hydrogen-induced crack growth kinetics is theoretically described. It is established that the critical value of the stress intensity factor is equal to:

$$K_{\mathrm{I}th}^{(\mathrm{H})} = \sqrt{AE\sigma_{0.2}\ln(1 - \psi(C))}, \tag{5.38}$$

where A is a parameter that depends on the structure of the material and the stress–strain state at the crack tip.

The proposed model makes it possible to construct a kinetic diagram of hydrogen-stress cracking by calculations. To do this, it is necessary to experimentally establish the dependence $\psi = \psi(C)$ of the relative narrowing of the pre-uniformly hydrogen-charged non-standard sample on the hydrogen concentration in its smallest cross-section at the time of fracture. This greatly complicates its widespread use.

Thus, within the diffusion model, it is possible to theoretically construct a typical kinetic diagram of the process (Fig. 5.2) with a certain choice of dependence (5.5), without using additional assumptions about the predominant role of such factors as changing the mechanisms controlling crack growth rate in different parts of the $V - K_{\mathrm{I}}$ curve (reaction of metal with hydrogen in section I and the process of hydrogen transport in section II), or crack branching, sometimes called the "universal cracking mechanism" (Romaniv et al. 1977), which provides stabilization of the crack growth rate in section II of the $V - K_{\mathrm{I}}$ curve. Note that experimentally branched cracks are common (Carter 1970; Speidel 1972), but not always met. Therefore, some authors consider the fact of crack propagation at a constant rate with increasing SIF to be a consequence rather than a cause (Speidel 1972). Here we speak about the main role of these factors, but we do not deny their importance as factors that can somehow change the dependence $V - K_{\mathrm{I}}$.

As mentioned above, mechanical stresses are not the only (besides the concentration gradient) cause of hydrogen redistribution in metals. This process may be caused by the influence of various defects in the crystal structure of metals, which are traps for hydrogen atoms. These include dislocations (Panasyuk et al. 1989; Fidelle et al. 1974; Gabidullin 1977), micropores (Chew 1972; Ellerbrock et al. 1972; Gabidullin and Yakushev 1973; Atten-Beolh and Hewitt 1974), vacancies (Heller 1956), and other defects. Consideration of these factors when describing the diffusion process in an undeformed metal leads to the equations in which the diffusion coefficient depends on the density of defects (Gabidullin 1977) and concentration (Speidel 1972; Ellerbrock et al. 1972; Gabidullin et al. 1973). Such equations are quite complex and can usually be solved only by numerical methods (Cushey and Pillinger 1975; Allen-Booth et al. 1975).

In the deformed metal at the notch (crack) tip, the interaction of the absorbed hydrogen with various types of defects can be qualitatively represented in the following way. Prior to deformation, hydrogen, concentrated in traps and dissolved in a regular crystal lattice, is in equilibrium state and, on average, uniformly distributed in the volume of the metal. During the sample deformation in the vicinity of the crack tip there is an area of plastically deformed material—the pre-fracture zone—in which the density of defects increases markedly compared to the rest of the sample. In this case, the balance in the system is disturbed, because the appearance of a new defect causes a hydrogen flow directed to its center (flow to the nucleus of dislocation or release of gas into the microcavity). An increase of the metal defectiveness in some area (in this case in the plastic zone at the crack tip) will increase the total amount of absorbed hydrogen in this area—both dissolved in the metal lattice and limited by defects located in this volume.

An attempt to quantitatively describe the process of hydrogen redistribution in metal due to changes in the defectiveness of the latter, and even more due to the combined action of mechanical stresses, diffusion migration of defects, the influence of hydrogen itself on the initiation and propagation of defects, encounters significant difficulties both when formulating the corresponding mathematical problem and when solving it.

At present there is only one known work (Ochiani et al. 1975), the authors of which considered, although quite approximately the problem of the effect of plastic deformation on the localization of hydrogen in the metal. Here, the problem of diffusion of preliminary uniformly distributed hydrogen in the metal is solved numerically by the finite element method for a plane sample with two side notches made of a special material (HT-60 steel). The curves of changes in the hydrogen concentration around the notch tip with time for two values of the initial concentration: C_B and $0.7\,C_B$ are constructed. Comparison of these curves with experimental diagrams of delayed fracture allows us to conclude that the critical value of the concentration C_{cr} for a given load does not depend on its initial value. This is a confirmation of the existence of a universal (under certain conditions) dependence of type (5.5). In addition, the authors of this paper show that the critical value C_{cr} of concentration decreases with increasing load ($C_{cr} = 9.7C_B$ for a load of 69.26 kg/mm^2 and $C_{cr} = 6.5C_B$ for 82.14 kg/mm^2), which corresponds to the previously presented ideas about the character of dependence (5.5).

The fracture kinetics has not been investigated within this approach. However, the results obtained are quite important, because they demonstrate the existence of the influence of changes in the defects of the material on the distribution of hydrogen in it, and hence on the character of the delayed fracture process.

Recently in the studies of the problem of hydrogen brittleness of metals, more importance is payed to the transfer of hydrogen inside the metal by dislocations (Allen-Booth et al. 1975). For the first time suggestion that hydrogen is transported in the form of Cottrell atmospheres by dislocations was proposed in 1951 (Bastien and Azou 1951). Experimentally the results of Louehan et al. (1972); Broudeur et al. 1872) are in favor of this idea. Hydrogen captured by dislocations, can settle on various structural defects of the metal (grain boundaries, inclusions, microcavities), which are an obstacle to the dislocations movement, which cause dangerous concentrations of hydrogen in these areas.

Quantitative influence of hydrogen transfer by dislocations on the process of its localization in the metal was studied in (Allen-Booth et al. 1975; Gabidullin et al. 1971). In (Gabidullin et al. 1971), the rate of hydrogen transport and the distances to which it can be transferred during plastic deformation were estimated. Assume that the corresponding values of the driving force of hydrogen transfer are the energy E_B gradient of its connection with the dislocation. Assuming the value E_B to be the effective energy in the region whose size is given by the maximum distance at which the dislocations interact with the impurity atom and which is assumed to be $30b$ (b is the Burgers vector), we can obtain the following estimate for the upper limits of the rate of hydrogen transfer by dislocations:

$$w_c = \frac{DE_B}{30bkT}, \tag{5.39}$$

where k is the Boltzmann constant.

Hence the maximum distance of hydrogen transport for time t_p of plastic deformation is:

$$x_c = w_c t_p = \frac{DE_B}{30bkT} t_p. \tag{5.40}$$

For the E_B values of an order of 0.1–0.5 eV, which agrees with the experimental data for many metals, for iron-based alloys ($D \sim 10^{-6}\ \mathrm{cm^2/s}$) it was determined from relation (5.39) that $w_c \sim 10\ \mathrm{cm/s}$ ($T = 300$ K, strain rate $\varepsilon_c = \lambda b w_c \sim 10^{-7}\ \mathrm{s^{-1}}$), which is much higher than the value typical of hydrogen diffusion in the metal lattice.

The mechanism of localization of hydrogen in the metal is described in Allen-Booth et al. (1975). As the dislocation and the associated atmosphere move, the latter can interact with inclusions, cavities, and other structural defects located along the dislocation path. In this case, part of the hydrogen is deposited either in the pre-existing cavities or in the pores formed near the inclusions. Since the rate of hydrogen atoms transport by dislocations is much higher than the rate by transferring ordinary diffusion, the amount and pressure of hydrogen released into the pores increases with time. To calculate the concentration C_v of hydrogen trapped in cavities and the pressure P_v of gas in pores, as well as the rates of change of these values the appropriate expressions were obtained in Allen-Booth et al. (1975) that may be useful to describe and explain various effects associated with hydrogen embrittlement.

Unfortunately, due to the locality, this theory cannot be applied to describe the process of embrittlement on a macro scale, when the development of the process must take into account the combined effects of hydrogen and a set of different defects. Therefore, it is important to develop on the basis of the above ideas a certain continual theory describing hydrogen transport in the metal by dislocations and the resulting redistribution of hydrogen, for the theoretical determination of various macroparameters (integral characteristics) of the embrittlement action of hydrogen on metal.

Thus, three mechanisms of hydrogen transport inside the metal are analyzed: diffusion in the stress field, hydrogen redistribution due to changes in metal defectiveness, and hydrogen transport by dislocations. Each of these factors has been shown to significantly affect the localization of hydrogen in the metal and, consequently, to create a critical state in some characteristic volume. Therefore, it can be expected that, taking into account the combined action of all three mechanisms of hydrogen transport for the embrittlement process, results will be found that differ significantly from those obtained during the study of each of the three types of transfer separately.

It is of great interest to study all these mechanisms of hydrogen transport in the relationship and interaction and, if possible, to identify among them the determining one under certain conditions. This problem is extremely difficult and attempts to solve it are currently unknown.

Thus, the transfer of hydrogen atoms within a metal is not the only process that can determine the kinetics of delayed fracture of metals. Although in the case of internal hydrogen embrittlement, assumptions about the diffusion mechanisms of subcritical crack growth process do not cause a fundamental objection. For the case of embrittlement in a hydrogen-containing medium, some authors (Williams 1970; Nelson 1974; Nelson et al. 1971) reject the dependence of crack growth on the rate of hydrogen movement inside the metal, considering surface phenomena—adsorption, dissociation, and chemisorption to be the main mechanism of hydrogen transport on the metal surface. This is based on the following experimentally established effects:

(1) the growth of the crack in the gaseous hydrogen begins almost immediately after loading the sample without any significant incubation period (Williams 1970; Nelson 1974);
(2) the activation energy of the crack growth process in region II of the $V - K_{\mathrm{I}}$ curve differs markedly from the hydrogen diffusion activation energy, and it can be compared with the activation energy of surface processes (Williams 1970);
(3) crack growth rate in atomic hydrogen is several orders of magnitude higher than its analogue in molecular gas (Nelson et al. 1971); since in this case such a degree of interaction as the dissociation of hydrogen molecules on the metal surface is excluded, it can be assumed that in the environment of gaseous hydrogen this parameter is controlled by a surface process—dissociation;
(4) when hydrogen impurities are introduced into the gaseous hydrogen, which activate (H_2S) or inhibit (SO_2) the process of surface interaction of the metal with hydrogen, the crack growth rate increases or decreases accordingly (Srikrishnan et al. 1975).

Comparing a rather high rate of the surface interaction process [the formation of a molecular layer of adsorbed hydrogen at room temperature requires about 10^{-6} s (Tion et al. 1975)] with the rate of diffusion movement of hydrogen in metal, which is much lower, the authors (Williams 1970) conclude that in the environment of gaseous hydrogen embrittlement takes place without noticeable penetration of hydrogen into the metal and is due only to surface phenomena. According to the above ideas about the rate of hydrogen transport by dislocations (Bucur 1977), which is compared with the rate of surface processes, this conclusion of the authors (Williams 1970) is not completely proved. In the gaseous hydrogen, the cause of embrittlement can be processes that take place inside the metal and are experimentally confirmed by the results of Snape (1969), Loginow and Pholps (1975), consistent with the ideas (Fidelle 1974) of the existence of a single mechanism of hydrogen embrittlement of metals—both "internal" and in the hydrogen-containing environments.

However, the assumption of the authors (Williams 1970) about the significant role of surface phenomena that control the kinetics (rather than the mechanism) of embrittlement in the hydrogen environment may be correct. Unfortunately, the differential equations of adsorption and sorption in general have not yet been fully established to create a mathematical model of the process of delayed fracture in a gaseous medium (Bucur 1977). The available expressions that determine the rate of chemisorption depending on the parameters of the environment—temperature, pressure (Williams 1970; Nelson 1974; Nelson et al. 1971; Bucur 1977), obviously, are insufficient for the theoretical construction of the $V - K_{\mathrm{I}}$ curve, because they are true for some average time and do not take into account rate variation during surface saturation with hydrogen. Without this, within a single approach, it is impossible to obtain a theoretical diagram $V - K_{\mathrm{I}}$, taking into account the critical value $K_{\mathrm{I}th}$ of SIF, the existence of which has been proved experimentally for the embrittlement in gaseous hydrogen (Chandler and Walter 1974).

Further development of theoretical studies of the kinetics of hydrogen-induced delayed fracture of metals should obviously be directed to the development of analytical models, considering the combined effect of different mechanisms of hydrogen delivery to dangerous areas of the metal.

As to experiments, it is important to reliably determine the characteristics of hydrogen transport in the metal and in the "metal–environment" system, depending on the various parameters of the metal, environment, research conditions, and dependence $C_{cr} - K_{\mathrm{I}}$, which is fundamental, because without it you cannot establish the reliability of other theoretical ideas.

Currently, the most effective is a simplified calculation model (Louehan et al. 1972), which is based on the generalization of the deformation strength criterion in the case of hydrogen effect. The determination of the hydrogen-induced crack growth rate is reduced to the following key points: the choice of the criterion of local microfracture of the material, based on the elastic–plastic situation in the pre-fracture zone, and establishing conditions for the elementary event of local fracture. An important element of the calculation model is the correct establishment of the distribution of hydrogen concentration in the pre-fracture zone with account of the elastic–plastic state, i.e., finding the dependence $C = C(x, t)$, taking into account the stress of the material at the crack tip. However, this model does not take into account the pre-hydrogen charging of the material, which is typical of the hydrogen-temperature fracture.

5.5 Methods and Means of Determining the Hydrogen Content in Structural Materials

Today in the field of non-destructive testing and technical diagnostics the problem of determining the concentration of hydrogen in metals and on this basis to assess the strength and life time of materials and structures remains actual (Skalskyi et al. 1999). An important aspect of determining the state of structural materials is the structure of metals and alloys, their electrical, magnetic properties, and parameters, which are manifested through the relevant physical phenomena and effects. To solve the problem of assessing the state of the "metal–hydrogen" system, ferromagnets and their properties are studied on the basis of modern approaches to the interpretation of known physical phenomena and effects. An important aspect in such studies is the dynamic processes of ferromagnets magnetization. What is special here is the phenomenon of magnetic aftereffect, which manifests itself in the form of accommodation (increase in time) and decomodation (decline in time) of magnetic permeability after switching on and cutting off the external magnetic field, respectively, (Vonsovsky 1971; Chikazumi 1987). The study of the temporal change of the parameters of ferromagnetic materials on the basis of magnetic aftereffect makes it possible to establish their relationship with the hydrogen concentration, and thus determine the state of the structural material.

The interaction of hydrogen with metals is analyzed by various methods: nuclear physics, Mesbauer spectroscopy, nuclear magnetic resonance, neutronography (Shvachko 1998). Hydrogen concentration is determined by the following methods: estimation of heat capacity; weighing; chromatography; vacuum extraction. The results of the study of the metal-hydrogen system to determine the heat capacity of hydrogen dissolved in metals at constant pressure C_p and at constant volume C_V are of particular interest. Under such conditions, the equilibrium hydrogen concentration is analytically determined depending on temperature and pressure (Smirnov 2006).

Hydrogen content can be determined using an evolograph with a gas sensitivity less than 1 mm^3. The method of gas chromatography (mass spectrometry) allows us to extract hydrogen from a sample of the test material under conditions of high-temperature heating in vacuum (Shved 1985; Weinman et al. 1990). The method of vacuum extraction allows us to determine the amount of unbound gaseous hydrogen using a mass spectrometer in a vacuum system, when it is heated to a temperature of 200–400 °C (Shved 1985).

In the field of non-destructive testing and technical diagnostics, devices and primary transducers for determining the hydrogen content in metals have been developed and introduced in practice. For example, a device of hydrogen content and its heavy isotopes in the surface layer of a solid without its failure is known (http://iki.cosmos.ru/innov/rus/isiv3.htm). The principle of operation of this device is based on spectroscopy of hydrogen nuclei, which are scattered forward during their elastic interaction with alpha particles. The effective thickness of the surface layer in which the hydrogen concentration is determined, is $\sim 5 - 20 \times 10^{-6}$ m and depends on the chemical composition of the test sample.

Some express analyzer is also designed to determine the hydrogen content in metals (http://www.horiba.com/int/scientific/applications/metallurgy/pages/hydrogen-analysis-in-steel-and-metals-solid-extraction-or-fusion/). The principle of its operation is based on the reductive melting of the sample in the flow of inert gas and quantitative analysis of melting products by thermoconductometric method.

In order to analyze the concentration of hydrogen in metals, the primary hydrogen transducers are also used with the following characteristics: operating temperature range 77–330 K; operating pressure range $1 - 5 \times 10^5$ Pa; measured hydrogen concentration in the range of 0.2–95% with an accuracy of $+0.1$ (https://www.te.com/usa-en/product-cat-ptt0036.html).

The devices of the American company "Leco" namely analyzers RH (RHEN)-600/602, RH-402 (LECO 2022) are important for determining the concentration of hydrogen, in metals and alloys. For example, the analyzer RH-402 is widely used for rapid analysis of hydrogen content in ferrous metals, steels, alloys. The principle of this analyzer operation is based on high-temperature extraction of hydrogen in the inert gas flow in a reducing or neutral medium in an induction furnace with a programming temperature up to 2800 °C. The released hydrogen, which is transported by the carrier gas (nitrogen) through a purification system to remove impurities, is moved to a highly sensitive thermal conductivity detector. The range of measured concentrations of hydrogen per mass of the test sample of 5 g is 0.001–400 million particles.

A method and apparatus for determining hydrogen in metals, based on the release of hydrogen into the gaseous phase by the influence of pulsed laser radiation on the test sample of the metal or a comparative sample with a given hydrogen content, is known (Glukhov et al. 2006). Hydrogen is recorded using a chemical transducer based on the metal–dielectric–semiconductor (MDS) structure, the gas-sensitive element of which is installed in the measuring chamber of the device with a pulsed laser, a system of focusing and beam guidance on the test sample.

There is also a method of determining the hydrogen concentration in ferromagnets by the parameter of the magnetic aftereffect under hydrogenetation and loading of the sample (Andreykiv et al. 2004). The method is based on measuring the time of relaxation processes in the ferromagnetic material. Non-hydrogen-charged and hydrogen-charged samples are magnetized by an external magnetic field. The residual magnetization time of each sample is measured and hydrogen concentration is analytically determined, taking into account material characteristics, hydrogen charging conditions, and external factors influencing the hydrogen charging process.

These mentioned methods of determining hydrogen concentration are associated with the use of vacuum systems and, accordingly, require high-precision metrological characteristics of the equipment. Generally speaking, the method of determining the concentration of hydrogen in the ferromagnetic material (sample) at the relevant factors that affect the experiment is quite complex. It is used mostly in the laboratory. Therefore, the scientific and technical problem is to involve known magnetic methods of non-destructive testing and to develop methods on the basis of which one can implement the procedure for measuring the concentration of hydrogen in ferromagnetic structural elements.

5.6 Application of AE to Fix the Hydrogen Degradation of Structural Materials

Successful application of the AE method to control the state of materials and products is possible only when reasonable methods and appropriate tools for their implementation are available. Today, despite the huge number of publications and undeniable progress in the development of the hardware, there are some difficulties in the correct choice and application of methodological foundations of this method.

The phenomenon of magnetoacoustic emission, as one of the new trends of development the methods and means of non-destructive testing, is increasingly being studied. It is especially effective in cases when it is impossible to additionally load or deform the structural material (ferromagnet) due to its degradation.

The first attempts to record elastic waves that occur during the remagnetization of ferromagnets and to explain the physical meaning of this phenomenon were made more than 50 years ago (Lomayev et al. 1984). However, these experiments have not been developed. Later, new MAE studies were conducted using better experimental equipment (Bertotti and Montorsi 1990). Recent studies of the MAE have revealed the dependence of the parameters of its signals on the modes of heat treatment, plastic deformation, and other factors in structural materials. All this indicates the possibility of using the phenomenon of MAE for NDT of ferromagnetic materials.

The influence of hydrogen on the mechanical properties of metals is well studied (Kolachev 1985; Skalskyi and Andreykiv 2006), in particular, the following sources of its diffusion into metal: casting, pressure treatment, welding, heat treatment, acid etching, electroplating, stress corrosion, aggressive working environments, etc. Hydrogen can also enter the metal from the air during storage of products in warehouses or during their operation. The presence of hydrogen in the metal changes the parameters of its crystal lattice, electrical resistance, magnetic, plastic, strength, structural, and other properties.

The most common methods of controlling the hydrogen content in the metal are the method of vacuum extraction (vacuum heating and vacuum melting) (Shved 1985). However, they are very time consuming and require sampling of metal from specific products or structural elements.

Non-destructive methods for estimating the hydrogen content in metals are also known. These are based on various physical principles: determining the work function of the electron, impedance, hysteresis loop parameters, and so on. All these methods proved to be practically unsuitable due to the high complexity, the need to use expensive and complex equipment.

In (Nechai and Moskvin 1975), an attempt was made to use the BE to study the effect of hydrogen on the magnetic structure of the metal during cyclic remagnetization of the sample. Hydrogen has been shown to increase the magnetic noise voltage in metals with positive magnetostriction and decrease it with negative one.

The authors of Migirenko et al. (1973) studied the effect of electrolytic hydrogen charging and mechanical stress on the manifestation of the BE. The obtained results show that the hydrogen charging of samples to significant (from 1 mm/100 g and above) concentrations by its action on the BE is similar to the action of mechanical stresses in the elastic region of deformations. It has been hypothesized that the effect of significant concentrations of hydrogen and external elastic stresses on the BE is associated with the restructuring of the magnetic structure, namely the change in its magnetization due to magnetostriction. Summarizing the above and taking into account the results of the review (Lomayev et al. 1984), it can be stated that the effect of hydrogen on the BE is insufficiently studied today.

References

Allen-Booth DM, Atkinson C, Bilbyu BA (1975) A numerical solution of the diffusion equation resulting from the void theory of the trapping of hydrogen in iron and steel. Acta Metall 23(3):371–376

Alymov VT (1976) Theory of crack growth in metals under the action of hydrogen. Mater Sci 11(6):628–630

Andreikiv AE, Panasyuk VV, Kharin VS (1978) Theoretical aspects of the kinetics of hydrogen embrittlement of metals. Mater Sci 14(3):227–244

Andreikiv OÉ, Lysak MV, Sergienko OM, Skal's'kii VR (1991) Acoustic emission in tests on materials in hydrogenous and corrosive media. Mater Sci 26(5):512–520

Andreykiv OY, Skalskyi VR, Hembara OV (1980) Metod otsinky vysokotemperaturnogo vodnevoho ruinuvannia bimetallevykh elementiv konstruktsii (A method of estimation of hydrogen fracture of bi-metallic structural elements). Fizyko-Khimichna Mekhanika Materiailv (Physicochemical Mechanics of Materials) 4:15–23

Andreykiv OYe (2003) Dovhovichnist metallichnykh materialiv u vodenvmisnykh seredovyshchakh. Progresyvni materialy i tekhnologii (Durability of metallic materials in hydrogen-containing environments. Advanced materials and technologies), vol 2. Publishing House of Akademperiodyka

Andreykiv OYe, Hembara OV (2008) Mekhanika ruinuvannia ta dovhovichnist metallevykh materialiv u voden-vmistkykh seredovyshchakh (Fracture mechanics of materials and durability of metalmaterials in hydrogen-containing media). Naukova dumka Publishing House

Andreykiv OYe, Panasyuk VV, Poliakov LP, Kharin VS (1987) Mekhanika vodorodnogo okhrupchivaniia metallov i raschet elementov konstruktsii na prochnost (Mechanics of hydrogen embrittlement and strength calculation of structural elements). Preprint PhMI Acad Sci Ukr. SSR, N 133

Andreykiv OYe, Nylytorchyn HM, Tkachov VI (2001) Mitsnist i ruinuvannia metalichnykh materialiv i elementiv konstruktsii u vodenvmistkykh seredovyshchakh (Strenghth and fracture of metallic materials and structural elements in hydrogen-containing environments). Phyzyko-Mekhanichnyi Instytut: postup i zdobutky (Physico-Mechanical Institute: Progress and achievements), Lviv

Andreykiv OYe, Ivanytskyi YaL, Chekurin VF, Hembara OV (2004) Sposib vyznachennia kontsentratsii vodniu v malovuglytsevykh staliakh za parametramy magnetnoi pisliadii (The method of determining the concentration of hydrogen in low-carbon steels by the parameters of the magnetic aftereffect) (Declaration patent №71821A UA, G01N17/00, 15 Dec 2004)

Archakov YuI (1985) Vodorodnaia koroziia (Hydrogen corrosion). Metallurgiya Publishing House

Atten-Beolh DM, Hewitt JA (1974) A mathematical model describing the effects of micro voids upon the diffusion of hydrogen in iron and steel. Acta Metall 22(2):171–175

Barth CF, Steigerwald EA (1970) Evaluation of hydrogen embrittlement mechanisms. Metall Trans 1(12):3451–3455

Bastien P, Azou P (1951) Influence l'ecrouissage sur le frottement inturieur du fer et de l'acier charges ou non en hydrogene. Comptes Rendus de l'académie des Sciences 232:1845–1848

Bernstein IM (1970) The role of hydrogen in the embrittlement of iron and steel. Mater Sci Eng 6:1–19

Bertotti G, Montorsi A (1990) Dependence of Barkhausen noise on grain size in ferromagnetic materials. J Magn Magn Mater 86(1):214–216

Bokstein SZ, Ginzburg SS (1978) Elektronnomikroskopicheskaia avtoradiografiia v metallovedenii (Electron microscopic autoradiography in metal science). Metallurgiya Publishing House

Broudeur R, Fidelle J-P, Auchere H (1872) Experience montrant le role des dislocations dans le transport de l'hydrogene. L'hydrogene dans les metaux 1:106–107

Bucur RV (1977) Adsorption and absorption of hydrogen by thin palladium layers. Surf Sci 62(2):519–535

Carter CS (1970) The effect of heat treatment on the fracture toughness and subcritical crack growth characteristics of a 350-grade maraging steel. Metall Trans 1(6):1551–1659

Cermak J, Kufudakis A (1966) Relation entre l'onde de diftuslon de 1'hydrogene se propageant dans une to1e metalllque et la deformation provoquce par cette onde. Memoires et etudes scientifique de la revue de metallurgie 63(9):767–772

Chandler WT, Walter RJ (1974) Testing to determine the effect of high-pressure hydrogen environments on the mechanical properties of metals Hydrogen embrittlement testing. ASTM STP 543:170–197

Cherepanov GP, Nelson HG (1973) On the theory of crack growth due to hydrogen embrittlement. Corrosion 29(8):305–309

Chew BA (1972) Void model for hydrogen diffusion in steel. Metal Sci J 5:195–200

Chikazumi S (1987) Fizika ferromagnrtizma. Magnitnyie kharakteristiki i prakticheskiie primeneniia (Physics of ferromagnetism. Magnetic characterristics and practical application). Mir Publishig House

Cotterill P (1963) Vodorodnaia khrupkost metallov (Hydrogen brittenes of metals). Metallurgizdat Publishing House

Cushey GR, Pillinger WL (1975) Effect of trapping and hydrogen permeation. Metall Trans A 5(3):467–470

Dal V, Anton V (eds) (1986) Staticheskaia prochost i mekhanika razrusheniia stalei (Static strength and fracture mechanics of steels). Metellutgiya Publishing House

Druchenko VA, Novakovskii VM, Chirva AK (1977) O mikroakustike korrozionnykh protsesov (On microacoustics of corrosion processes). Zashchita Metallov (Metals Protection) 13(3):281–286

Fidelle JP (1974) Closing commentary—1HE–HEE: are they the same? Hydrogen embrittlement testing. ASTM STP 543:267–272

Fidelle JP, Bernardi R, Broudeur R, Roux C, Rapinet M (1974) Disk pressure testing of hydrogen environment embrittlement. Hydrogen Embrittlement Testing ASTM STP 543

Gabidullin RM (1977) Effect of dislocations on kinetics of metal degassing. Sov Mater Sci 12(1):42–45

Gabidullin RM, Kolachev B, Drozdov P (1971) Otsenka uslovii proyavleniia obratimoi vodorodnoi khrupkosti metallov (Evaluation of the conditions for the manifestation of reversible hydrogen embrittlement of metals). Problemy Prochnosti (Strength Problems) 12:36–40

Gabidullin RM, Yakushev VA (1973) O termodinamike raspredeleniia vodoroda v metallakh (On the thermodynamics of the distribution of hydrogen in metals). Izvestiia vysshykh uchebnykh zavedenii, tsvetnaia metallurgiia (Reports of higher educational establishments, non- ferrous metallurgy) 2:40–43

Gabidullin RM, Soyeskov AI, Dashkova LA (1973) Vliyaniie poristosti na kinetiku degazatsii i velichinu kazhushchikhsia koeffitsiyentov diffuzii vodoroda v aluiminii (Influence of porosity on degassing kinetics and apparent diffusion coefficients of hydrogen in aluminum). Tekhnologiya legkikh splavov: Nauchno-tekhnicheskii buleten VILSA (Technology of light alloys: Scientific and Technical Bulletin of All-Union Institute of Light Alloys) 2:9–13

Geld PA, Riabov RA (1974) Vodorod v metallakh i splavakh (Hydrogen in metals and alloys). Metallurgiya Publishing House

Geld PV, Riabov RA, Kodes ES (1979) Vodorod i nesovershenstva struktury metalla (Hydrogen and metal structure imperfections). Metallurgiya Publishing House

Gerberich WW, Chen YT, John CS (1975) A short-time diffusion correlation for hydrogen-induced crack growth kinetics. Metall Trans A 6(8):1485–1498

Gerberich WW, Chen YT (1975) Hydrogen-controlled cracking. An approach to threshold stress intensity. Metall Trans A6(2):271–278

Glukhov NP, Kaliteevsky SD, Lazarev SD et al (2006) Sposob opredeleniia vodoroda v metallakh i ustroistvo dla yego realizatsii (A method for determining hydrogen in metals and a device for its implementation) (RU Patent 2282182 (13) C1, G01N17/00, 20 Aug 2006)

Goldshtein RV, Entov VM, Pavlovskii BR (1977) Model razvitiia vodorodnykh treshchin v metalle (Model of hydrogen crack propagation in metal). Doklady Akademii Nauk SSR (Reports of the Academy of Sciences of USSR) 237(4):828–832

Hahn GT, Rosenfield AR (1966) Experimental determination of plastic constraint ahead of a sharp crack under plain-strain conditions. Trans Am Soc Metals 59:909–919

Heller WR (1956) Hydrogen in iron and its alloys. Stress corrosion cracking and embrittlement. John Wiley

Hirth JP (1980) Effect of hydrogen on the properties of iron and steel. Metall Trans 11A(6):861–890

Hydrogen Pressure Transducer. Retrieved 12 Feb 2023 from https://www.te.com/usa-en/product-cat-ptt0036.html

Hydrogen Analysis in steel and metals: Solid Extraction or Fusion. Retrieved 15 Oktober 2022 from http://www.horiba.com/int/scientific/applications/metallurgy/pages/hydrogen-analysis-in-steel-and-metals-solid-extraction-or-fusion/

Izmeritel soderzhaniia isotopov vodoroda (Hydrogen isotop meter). Retrieved 25 Mar 2021 from http://iki.cosmos.ru/innov/rus/isiv3.htm

Johnson HH, Morlett JG, Troiano AR (1958) Hydrogen, crack initiation and delayed failure in steel. Trans Metall Soc AIME 212:528–538

Karpenko HV, Krypyakevych RI (1962) Vliyaniie vodoroda na svoistva stali (Influence of hydrogen on steel properties). Melallurgizdat Publishing House

Kasatkin BS, Smiyan OD, Mikhailov VE (1986) Vliianiie vodoroda na sklonnost k obrazovaniiu treshchin v zone termicheskogo vliianiia s kontsentratorom napriazhenii (The effect of hydrogen on the tendency to form cracks in HAZ with stress concentrator). Avtomaticheskata Svarka (Automatic Welding) 11:20–23

Kikuta EI (1976) Izucheniie vodorodnoi khrupkosti i rol vodoroda v mikrostrukture (Study of hydrogen brittleness and role of hydrogen in microstructure). Yese Gakkaisi 45(2):1016–1021

Kolachev BA (1985) Vodorodnaia khrupkost metallov (Hydrogen brittleness of metals). Metallurgiya Publishing House

Kott D (1978) Mikromechanizmy razrusheniia i treshchinostoikost konstruktsionnykh splavov (Fracture micromechanisms and crack growth resistance of structural alloys). In: Taplin D (ed) Razrusheniye materialow (Fracture of Materials). Mir Publishing House

Kronshtal' OV, Kharin VS (1992) Effect of heterogeneity of materials and heat cycles on diffusion of hydrogen as a factor of the risk of development of hydrogen degradation of metals. Mater Sci 28(6):475–486

LECO® Analyzers. Retrieved 10 May 2022 from http://www.alpharesources.com/ohn-inorganic-analyzers.php

Liu S (1972) Korrozionnoie rastreskivaniie i vzaimodeistviie mezhdu polem napriazhenii u konchika treshchiny i rastvorennymi atomami (Corrosion cracking and the interaction between the stress field at the crack tip and dissolved atoms). Teoreticheskiie osnovy inzhenernykh raschetov. Trudy AOИM) (Theoretical bases of engineering calculations. Proceedings of ASTM), Ser, D 92(2):219–225

Loginow AW, Pholps EH (1975) Steels for seamless hydrogen pressure vessels. Corrosion 81(11):404–412

Lomayev GV, Malyshev VS, Degtiarev AP (1984) Obzor primeneniia effekta Barkhausena v nerazrushayushchem kontrole (Overview of the application of the Barkhausen effect in non-destructive testing). Defektosopiya (Defectoscopy) 3:54–71

Louehan MR, Caskey GR, Donovan JA, Rawt DE (1972) Hydrogen emibrittlement of metals. Mater Sci Eng 10(6):387–388

McMeeking RM (1977) Finite deformation analysis of crack tip opening in elastic-plastic materials and implication for fracture. J Mech Phys Soids 25(5):357–381

Migirenko GA, Moskvin VN, Nechai EP (1973) Priminieniie metoda magnitnykh shumov dlia issedovaniia navodorazhyvaniia stali (Applications of the magnetic noise method to study the hydrogenation of steel). Metody opredelieniia i issledovaniia sostoianiia gazov v metallakh (Methods of determination and study of the state of gases in metals). MDNTP Part 1:88–93

Mnushkin OS, Kopelman LA (1980) O mekhanizme vodorodnoi khrupkosti stali (On the mechanism of hydrogen brittleness of steel). Izvestiya AN USSR. Metally (Reports of the Academy of Sciences of the USSR. Metals) 2:154–160

Moroz LS, Chechulin BB (1967) Vodorodnaia khrupkost metallov (Hydrogen brittleness of metals). Metallurgiya Publishing House

Nechai EP, Moskvin VN (1975) Ob ispolzovanii effecta Barkhauzena dla kontrolia stepeni navordorozhyvaniia metallov (On the use of the Barkhausen effect to control the degree of hydrogenation of metals). Tezisy dokladov II vsesoyuznoi mezhvuzovskoi konf. po el./mag. metodam kontrolia kachestva materialov (Abstracts II All-Union. interuniversity conference on el./mag. methods of quality control of materials and products). Part II

Neimitz A (1998) Mechanika pękania. Wydawnictwo Naukowe PWN SA

Nelson HG (1974) Testing for hydrogen environment embrittlement: primary and secondary influences. Hydrogen embrittlement testing. ASTM STP 513:152–169

Nelson HG, Williams DP, Tetelman AS (1971) Embrittlement of ferrous alloy in a partially dissociated hydrogen environment. Metall Trans 2(4):953–959

Nykyforchyn H, Skrypnyk I, Lutchyn V (1998) Modeli dlia dyfuziinogo zernogranychnogo rostu mikropor pry vysokykh temperatutakh (Models of diffusion grain boundary microvoids propagation at high temperatures). Mashynoznavstvo (Machine science). Publishing House LP, 4/5

Ochiani S, Yoshinaga S, Kikuta Y (1975) Formulation of stress (strain)—induced diffusion of hydrogen and its solution by computer-aided finite element method. Trans Iron Steel Inst Jpn 15(10):503–507

Paisl G (1981) Deformatsiia reshetki metalla, svyazannyie s vodorodom (Deformations of metal lattice due to hydrogen). In: Alefeld G, Felkel I (eds) Vodorod v Metallakh. T. 1: Ostovnyie svoistva (Hydrogen in metals, vol 1: Basic propertiers). Mir Publishing House

Panasyuk VV, Andreikiv AE, Kharin VS (1982a) Theoretical analysis of crack growth in metals under the action of hydrogen. Mater Sci 17(4):340–352

Panasyuk VV, Andreykiv OY, Gembara OV (2000) Hydrogen degradation of materials under long-term operation of technological equipment. Int J Hydrogen Energy 25:67–74

Panasyuk VV, Andreykiv OY, Ritchie RO, Darchuk OI (2001) Estimation of the effects of plasticity and resulting crack closure during small fatigue crack growth. Int J Fract 107:99–115

Panasyuk VV, Andreikiv AE, Kharin VS (1982) Theoretical analysis of crack growth in metals under the action of hydrogen. Mater Sci 17(4):340–352

Panasyuk VV, Andreykiv OYe, Kharin VS (1984) Crack growth in metals affected by hydrogen. In: Advances in fracture research: proceedings 6th international conference on fracture (ICF 6). Pergamon Press

Panasyuk VV, Andreykiv Aye, Parton VZ (1989) Osnovy mekhaniki khrupkogo razrusheniia (Bases of brittle fracture mechanics). Naukova dumka Publishing House

Panasyuk VV, Andreykiv OYe, Darchuk OI, Kuznyak NV (1994) Influence of hydrogen-containing environments on fatigue crack extension resistance of metals. Handbook of fracture crack propagation in metallic structures, vol 2. Elsevier

Peth NJ (1956) The lowering of fracture—stress due to surface adsorption. Phillos Mag 1(4):331–337

Pokhmurskii VI, Fedorov VV (1998) Vplyv vodniu na dyfuziini protsesy v metalakh (Influence of hydrogen on diffusion processes in metals). Lviv

Pokhodnia IK (1972) Gazy v svarnykh shvakh (Gases in welds). Mashynostroyeniye Publishing House

Pokhodnia IK, Demchenko LI, Shlepkov VN (1978) O mekhanizme obrazovaniia por v svarnykh soyedineniyakh (On the mechanism of pores formation in welded joints). Avtomaticheskaia Svarka (Automatic Welding) 6:1–5

Pokhodnia IK (1998) Problemy svarki vysokoprochnykh nizkolegirovannykh salei. Suchasne materialoznawstwo XXI storichcha (Problems of welding of high-strength low-alloy steels. In: Modern materials science of XXI century). Naukova dumka Publishing House

Pokhodnia IK, Yavdoshin NR, Paltsevych AP, Pokhodnia IR, Yavdoshin AP, Shvachko AS, Kotelchuk AS (2004) Metallurgiia dugovoi svarki. Vzaimodeistviie metalla s gazami (Metallaurgy of arc welding. Interaction of metals with gases). Naukova dumka Publishing House
Pokhodnya IK, Shvachko VI (2001) Nature of hydrogen brittleness of structural steels. Mater Sci 37(2):241–251
Pokhodnya IK, Shvachko VI, Utkin SV (2002) Influence of hydrogen on the equilibrium of a dislocation submicrocrack in α-Iron. Mater Sci 38(1):1–10
Roldugin VI, Martynov GA (1988) Raschet elektronnoi plotnosti vblizi poverkhnosti metalla metodom kvantovykh funktsii raspredeleniya dlia granitsy metal-vakuum (Calculation of the electron density near the metal surface by the method of quantum distribution functions for the metal-vacuum interphase). Poverkhnost (Surface) 2:19–27
Romaniv ON, Nikiforchin GN, Deev NA (1977) Kinetic effects in the mechanics of delayed fracture of high-strength alloys. Sov Mater Sci 12(4):347–360
Schnitt-Thomas KG, Stengel W (1983) Möglichkeiten zur Früherkennung von Wasserstoffschädigungen in metallischen Werkstoffen durch Anwendung der Schallemissionanalyse. Werkst Korros 34:7–13
Shapovalov VI, Trofimenko VV (1987) Flokeny i control vodoroda v stali (Flakes and hydrogen control in steel). Metallurgizdat Publishing House
Shewmon P (1974) *Diffuziia v tviordykh telakh (Diffusion in solids)*. Metallurgiya Publishing House
Shewmon PG (1976) Hydrogen attack of carbon steel. Metall Trans A 7(2):279–286
Shober T, Vepul H (1981) Sistemy Nb–H(D); Ta–H(D), V–H(D): structura, diagrammy, morfologiia, metody prigotovleniia. Vodorod v metallakh (Systems Nb–H(D); Ta–H(D), V–H(D): structure, diagrams, methods of preparation. Hydrogen in metals). Mir Publishing House
Shvachko VI (1998) Analysis and investigation of hydrogen in steels by the mass-spectral method. Mater Sci 34(4):544–558
Shved MM (1985) Izmeneniie ekspluatatsionnykh svoistv zheleza i stali pod vliianiem vodoroda (Changes in the operational properties of iron and steel under the influence of hydrogen). Naukova dumka Publishing House
Skalskyi VR, Andreykiv OYe (2006) Otsinka obiemnoi poshkodzhennosti materialiv metodom akustychnoi emisii (Evaluation of volumetric damage of materials by the method of acoustic emission). Publishig House of I. Franko LNU
Skalskyi VR, Koval PM (2007) Some methodological aspects of application of acoustic emission. Spolom Publishing House
Skalskyi VR, Andreykiv OYe, Serhiyenko OM (1999) Otsinka vodnevoi poshkodzhenosti materialiv za amlitudamy sygnaliv akustychnoi emisii (Assessment of hydrogen damage of materials by amplitudes of acoustic emission signals). Tekhnicheskaia diagnostika i nerazrushayushchii control (Technical diagnostics and non-destructive testing) 1:17–27
Smirnov LI (2006) Teploiemkost vodorodnoi podsistemy v sistemakh metall-vodorod (Heat capacity of hydrogen sub-system in systems metal-hydroghen). Metallofizilka i noveishiye tekhnologii (Metallophysics and the Latest Technologies) 28:295–330
Snape E (1969) Stress-induced failure of high-strength steels in environment containing hydrogen sulphide. Br Corros J 4(5):253–259
Speidel MO (1972) Branching of subcritical cracks in metals. L'hydrogene dans los metrix 2:358–362
Srikrishnan V, Liu HW, Ficalora PJ (1975) Selective chemisorption and hydrogen embrittlement. Scr Metall 9(12):1341–1344
Steigerwald EA, Schaller FW, Troiano AR (1959) Discontinuous crack growth in hydrogenated steel. Trans Metall Soc AIME 215:1048–1052
Steklov OI (1992) Ispytaniie stalei i svarnykh soiedinenii v navodorozhyvaiushchikh sredakh (Testing of steels and welded joints in gydrogenating environments). Mashinostroyeniye Publishing House
Tion JK, Richards RJ, Buck O, Marcus HL (1975) Model of dislocation sweep-in of hydrogen during fatigue crack growth. Scr Metall 9(10):1097–1101

Tkachov VI (1999) Mechanism of reversible effect of hydrogen on mechanical properties of steel. Mater Sci 35(4):477–484
Troiano AR (1959) Delayed failure of high-strength steels. Corrosion 15(4):57–62
Van Ellerbrock H-G, Vibrans G, Stuwc H-P (1972) Diffusion von Wasserstoff in Stahl mit inneren Hohlraumen. Acta Metall 20(1):53–60
Van Leeuwen H-P (1974a) The kinetics of hydrogen embrittlement: a quantitative diffusion model. Eng Fract Mech 6(1):141–161
Van Leeuwen H-P (1974b) Analyse quantitative de la fragilisation par l'hydrogene. Memoires et etudes scientifique de la revue de metallurgie 71(9):509–525
Van Leeuwen H-P (1975) Plateau velocity of SCC in high-strength steel—a quantitative treatment. Corrosion 31(2):42–50
Van Leeuwen HP (1985) Fugacity of gaseous hydrogen. Hydrogen degradation of ferrous alloys. Noyes Publishing, Park Ridge
Vavrukh MV, Solov'yan VB (1986) Localization of hydrogen impurities in metal. Mater Sci 21(4):317–320
Volchok IP (1993) Soprotivleniie razrusheniiu stali i chuguna (Fracture resistance of steel and cast iron). Melallurgiya Publishing House
Vonsovsky SV (1971) Magnetizm. Magnitnyie svoistva dia-, para-, ferro-, antiferro i ferromagnetikov (Magnetism. Magnetic properties of dia-, para-, ferro-, antiferro and ferromagntics). Nauka Publishinhg House
Weinman AB, Melekhov RK, Smiyan OD (1990) Vodorodnoie okhrupchivaniie elementov kotlov vysokogo davleniia (Hydrogen embrittlement of high-pressure boiler elements). Naukova dumka Publishing House
Williams DP (1970) Embrittlement of 4130 Steel by low pressure gaseous hydrogen. Metall Trans 1(1):63–68
Yousscj A, Jaeger LG (1974) The rote of finite deformation analysis in plane stress and strain fracture. In: International conference on vehicle structures, pp 164–172
Yuyama S (1986) Fundamental aspects of acoustic emission applications to the problem caused by corrosion. Corrosion monitoring in industrial plants using nondestructive testing and electrochemical methods. ASTM STP 908, Philadelphia
Zakharov AP (1987) Vzaimodeistviie vodoroda s metallami (Interaction of hydrogen with metals). Nauka Publishing House
Zakkay VF, Gerberich UU, Parker ZR (1973) Strukturnyie tipy razrusheniia. Razrusheniie (Structural types of failure. Fracture). Mir Publishing House

Chapter 6
Determination of Magnetic Ductility and Residual Magnetization of Steels

The phenomenon of magnetic ductility (magnetic aftereffect of the material) is a certain delay in the magnetization following the change of the magnetic field applied to the ferromagnet. Speaking about the magnetic aftereffect, we usually exclude from this concept the phenomena associated with changes in magnetic properties because of changes in the material itself under the action of natural diffusion or metallurgical processing, e.g., that causes the separation of phases.

It was found in the papers by Vonsovsky (1948), Tikazumi, Mishina, and others (Vonsovsky 1971; Tikadzumi 1987; Mishin 1991; Tikadzumi 1983) that the change in magnetization with time depends on the thermodynamic state of the body—its phase composition, temperature, deformation, and the concentration of impurities, including hydrogen. Thus, the measurement of magnetic aftereffect in relation to other physical processes is of some interest to be used when developing new non-destructive testing methods for assessing the state of ferromagnetic materials in products and structural elements operating in different environments, especially in hydrogen-containing ones.

This chapter presents the results of the evaluation of the magnetic ductility and residual magnetization of ferromagnets, obtained on samples of various structural steels. Some of them are published in (Mykhalchuk and Plakhtii 2005; Skalskyi et al. 2006, 2009).

6.1 Physical Essence of Magnetic Ductility

As a result of magnetic ductility, the equilibrium magnetization is established after time τ, which can vary from 10^{-4} sec to several hours (Baryakhtar 1996). At the same time magnetic relaxation takes place. This is the process of establishing balance in the force subsystem of the body. For example, the relaxation of the magnetic susceptibility χ of ferromagnets to its equilibrium value is caused by the action of

V. Skalskyi and Z. Nazarchuk, *Magnetoelastic Acoustic Emission*, Springer-AAS Acoustics Series, https://doi.org/10.1007/978-981-99-4032-5_6

magnetic domain walls in them with defects that are reoriented or diffuse. In the first case, because the defects jump from one position in the crystal lattice to another, the energy of magnetic anisotropy changes due to changes in the symmetry of the defect's local environment. In addition, the field of mechanical stresses of defects interacts with magnetization, thus causing the magnetostriction—deformation of bodies during their magnetization. As a result of magnetostriction, the geometric dimensions of the body and its volume change. Finally, the presence of a defect leads to changes in the exchange integrals and, as a consequence, to the inhomogeneity of magnetization.

In the second case, the interaction involves defects that diffuse in the lattice (at a distance commensurable with the thickness of the domain wall). The relaxation time τ, specified by the diffusion rate, significantly depends on the type of diffusion (electron, ionic, etc.) and increases sharply with decreasing temperature. For example, in carbonyl iron at $-12\,^{\circ}\mathrm{C}\ \tau \sim 1000\,\mathrm{s}$ and at $+100\,^{\circ}\mathrm{C}\ \tau \sim 0.02\,\mathrm{s}$.

Magnetic ductility is described quantitatively by introducing the corresponding components of domain wall delay into the equation of domain wall dynamics.

In magnetic hard materials, which are an ensemble of interacting single-domain particles (grains in a polycrystalline alloy), the magnetic ductility is particularly high and is caused by the irreversible rotation of the particle's magnetization due to thermal fluctuations. In some materials, a whole spectrum of time τ constants is observed, which can be determined by the number and location of peaks in the frequency dependence χ during magnetization by a monochromatic external magnetic field.

6.2 Models of Magnetic Ductility of Ferromagnets

Some theoretical considerations. Let field $H = H_1$ be applied to the magnet, which at time $t = 0$ suddenly changes to the value of $H = H_2$. As shown in Fig. 6.1, the change in magnetization consists of a partial change J_i, which occurs immediately, and a subsequent change J_n, which begins with a delay after the field jump.

In the general case, the change of J_n with time can be represented as a function

$$J_n = J_n(t). \tag{6.1}$$

The value of J_n depends both on the value of the instantaneous change in the magnetization J_n at the first moment and on the state of magnetization, created in the field $H = H_2$, when the initial change of J_i ends. If, for example, this point is in the region of the reverse rotation of magnetization, a very small change in the magnetization J_n is observed. When it is in the region of irreversible magnetization (as an example we can mention the state of residual magnetization or demagnetized state corresponding to the coercive field), the value of J_n is quite large.

In the simple case, the form of the function $J_n(t)$ is determined by a single relaxation time τ

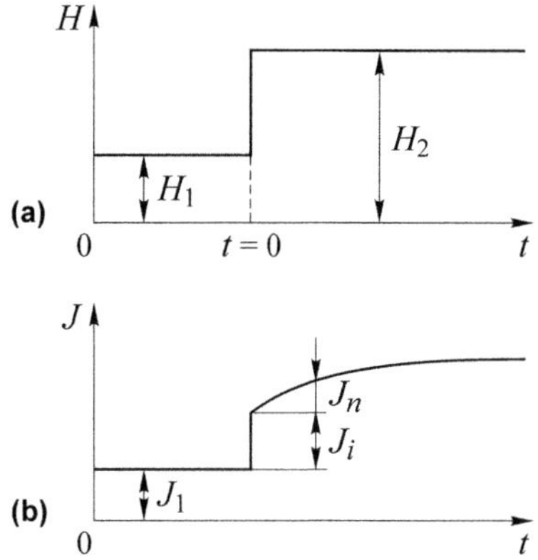

Fig. 6.1 Magnetic field jump and the magnetic aftereffect that accompanies this change

$$J_n(t) = J_{n0}(1 - e^{-t/\tau}), \tag{6.2}$$

where J_{n0} is the change in magnetization in the time interval from $t = 0$ to $t = \infty$.

Figure 6.2 presents as an example, the results of Tomono's experiments (Tikadzumi 1987) on the measurement of magnetic aftereffects in pure iron with low carbon content. A strictly linear graph of the dependence of lg J_n on t indicates the correctness of relation (6.2). In this case, the change of the magnetic field from $H_1 = 1.2$ A/m to $H_2 = 0$ is embedded in the region of the initial magnetic permeability. In this case, the ratio $J_{n0}/J_i = 30\%$. Denote this value by ξ. Then, taking into account the jump-like change in magnetization, we obtain the expression

$$J = \chi_a H\{1 + \xi(1 - e^{-t/\tau})\}. \tag{6.3}$$

The phenomenon of magnetic aftereffect causes a lag in the change of magnetization and induction during the ferromagnet magnetization in a variable field. To

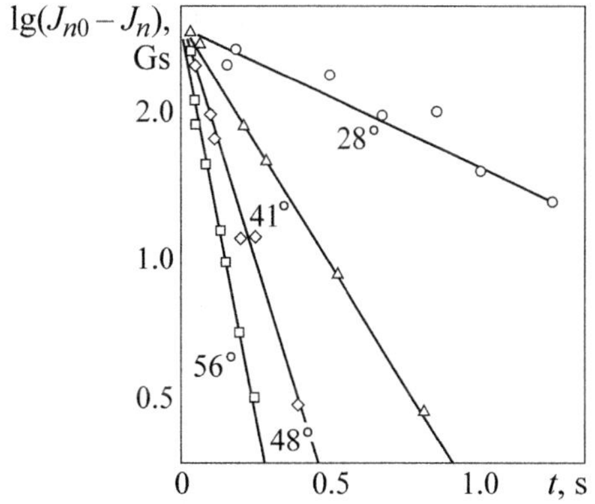

Fig. 6.2 Magnetic aftereffect in low-carbon steel (numbers near the lines indicate the heating temperature)

analyze this case, let us write the differential equation for magnetization in a variable field

$$\frac{d}{dt}(J - \chi_a H) = -\frac{1}{\tau}\{J - \chi_a H(1 + \xi)\}. \tag{6.4}$$

Obviously, it is like equation (6.3), which specifies the magnetization change in a constant field. Consider the variable magnetic field to be harmonic and monochromatic

$$H = H_0 e^{i\omega t}. \tag{6.5}$$

The oscillations of the field will cause the oscillations of the magnetization J, but due to the aftereffect the change of J is delayed with respect to the field, so we can write the following equality:

$$J = J_0 e^{i(\omega t - \delta)}, \tag{6.6}$$

where δ is the phase delay.

To determine δ and J_0, we substitute (6.5) and (6.6) in (6.4). As a result we obtain

$$\mathrm{tg}\delta = \frac{\xi\omega\tau}{(1 + \xi) + \omega^2\tau^2}, \tag{6.7}$$

$$J_0 = \frac{\omega\tau}{\omega\tau\cos\delta - \sin\delta}\chi_a H. \tag{6.8}$$

Since the delay in phase during alternating magnetization is accompanied by energy losses, the value of δ is usually called the angle of magnetic loss, and tgδ is the loss factor.

Figure 6.3 shows the curves of temperature dependence of the loss factor in variable fields of different frequency ω, obtained by Tomono on the same material (Tikadzumi 1987). It is seen from the figure that the loss factor at each frequency reaches a maximum at a certain temperature. This is explained by the fact that the relaxation time τ changes with temperature.

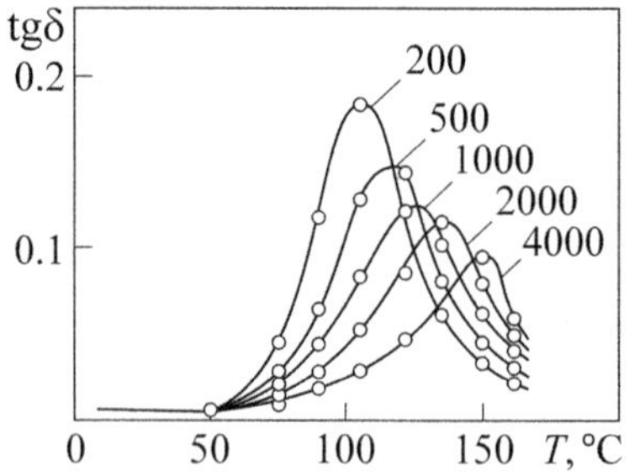

Fig. 6.3 Temperature dependence of the loss factor in low-carbon steel at different frequencies of the variable magnetic field (numbers near curves indicate frequency in hertz)

Fig. 6.4 Dependence of lgτ on $1/T$, constructed by points obtained by quasi-static measurements and measurements in an alternating magnetic field

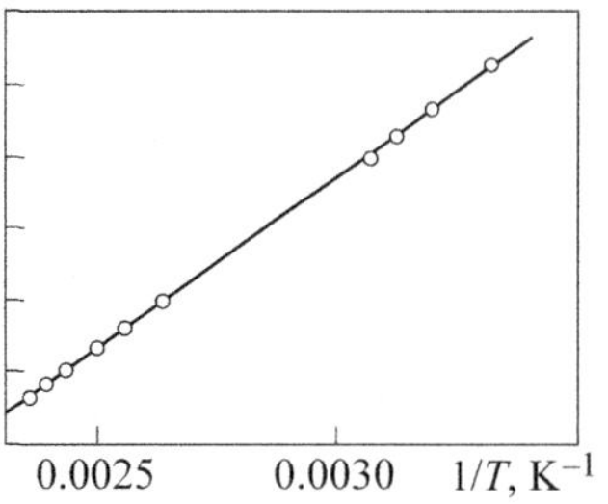

Considering (6.7) as a function of τ, we can say that at sufficiently large relaxation times the denominator of the fraction grows much faster than the numerator and $\text{tg}\delta \to 0$. At sufficiently small τ the numerator begins to play a major role. Thus, we obtain that $\text{tg}\delta \to 0$. Therefore, for some τ the loss factor $\text{tg}\delta$ reaches a maximum. The maximum relaxation time is determined by the formula

$$\tau = \frac{\sqrt{1+\xi}}{\omega}. \tag{6.9}$$

Therefore, time τ can be found by measuring losses. In Fig. 6.4, the dependence of $\lg \tau$ on $1/T$ is constructed according to the values of τ obtained as a result of measurements in quasi-static (Fig. 6.2) and variable (Fig. 6.3) fields. In both cases, all points fall on the same line, thus indicating the common cause of both effects.

The widespread models of magnetic aftereffect. To clearly describe the phenomenon of magnetic aftereffect, it is convenient to use the following model (Tikadzumi 1987). Imagine a concave asphalt path covered with a layer of mud of a certain thickness. A heavy metal ball is placed on the path (Fig. 6.5). Under the action of gravity, the ball moves, sinking in the mud layer, until a good balance of applied forces is reached, after which it, immersed in the mud, gradually shifts further. This picture corresponds to the phenomenon of static aftereffect. The case of a variable magnetic field corresponds to a situation where a variable force is applied to the ball, under the action of which it performs oscillating movements, moving from left to right. At low temperatures, the mud hardens, the ball can roll freely on its surface, and no loss occurs. When the temperature is high and the ductility of the mud is low enough, the ball rolls on the asphalt surface, and its movement becomes free again. At the intermediate temperature, when the relaxation time of the ductility exactly coincides with the period of the ball oscillation, its motion meets the greatest resistance. The described situation corresponds to the case of variable magnetization at a temperature when the loss becomes maximum.

Below consider the specific mechanisms of the magnetic aftereffect. It should be noted that not in all cases it is possible to describe this phenomenon, setting a single time τ, as it is done above. In general case, the magnetization of the ferromagnet occurs due to the jump-like displacement of the domain walls and due to irreversible rotation, and the ratio of the contributions of both mechanisms may

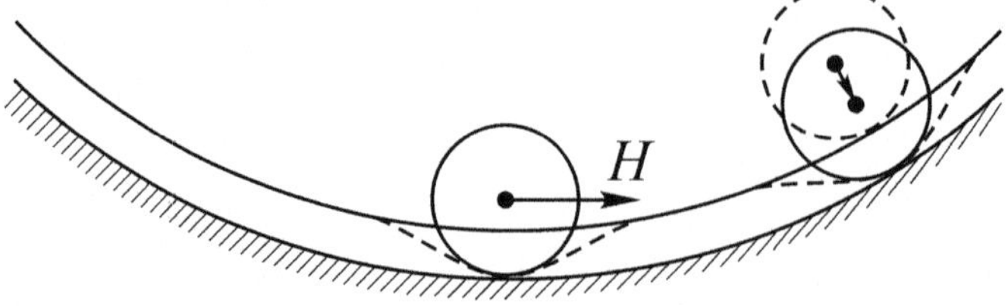

Fig. 6.5 Snook's model to describe the magnetic aftereffect

change. Therefore, it is more correct to assume that the relaxation time of the aftereffect associated with these processes acquires values that are in some interval τ. The dependence of lg J_n on t in this case will deviate from the strictly linear one, which is presented in Fig. 6.2. Assume that the distribution τ is given by the function In τ and that the part of areas where the relaxation time takes values corresponding to this distribution of the interval from In τ to In τ + d(In τ) is equal to $g(\tau)\,\mathrm{d}(\ln\tau)$. Since $g(\tau)\,\mathrm{d}(\ln\,\tau) = [g(\tau)/\tau]\,\mathrm{d}\tau$, then normalizing $g(\tau)$

$$\int_0^\infty \frac{g(\tau)}{\tau}\mathrm{d}\tau = 1, \tag{6.10}$$

we obtain that the total time change of the magnetization can be represented as

$$J_n(t) = J_{n0}(1 - \int_0^\infty \frac{g(\tau)}{\tau}e^{-t/\tau}\mathrm{d}\tau). \tag{6.11}$$

For simplicity, let the distribution function $g(\tau)$ in the interval from τ_1 to τ_2 be constant, and outside this interval it moves to zero, as in Fig. 6.6 (Tikadzumi 1987).

Then from the relation (6.10) we find

$$g = \frac{1}{\ln\frac{\tau_2}{\tau_1}} \text{ at } \tau_1 < \tau < \tau_2. \tag{6.12}$$

Fig. 6.6 Richter-type magnetic aftereffect

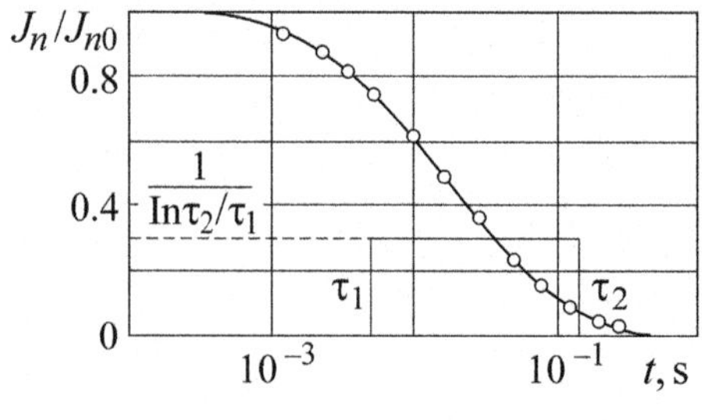

Denoting $t/\tau = y$, the second term in formula (6.11) can be written as follows:

$$\Delta J_n = J_{n0} - J_n = \frac{J_{n0}}{\ln\frac{\tau_2}{\tau_1}} \int\limits_{\tau_1}^{\tau_2} \frac{e^{-t/\tau}}{\tau} \mathrm{d}\tau = \frac{J_{n0}}{\ln\frac{\tau_2}{\tau_1}} \int\limits_{t/\tau_2}^{t/\tau_1} \frac{e-y}{\tau} \mathrm{d}y. \tag{6.13}$$

Introducing the notation

$$N(\alpha) = \int\limits_{\alpha}^{\infty} \frac{e^{-y}}{y} \mathrm{d}y, \tag{6.14}$$

we write formula (6.13) as

$$\Delta J_n = \frac{J_{n0}}{\ln\frac{\tau_2}{\tau_1}} \left[N\left(\frac{t}{\tau_2}\right) - N\left(\frac{t}{\tau_1}\right) \right]. \tag{6.15}$$

Function $N(\alpha)$ looks like

$$\begin{aligned} &N(\alpha) = -0.577 - \ln\alpha + \alpha - \frac{1}{2}\frac{\alpha^2}{2!} + \frac{1}{3}\frac{\alpha^3}{3!} + \ldots \text{ at } \alpha << 1, \\ &N(\alpha) = \frac{e^{-\alpha}}{\alpha}(1 - \frac{1!}{\alpha} + \frac{2!}{\alpha^2} - \ldots), \text{ at } \alpha >> 1, \\ &N(1) = 0.219 \text{ at } \alpha = 1. \end{aligned} \tag{6.16}$$

Using these approximate formulas, we obtain that at the very beginning ($t << \tau_1 < \tau_2$) the ΔJ_n value decreases with time almost linearly

$$\Delta J_n \cong J_{n0} \left(1 - \frac{\frac{1}{\tau_1} - \frac{1}{\tau_2}}{\ln\frac{\tau_2}{\tau_1}} \right), \tag{6.17}$$

in the middle ($\tau_1 << t << \tau_2$) it is proportional to $\ln t$

$$\Delta J_n \cong \frac{J_{n0}}{\ln\frac{\tau_2}{\tau_1}} (\ln\tau_2 - 0.577 - \ln t), \tag{6.18}$$

and at the very end, when enough time has passed ($\tau_1 < \tau_2 << t$), the ΔJ_n value tends to zero according to the expression

$$\Delta J_n \cong J_{n0} \frac{\tau_2}{\ln\frac{\tau_2}{\tau_1}} \frac{1}{t} e^{-t/\tau}. \tag{6.19}$$

Figure 6.6 shows the dependence of J_n/J_{n0} on t, obtained as a result of experimental studies of the magnetic aftereffect in carbonyl iron annealed at 55 °C. Solid line indicates the plot of function (6.15) at $\tau_1 = 0.0048$ s and $\tau_2 = 0.14$ s. One can see that the experimental points agree very well with the continuous theoretical curve. During variable magnetization of carbonyl iron at the frequency ω, satisfying the relation $\tau_1 < 1/\omega < \tau_2$, under the same conditions, the maximum loss factor $\mathrm{tg}\,\delta$ is observed. Similar phenomenon is called a magnetic aftereffect of the Richter type in honor of Richter, who studied this case in detail.

According to formula (6.7), at sufficiently low temperatures, when the relaxation time τ becomes very large, the loss factor $\mathrm{tg}\,\delta$ should tend to zero; however, as can be seen from Fig. 6.3, at low temperatures there remains an almost constant nonzero value of $\mathrm{tg}\,\delta$, which also does not depend on ω. This phenomenon is called the magnetic aftereffect of the Jordan type. Apparently, this effect is caused by a very wide distribution range of τ. The matter is that when the value of τ_1 is small enough and τ_2 is large enough, then under variable ferromagnet magnetization at almost all frequencies ω the value of $1/\omega$ falls within the specified interval, and therefore, the aftereffect is independent of ω.

Since for this $\tau_1 << t << \tau_2$, the magnetization $\Delta J_n(t)$ is expressed by formula (6.18), and its relative change in time is described by the relation

$$\frac{\Delta J_n}{J_n} = \text{const} - \beta \ln t. \tag{6.20}$$

Such a magnetic aftereffect is often observed in materials for permanent magnets. Figure 6.7 illustrates the curves of the magnetic aftereffect in the alnico 5 alloy (Tikadzumi 1987). In all cases, the experimental points are well located on straight lines corresponding to the dependence of ΔJ on $\lg t$.

The specific mechanism of the magnetic aftereffect in low-carbon steel presented above was proposed by Snook, and then Neil made some clarifications to this mechanism (Tikadzumi 1987). According to this theory, the magnetic aftereffect in iron is associated with a change in the position of the carbon or nitrogen atoms integrated in the lattice. As it is shown in Fig. 6.8 iron forms body-centered cubic lattices (bcc), and since the radii of carbon and nitrogen atoms are much smaller than the radii of iron atoms, it can be assumed that C or N atoms represented by small circles are

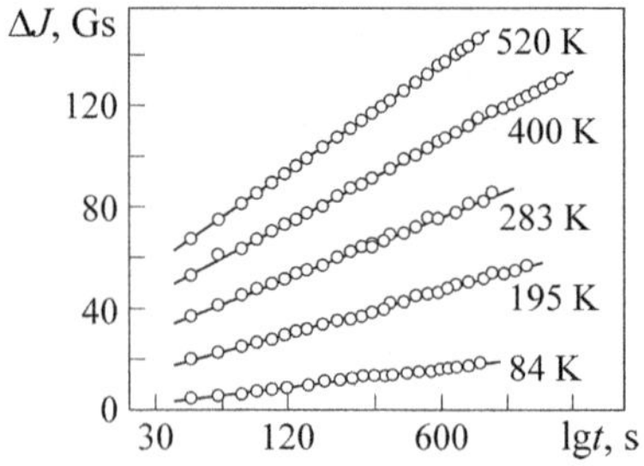

Fig. 6.7 Magnetic aftereffect in alnico 5 alloy

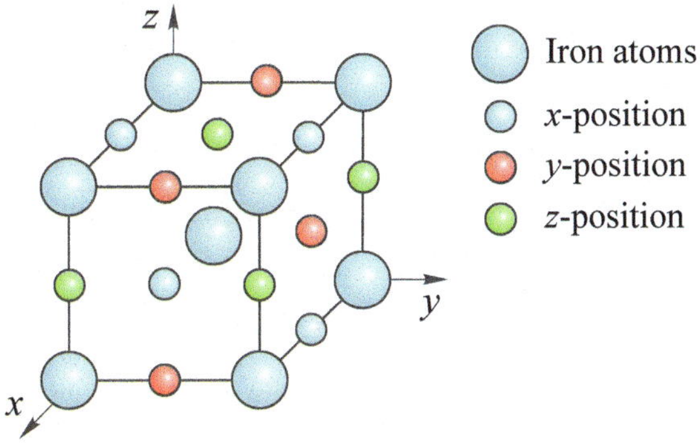

Fig. 6.8 Three types of positions that can be occupied by integrated carbon atoms inside the bcc lattice of iron

integrated into lattices. As can be seen from Fig. 6.8 the integrated atoms can occupy *x*-, *y*-, and *z*-positions. Let the carbon atoms be predominantly in the *x*-positions. In this case, there is a change in the magnetodipole interaction of atomic pairs whose axes are oriented along the *x*-direction, and uniaxial magnetic anisotropy is induced.

Neel model. In addition to the diffusion mechanism, Neel proposed one more mechanism that is fundamentally different from the type of magnetic aftereffect described above. It is called the aftereffect of thermal fluctuations. It occurs due to thermal oscillations of the spins themselves, which determine the ferromagnetic properties. Consider a simple example that is an elongated microparticle with a single-domain structure. Suppose that a magnetic field $\vec{H}$, directed in the negative direction is applied to such a particle, magnetized along the longitudinal axis in the positive direction. If H is less than the threshold field H_0, then the direction of the vector $\vec{J_s}$ remains positive, and if H is greater than H_0, the magnetization vector is inverted and its direction becomes negative. The energy of a particle with volume V for positive vector orientation is equal to

$$U_+ = V(-K_u + J_s H), \tag{6.21}$$

and for negative orientation

$$U_- = V(-K_u - J_s H). \tag{6.22}$$

At absolute temperature T the thermal oscillations occur. The energy of such oscillations is $kT/2$ per one degree of freedom of each spin. This energy distribution is maintained when all the spins of the particle rotate simultaneously, while remaining parallel to each other. The energy of the thermal oscillations is usually small compared to the height of the potential barrier, so the simultaneous rotation of the spins can be ignored. However, if the particle volume is small enough, it can be expected that at $H \approx 0$ the potential barrier VK_u will be equal to $kT/2$. At a temperature $T = 273$ K, the energy of thermal oscillations $kT = 3.77 \times 10^{-21}$ J, and therefore at $K_u = 10^5$ J/m^3 for the critical volume V_0 we find

$$V_0 K_u \sim \frac{kT}{2},$$

where

$$V_0 \approx \frac{kT}{2K_u} = \frac{3.77 \times 10^{-21}}{2 \times 10^5} = 1.9 \times 10^{-26}\,\mathrm{m}^3. \tag{6.23}$$

If the particle is of almost a spherical shape, then its critical radius is

$$r_0 = \left(\frac{3}{4\pi} V_0\right)^{1/3} = 1.7 \times 10^{-9}\,\mathrm{m} = 17\,\text{Å} \tag{6.24}$$

In such a particle, thermal oscillations with an orientation of spontaneous magnetization that resembles the Langevin paramagnetism are constantly excited. For this reason, these oscillations are called superparamagnetic.

In the presence of the field H, the height of barrier $U_{\max}$ is equal to $V(2K_u - J_s H)^2/4K_u$ relative to the level U_+ and $V(2K_u + J_s H)^2/4K_u$ relative to the level U_-. Therefore, the number of particles in which due to thermal excitation the magnetization rotates from the positive to negative direction is greater than the number of particles in which the reverse process occurs.

Even in the fields smaller than the threshold one, there is a gradual rotation of magnetization. It should be noted that if the field H differs little from H_0, the described phenomenon occurs even when the volume of particles V significantly exceeds the superparamagnetic critical value V_0, which is determined by formula (6.23). Neel, having analyzed in detail the mechanism of magnetization rotation, showed that at room temperature for particles with a volume of $1 \times 10^{-24}\,\mathrm{m}^3$ time is $\tau \approx 10^{-1}$ s, and for the particles with a volume of $2 \times 10^{-24}\,\mathrm{m}^3$ it reaches 10^9 s (several decades). Thus, the times τ_1 and τ_2 of the magnetic Richter-type aftereffect in this case are separated from each other by a rather large interval, and therefore the change in magnetization is proportional to $\ln t$ in accordance with (6.18). Since the size and anisotropy constants of the particles actually have some distribution, we can assume that the magnetic aftereffect of this type is proportional to $\ln t$.

Obviously, we can speak about the aftereffect caused by thermal fluctuations not only in the case of magnetization rotation, but also in the case of the domain walls shift. Under minor action on the domain boundaries, the walls fixed on any obstacles may begin to shift. Neel (Tikadzumi 1987) performed detailed calculations of the magnetic field, which plays the role of such an excitation factor and which occurs at the domain walls caused by the violation of the spins parallelism due to thermal oscillations.

The fundamental difference between the phenomena of diffusion magnetic aftereffect and aftereffect caused by the thermal fluctuations is that in the first case, even after changes in magnetic properties, the material "remembers" the former states that affect the distribution of carbon atoms. In other words, if, changing the magnetization, the material is transferred from state A to state B, and then after some time

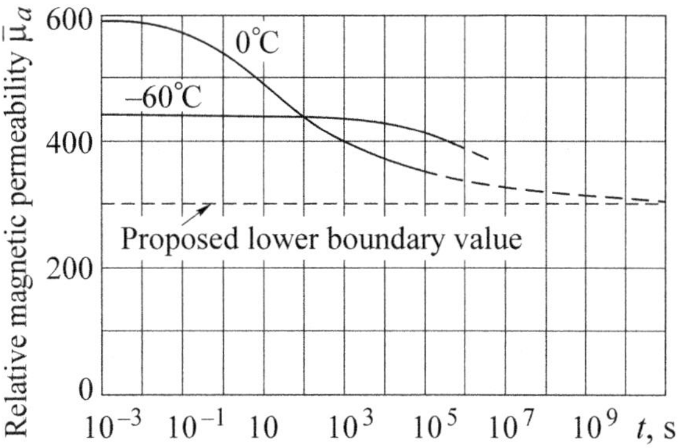

Fig. 6.9 Desaccomodation of magnetic permeability in Mn–Zn ferrite

to state C, then usually together with the magnetic aftereffect corresponding to the transition from B to C, an aftereffect imposed on the first one and corresponds to the transition from A to C will be observed. In this case, we usually say about the principle of superposition. If the magnetic aftereffect is related to the thermal fluctuations, then after the change of the magnetic state, the "memory" of the material about the previous states is completely erased. Thus, in this case the principle of superposition is not fulfilled. In this sense, Neil called the magnetic aftereffect of the first type as invertible and the second type as irreversible.

The phenomenon of magnetic aftereffect can be also manifested in the form of desaccomodation of magnetic permeability. Desaccomodation is found by the resonant frequency change of the *LC* circuit as a result of decreasing with time magnetic permeability of ferrite used as the core of the inductance *L*. Desaccommodation of magnetic permeability in Mn–Zn ferrite significantly depends on temperature. This phenomenon is explained by the Snook model (Fig. 6.5). A ball is placed on the surface of a sufficiently solid layer of mud, which oscillates under the action of a variable force. After some time, the ball, immersed in the mud, sinks and the amplitude of oscillations becomes small, which corresponds to the appearance on the dashed line in Fig. 6.9 (Tikadzumi 1987). Assume that the ball corresponds to spontaneous magnetization, and its motion corresponds to the rotation of the magnetization vector. In the case of diffusion magnetic aftereffect, the concavity of the asphalt surface corresponds to the initial magnetocrystalline anisotropy, and the hollow in which the ball that sinks corresponds to the induced magnetic anisotropy, caused by the anisotropic distribution of carbon atoms.

6.3 Losses During Remagnetization of Ferromagnets

The phenomenon of eddy current losses is due to the fact that when the magnetization changes inside the sample, according to the law of electromagnetic induction, the currents, leading to retardation of the magnetization change are induced. For

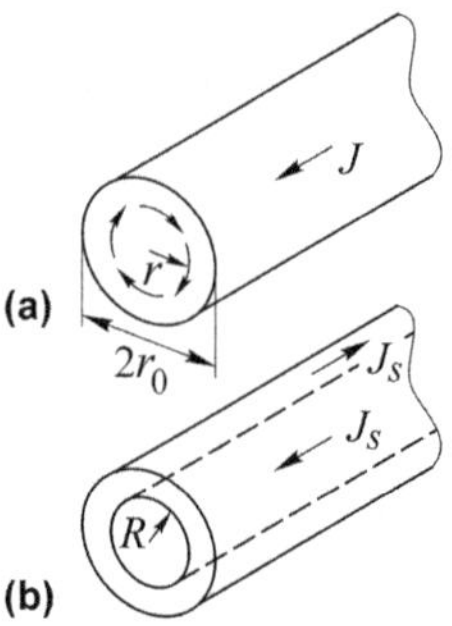

Fig. 6.10 Eddy currents in a cylindrical sample in the case of homogeneous magnetization (**a**) and shear of a domain walls (**b**)

simplicity, take a cylinder of radius r_0, which is magnetized along the longitudinal axis at a constant speed (Fig. 6.10). As a result of the magnetic field change, an electric field $\vec{E}$ appears inside the sample. The force lines of this field form circles with centers lying on the cylinder axis (in Fig. 6.10 the direction of the field $\vec{E}$ is shown by arrows).

In the general case, for any circuit, according to the law of electromagnetic induction rot $\overrightarrow{E} = -\partial \overrightarrow{B}/\partial t$, the following relation is fulfilled:

$$\int E_s d_s = -\iint \frac{dB_n}{dt} dS. \tag{6.25}$$

In this case, the distribution of field $\vec{E}$ has a cylindrical symmetry, so taking a circle of radius r for such a contour, we find from (6.25)

$$2\pi r E(r) = -\pi r^2 \frac{dJ}{dt},$$
$$E(r) = -\frac{r}{2}\frac{dJ}{dt}. \tag{6.26}$$

Denoting the resistivity of the metal from which the cylinder is made, by ρ, we obtain the following expression for the current density:

$$i(r) = -\frac{r}{2\rho}\frac{dJ}{dt}. \tag{6.27}$$

Thus, the power loss per volume unit is

$$P = \frac{1}{\pi r_0^2}\int_0^{r_0} 2\pi E(r) i(r) r dr = \frac{1}{2\rho r_0^2}\left(\frac{dJ}{dt}\right)^2 \int_0^{r_0} r^3 dr = \frac{r_0^2}{8\rho}\left(\frac{dJ}{dt}\right)^2. \tag{6.28}$$

As can be seen from this formula, the eddy current losses are proportional to the square of the rate of magnetization change. Therefore, under variable magnetization, the losses increase rapidly with increasing frequency of the magnetic field. The proportionality of the losses of the r_0^2 value suggests that they can be reduced by grinding the material. In addition, the losses are inversely proportional to the resistivity ρ, which means that the higher ρ of the material, the smaller they are.

So far, we have talked about the phenomena caused by changes in magnetization with time. There is also the opposite effect—with increasing $\mathrm{d}J/\mathrm{d}t$ eddy currents increase, which create a magnetic field that reduces the rate of change of magnetization. The distribution of such a magnetic field is similar to the field distribution in the solenoid, so eddy currents flowing closer to the center of a circle of radius r do not affect the area outside this circle. Therefore, when approaching the axis of the cylinder, the magnetic field of eddy currents increases, and when approaching the surface of the sample it weakens and on the surface tends to zero. Therefore, the change in magnetization on the surface occurs with the change in the magnetic field, and with deepening inside the cylinder, the change in magnetization increasingly lags behind the change in the field. It is known that in the magnetic field, variable with an angular frequency ω, the value of the magnetization weakens e times at depth (Skalsky et al. 2009)

$$s = \sqrt{\frac{2\rho}{\omega\mu_0\mu}}. \tag{6.29}$$

The value of s is called the depth of the skin layer. Its value remains almost unchanged during transition to the plane geometry of the sample, if the value of s is small compared to its size. For iron, in which $\mu = 500$, $\rho = 1 \times 10^7\ \Omega\,\mathrm{m}$, during magnetization in a harmonic field with a frequency of 50 Hz, the depth of the skin layer has the value

$$s = \sqrt{\frac{2 \cdot 1 \times 10^{-7}}{2\pi \cdot 50 \cdot 500 \cdot 4\pi \times 10^{-7}}} \approx 1.0 \times 10^{-3}(\mathrm{m}) = 1\,\mathrm{mm}. \tag{6.30}$$

Cores of transformers are usually made of thin plates. This is because when the plate thickness is less than s, the core is effectively magnetized throughout its full depth.

It was assumed above that the change in magnetization occurs equally in all parts of the magnet. However, in many cases the real magnets are magnetized because of the shift of the domain walls. Let us calculate the losses for a cylinder divided into two domains by a cylindrical domain wall of radius R (Fig. 6.10b). In this case, in the areas within the region $r < R$, there is no change in magnetization, so there

$$E(r) = 0 \quad (r \leq R). \tag{6.31}$$

For $r > R$, according to (6.25), we have

$$2\pi r E(r) = -4\pi J_s R \frac{dR}{dt},$$

$$E(r) = -2J_s R \frac{dR}{dt} \frac{1}{r} \ (r > R). \tag{6.32}$$

Thus, the losses per unit volume are

$$P = \frac{1}{\pi r_0^2} \int_R^{r_0} \frac{E^2(r)}{\rho} 2\pi r dr = \frac{8J_s^2 R^2}{\rho r_0^2} \left(\frac{dR}{dt}\right)^2 \ln\left(\frac{r_0}{R}\right). \tag{6.33}$$

To be able to compare this value with (6.28), we use the next expression for the resulting change in magnetization

$$\frac{dJ}{dt} = \frac{4J_s}{r_0^2} R \frac{dR}{dt}. \tag{6.34}$$

Then (6.33) will take the form

$$P = \frac{r_0^2}{2\rho} \ln\left(\frac{r_0}{R}\right) \left(\frac{dJ}{dt}\right)^2. \tag{6.35}$$

The losses of P in this case depend on the position of the domain wall R. Let us average this value by R. Since

$$\overline{\ln\left(\frac{r_0}{R}\right)} = \frac{1}{r_0} \int_0^{r_0} \ln\left(\frac{r_0}{R}\right) dR = 1, \tag{6.36}$$

the average loss value will be equal to

$$\overline{P} = \frac{r_0^2}{2\rho} \left(\frac{dJ}{dt}\right)^2. \tag{6.37}$$

Comparing (6.37) with (6.28), we see that in the presence of a domain wall, the losses become 4 times greater. This is explained by the fact that the change in magnetization in the case of domain walls shift is heterogeneous in volume, and for the same resulting change in dJ/dt the local changes in magnetization in some areas reach very large values. The distribution of eddy currents acquires as a result the same heterogeneous character, and since the losses are proportional to i^2, the appearance of such currents is caused by quite high values of $\overline{P}$.

Williams, Shockley, and Kittel (Tikadzumi 1987) were the first who confirm this experimentally. They studied the motion of domain walls in frames cut out from a single crystal of silicious iron so that their sides were oriented along the axes [100] and [010]. The experiments on the observation of domain images have shown that on each side of the frame only one wall is formed in this case. Besides, it has been found that during the magnetic field increase, the rate of the domain walls shift increases with increasing H linearly

$$v = c(H - H_0), \tag{6.38}$$

where H_0 is the threshold field required to separate the domain walls from the obstacles to their movement.

To calculate accurately the eddy currents in a real magnet with many domain walls is an extremely difficult task. This is related to the need to take into account many factors—starting from the walls geometry, the distance between them, the distribution, and the method of their deformation and ending with the shape and size of the sample. Let us find out what will be the error, for example, if you do not take into account the domain structure in the calculations. Compare the eddy currents distribution in the cases shown in Fig. 6.10a, b. As can be seen from Fig. 6.11, in the second case, contrary to the first one, there is a jump-like change in eddy currents in the place of the domain wall location. The value of the change of Δi is easy to calculate by substituting the value of $r = R$ in (6.32):

$$\Delta i = \frac{2J_s}{\rho} v. \tag{6.39}$$

The field of formula (6.39) application is not limited to this case. The obtained result is almost always true when there is a shift of the 180° domain walls with velocity v. This means that the eddy current distribution curve, calculated under the assumption that the change in magnetization is uniform (Fig. 6.12, curve *1*), is actually superimposed by sharp jumps caused by current changes in the places of the domain walls location (curve *2*). In the first case it is accepted to speak about macroscopic eddy currents, and in the second about microscopic ones. Let us denote these currents by i_{ma} and i_{mi}. Then the total loss can be written as

$$P = \rho \int (i_{ma} + i_{mi})^2 \mathrm{d}v = \rho \int i_{ma}^2 \mathrm{d}v + \rho \int i_{mi}^2 \mathrm{d}v + 2\rho \int i_{ma} i_{mi} \mathrm{d}v. \tag{6.40}$$

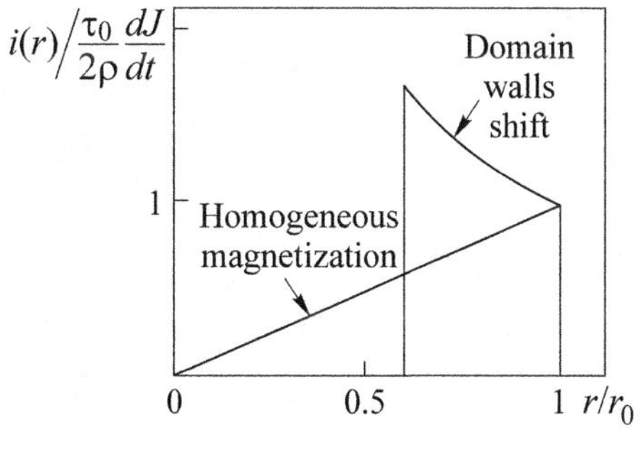

Fig. 6.11 Distribution of eddy currents in a cylindrical sample (in the case of homogeneous magnetization and in the case of domain walls shift)

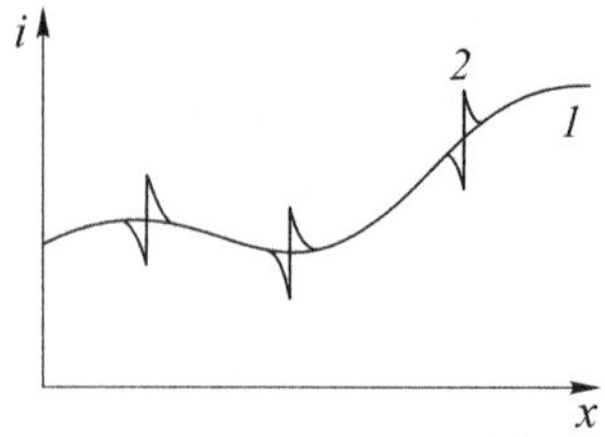

Fig. 6.12 Distribution of macroscopic (*1*) and microscopic (*2*) eddy currents

Magnet unit volume integration is performed in (6.40). If the distribution of macroscopic eddy currents in comparison with the distribution of the microscopic ones is smooth, as in Fig. 6.12, the third summand in (6.40) disappears because the positive values of i_{mi} are balanced by the negative ones. Therefore, when calculating the total losses, it is enough to find the losses for macroscopic eddy currents and separately for microscopic ones and then add them. If the sample contains a large number of leading impurities or if the size of the domains is commensurable with the size of the sample, the third summand does not disappear. Then the calculations are extremely complicated, and in addition, it becomes difficult to distinguish between macroscopic and microscopic eddy currents.

However, if the number of domain walls is large enough, and the velocity v of one wall shift is small, then in all the cases, as can be seen from (6.39), the microscopic eddy currents are small, and approximate calculations can be limited to the first summand, which includes the macroscopic eddy currents. This conclusion will be clear if we consider that the presence of a large number of domain walls corresponds to a state close to uniform magnetization.

6.4 Model of Magnetic Aftereffect in Hydrogenated Ferromagnets

Most often, methods based on the measurement of resistivity, nuclear magnetic resonance, neutron diffraction, internal friction, etc. are used to create the non-destructive methods for assessment of sufficiently high concentrations of hydrogen in real structural elements (Williams et al. 1952; Kolsky 1964; Lynnworth 1975; Goodwin et al. 2003). As it turned out, they are not efficient and time-consuming for engineering practice. So, it is necessary to create new methods that would determine the small and medium concentrations of the dissolved hydrogen.

It has been experimentally established (Tikadzumi 1987; Mishin 1991; Vonsovsky 1984; Kronmuller 1981) that the change in magnetization with time depends on the thermodynamic state of the body—its phase composition, temperature T, deformation ε_{ij}, and concentration of impurities, in particular, hydrogen. Therefore, the measurement of magnetic aftereffect in interrelation with other physico-mechanical processes is interesting when developing new methods of non-destructive testing of the state of materials and structural elements.

The model of relaxation of the ferromagnet magnetization. As noted above, the magnetic aftereffect in a ferromagnet consists in the lag of its magnetization from the external magnetic field change. If to change suddenly the magnetic field $\vec{H}$ by a value $\Delta\vec{H}$, the increment of magnetization $\Delta\vec{M} = \vec{M} - \vec{M}_{eq}$ ($\vec{M}_{eq} = \kappa\vec{H}$ is equilibrium magnetization, which corresponds to the instantaneous value of the magnetic field, and κ is equilibrium magnetic susceptibility of the material) can be given as the sum

$$\Delta\vec{M} = \Delta\vec{M}_i + \Delta\vec{M}_n \tag{6.41}$$

of the component $\Delta\vec{M}_i$, proportional to $\Delta\vec{H}$ (instantaneous), and the summand $\Delta\vec{M}_n$, that changes with time: $\Delta\vec{M}_n = \Delta\vec{M}_n(t)$ (Vonsovsky 1984). The dependence $\Delta\vec{M}_n(t)$ is relaxation, since the spin subsystem of the ferromagnet changes gradually with the change of the external magnetic field. The aftereffect at low magnetic field intensities is mainly caused by the domain walls shift, and in strong magnetic fields the rotation of the magnetization vector of the domains dominates. Both of these processes change the macroscopic magnetization and can occur by both diffusion and thermofluctuation mechanisms (Tikadzumi 1987).

Consider a ferromagnet that is placed in an external alternating magnetic field, the magnetic field vector $\vec{H}$ of which has a fixed direction (Andreikiv et al. 2002). To take into account the non-instantaneous character of the ferromagnet remagnetization, we assume that the rate of the component $\left(\vec{M} - \vec{M}_i\right)$ change is proportional to the deviation of the magnetization from the equilibrium value

$$\partial\left(\vec{M} - \vec{M}_i\right)\Big/\partial t = -\alpha\left(\vec{M} - \vec{M}_{eq}\right), \tag{6.42}$$

where $\vec{M}_i = \kappa_i\vec{H}$ is the instantaneous component of magnetization and κ_i is instantaneous magnetic susceptibility of the material.

In general, taking into account the influence of the thermodynamic state of the body on magnetic relaxation, the coefficient α will be considered as a function of the state parameters—temperature T, component of the strain tensor ε_{ij}, magnetization $\vec{M}$, and also time t

$$\alpha = \alpha\left(t, T, \varepsilon_{ij}, \vec{M}\right). \tag{6.43}$$

The explicit dependence on time in this formula allows us to take into account different mechanisms of domain structure rearrangements—diffusion, fluctuation, etc., which cause the magnetic aftereffect with different relaxation rates (Tikadzumi 1987; Mishin 1991; Kronmuller 1981).

Neglecting for weak magnetic fields the dependence on $\vec{M}$ in formula (6.43) and integrating equation (6.42) under the initial condition $\vec{M}|_{t=0} = \vec{M}_0$, we obtain

$$\vec{M}(t) = \kappa_i \vec{H}(t) + \left[(\kappa - \kappa_i) \int_0^t \alpha(\varsigma) \vec{H}(\varsigma) \exp\left(\int_0^{\varsigma} \alpha(\xi) \mathrm{d}\xi \right) \mathrm{d}\varsigma + \vec{M}_0 \right] \times \exp\left(- \int_0^t \alpha(\varsigma) \mathrm{d}\varsigma \right). \tag{6.44}$$

If the external magnetic field changes by a jump: $\vec{H}(t) = \theta(t)\vec{H}_0$, where $\theta(t)$ is the Heaviside unit function, then from the relation (6.44) it follows

$$\vec{M}(t) = \kappa_i \vec{H}_0 \theta(t) + \left[(\kappa - \kappa_i) \vec{H}_0 \int_0^t \alpha(\varsigma) \exp\left(\int_0^{\varsigma} \alpha(\xi) \mathrm{d}\xi \right) \mathrm{d}\varsigma + \vec{M}_0 \right] \times \exp\left(- \int_0^t \alpha(\varsigma) \mathrm{d}\varsigma \right). \tag{6.45}$$

When $\alpha(t) = \alpha = \text{const}$, relation (6.45) passes into the known law of relaxation (Tikadzumi 1987)

$$\vec{M}_n(t) = \vec{M}_0(1 - \exp(-\alpha t)). \tag{6.46}$$

As we can see from Fig. 6.13, already at the temperature of 28°C dependence (6.45) (curve *5*) gives better agreement with the experimental data than dependence (6.46) (curve *4*).

When there are two times of relaxation

$$\alpha(t) = \begin{cases} \alpha_1, 0 \le t \le t_1 \\ \alpha_2, t_1 < t \end{cases}, \tag{6.47}$$

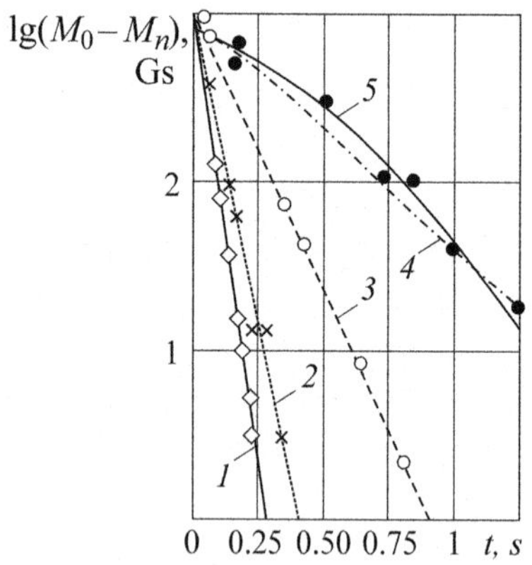

Fig. 6.13 Relaxation of magnetization of pure iron with low carbon content:◇—$T = 54\,°\text{C}$; × —48 °C; o—41 °C; •—28 °C [experimental data (Tikadzumi 1987)]; curves *1*–*4* are approximation by dependences (6) at $\alpha = 10.7;\ 7.5;\ 3.3;\ 1.4$; curve *5* is approximation of data obtained at $T = 28\,°\text{C}$ by dependence (6.45) at $\alpha(t) = 0.5432 + 1.4t$

we pass to the law of demagnetization.

Mathematical model of magnetic aftereffect in a ferromagnetic electrically conductive body. Equations (6.42) and (6.44) describe the change in the body magnetization with a change in the given external magnetic field under homogeneous conditions. With the change of the latter in the electrically conductive body an electric current appears, which, in turn, causes its own magnetic field. Therefore, to study the magnetic aftereffect in bounded electrically conductive bodies, equation (6.44) should be supplemented by a system of equations of macroscopic electrodynamics (Solodyak 2000; Landau and Lifshitz 1957)

$$\mathrm{rot}\vec{H} = j + \partial\vec{D}\Big/\partial t, \quad \mathrm{rot}\vec{E} = -\partial\vec{B}\Big/\partial t, \quad \mathrm{div}\vec{B} = 0, \quad \mathrm{div}\vec{D} = 0, \tag{6.48}$$

with relations connecting the vectors of induction and intensity of electric and magnetic fields $\vec{E}$ and $\vec{D}$ and $\vec{B}$ and $\vec{H}$

$$\vec{D} = \varepsilon_0\vec{E}, \quad \vec{B} = \mu_0\left(\vec{H} + \vec{M}\right), \tag{6.49}$$

and some formulas that determine the relationship between the current and electric field

$$\vec{j} = \sigma\vec{E}.$$

Here σ is the specific conductivity; ε_0 and μ_0 are electric and magnetic constants of vacuum.

Consider the quasi-stationary case when the displacement current can be neglected, i.e.,

$$\partial\vec{D}\Big/\partial t << \vec{j}, \quad \partial\vec{D}\Big/\partial t \approx 0. \tag{6.50}$$

Assuming that the temperature and deformation are homogeneous, we obtain from formulas (6.49), (6.50)

$$\Delta\vec{H}(\vec{r}, t) - \mathrm{grad}\,\mathrm{div}\vec{H}(\vec{r}, t) = \sigma\mu_0\left[\partial\vec{H}\Big/\partial t + \partial\vec{M}\Big/\partial t\right]. \tag{6.51}$$

The system of equations (6.42), (6.51) describes the change in the magnetization and magnetic field of an electrically conductive ferromagnetic body taking into account the magnetic flux relaxation. Restricting ourselves by weak magnetic fields, when parameter α does not depend on $\vec{M}$, and hence on the spatial coordinates, exclude from this system by means of relation (6.44) an unknown function $\vec{M}$. As a result, we obtain the integro-differential equation

$$\Delta\vec{H}(\vec{r},t) - \operatorname{grad}\operatorname{div}\vec{H}(\vec{r},t)$$
$$= \sigma\mu_0\left\{ (1+\kappa_i)\partial\vec{H}(\vec{r},t)\Big/\partial t + (\kappa-\kappa_i)\alpha(t)\vec{H}(\vec{r},t)\right.$$
$$- \kappa\alpha(t)\vec{H}_0\exp(-A(t))$$
$$\left. - \alpha(t)(\kappa-\kappa_i)\int\limits_0^t \alpha(\varsigma)\vec{H}(\vec{r},\varsigma)\exp(A(\varsigma)-A(t))\mathrm{d}\varsigma \right\}, \tag{6.52}$$

where $A(t) = \int_0^t \alpha(\varsigma)\,\mathrm{d}\varsigma$; $A(\varsigma) = \int_0^\varsigma \alpha(\xi)\,\mathrm{d}\xi$.

When α is independent of time, this equation becomes simpler

$$\Delta\vec{H}(\vec{r},t) - \operatorname{grad}\operatorname{div}\vec{H}(\vec{r},t)$$
$$= \sigma\mu_0\left[(1+\kappa_i)\partial\vec{H}(\vec{r},t)\Big/\partial t + \alpha(\kappa-\kappa_i)\int\limits_0^t \partial\vec{H}(\vec{r},\varsigma)\Big/\partial\varsigma\exp(-\alpha(t-\varsigma))\mathrm{d}\varsigma \right]. \tag{6.53}$$

If $\alpha \to \infty$ (magnetic relaxation is absent), Eq. (6.53) passes into the known equation of diffusion of the magnetic field into a conductor (Landau and Lifshitz 1957)

$$\Delta\vec{H}(\vec{r},t) = \sigma\mu\partial\vec{H}(\vec{r},t)\Big/\partial t. \tag{6.54}$$

Here $\mu = \mu_0\mu_r = \mu_0(1+\kappa)$ is the absolute magnetic permeability of the material.

Magnetic aftereffect in a long cylinder. Let us study the effect of the finite rate of relaxation of the magnetic flux on the magnetic aftereffect of the electrically conductive body. It is convenient to use for experiments the cylindrical samples. Therefore, consider a body in the form of a sufficiently long cylinder, which is placed in an external homogeneous magnetic field, the stress vector of which is oriented along the axis of the cylinder: $\vec{H} = \{0, 0, H_0\}$. The body is in the state of magnetic equilibrium, i.e., its magnetization is determined as $\vec{M}_0 = \{0, 0, \kappa H_0\}$. At time $t = 0$, the external magnetic field disappears, and the relaxation of the magnetization begins in the body. The change in the magnetic field within the proposed mathematical model is described by equation (6.52), which under the conditions of axial symmetry $\vec{H} = \{0, 0, H(r)\}$, where r is the radial coordinate, takes the form

$$\frac{\partial^2 H(r,t)}{\partial r^2} + \frac{1}{r}\frac{\partial H(r,t)}{\partial r} = \sigma\mu_0\left(\frac{\partial H(r,t)}{\partial t} + \alpha\kappa\int_0^t \frac{\partial H(r,\varsigma)}{\partial\varsigma}\exp(-\alpha(t-\varsigma))d\varsigma\right). \tag{6.55}$$

The solution of this equation must be subjected to the boundary conditions

$$H(r)|_{t=0} = H_0, \quad H(t)|_{r=R} = 0, \quad t > 0,$$
$$\partial H(t)/\partial r\,|_{r=0} = 0, \quad H(t)|_{r=0} < \infty. \tag{6.56}$$

Applying the Laplace transform, we arrive at the differential equation for the image $H_L(r, p)$

$$H_L'' + \frac{1}{r}H_L' - \sigma\mu_0 p\left(1 + \frac{\alpha\kappa}{p+\alpha}\right)\left[H_L - \frac{H_0}{p}\right] = 0. \tag{6.57}$$

The bounded solution of this equation is written as

$$H_L(r,p) - H_0/p = AI_0\left(\sqrt{\sigma\mu_0 p(1 + (\alpha\kappa)/(p+\alpha))}\, r\right), \tag{6.58}$$

where $I_0(x)$ is a zero-order Bessel function.

The integration constant A is calculated from the boundary condition (6.56):

$$A = -\frac{H_0}{pJ_0\left(\sqrt{\sigma\mu_0 p\left(1 + \frac{\alpha\kappa}{p+\alpha}\right)}R\right)}. \tag{6.59}$$

Using the expansion theorem (Lykov 1967), we find the following solution of the problem (6.55), (6.56):

$$H(r,t) = H_0\sum_{n=1}^{\infty}\frac{2a_n J_0\left(a_n\frac{r}{R}\right)}{t_0 J_1(a_n)} \times\left(\frac{\exp(-p_n^+ t)}{p_n^+\left(1 + \frac{\alpha^2\kappa}{(-p_n^+ + \alpha)^2}\right)} + \frac{\exp(-p_n^- t)}{p_n^-\left(1 + \frac{\alpha^2\kappa}{(-p_n^- + \alpha)^2}\right)}\right), \tag{6.60}$$

where $t_0 = \sigma\mu_0 R^2$; $J_1(a_n)$ is Bessel first-order function; a_n are the roots of the Bessel zero-order function $J_0(a_n)$; and R is radius of the cylinder.

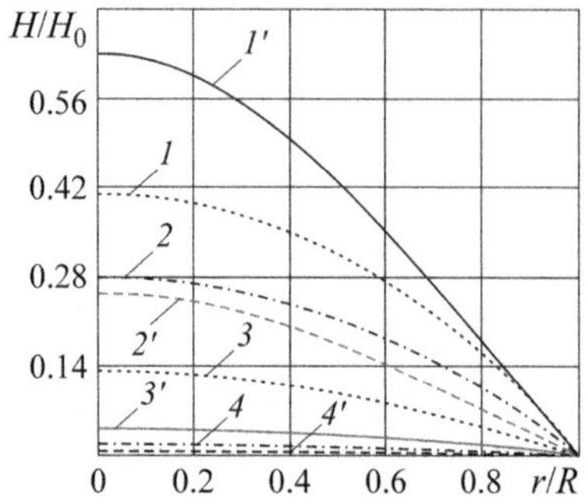

Fig. 6.14 Distribution of the magnetic field in the cylinder at time instants: $t = 0.5$ s (curves *1*, *1′*), 1 sec (curves 2, 2′), 2 s (curves *3*, *3′*), 5 s (curves *4*, *4′*) at $R = 0.005$ m, $k = 10^4$, $\mu_0 = 4\pi \times 10^{-7}$, $\sigma = 10$ MCm/m, which corresponds to iron (Tikadzumi 1987) (curves *1*–*4* are calculated for $\alpha =$ 1.4; curves *1′*–*4′* for $\alpha \to \infty$)

$$p_n^{\pm} = 0.5\left[\left(\alpha\mu_r + \frac{a_n^2}{t_0}\right) \pm \sqrt{\left(\alpha\mu_r + \frac{a_n^2}{t_0}\right)^2 - 4\alpha\frac{a_n^2}{t_0}}\right]. \tag{6.61}$$

In particular, when $\alpha \to \infty$(the magnetic relaxation is absent), relation (6.60) takes the form

$$H(r,t) = H_0 \sum_{n=1}^{\infty} \frac{2J_0(a_n r/R)}{a_n J_1(a_n)} \exp(-p_n t), \tag{6.62}$$

where $p_n = \left(a_n^2\right)/\left(t_0\mu_r\right)$.

The obtained expression (6.62) in the case of axial symmetry satisfies equation (6.54) and condition (6.56).

The graphs shown in Fig. 6.14 illustrate the effect of the finite rate of the magnetic flux relaxation on the magnetic field in the cylinder.

For the experimental study of the magnetic aftereffect, it is convenient to measure the electromotive force of electromagnetic induction E, which arise in the circuit around the cylinder.

The electromotive force is determined by the rate of change of the magnetic flux Φ_m in the cross-section of the body

$$E = -(d\Phi_m)/dt. \tag{6.63}$$

Considering that

$$\Phi_m = \int_S B\mathrm{d}S = 2\pi\mu_0 \int_0^R (H(r,t) + M(r,t))\,r\mathrm{d}r, \tag{6.64}$$

we obtain from relation (6.63)

$$E = 4\pi\mu_0 \sum_{n=1}^{\infty} \left\{ \exp(-\alpha t) \left[\frac{\alpha\kappa}{2} + \frac{\alpha^2\kappa\left(-p_n^+ + \alpha\right)}{p_n^+ t_0 \left[\left(-p_n^+ + \alpha\right)^2 + \alpha^2\kappa\right]} + \frac{\alpha^2\kappa\left(-p_n^- + \alpha\right)}{p_n^- t_0 \left[\left(-p_n^- + \alpha\right)^2 + \alpha^2\kappa\right]} \right] + \exp\left(-p_n^+ t\right) \left[\frac{\left[\alpha\kappa - p_n^+\left(-p_n^+ + \alpha\right)\right]\left(-p_n^+ + \alpha\right)}{p_n^+ t_0 \left[\left(-p_n^+ + \alpha\right)^2 + \alpha^2\kappa\right]} \right] - \exp\left(-p_n^- t\right) \left[\frac{\left[\alpha\kappa - p_n^-\left(-p_n^- + \alpha\right)\right]\left(-p_n^- + \alpha\right)}{p_n^- t_0 \left[\left(-p_n^- + \alpha\right)^2 + \alpha^2\kappa\right]} \right] \right\}. \quad (6.65)$$

When $\alpha \to \infty$ formula (6.65) takes the form

$$E = 4\pi\mu_0 \sum_{n=1}^{\infty} \frac{p_n}{a_n^2} \exp(-p_n t). \quad (6.66)$$

As we can see from Fig. 6.15, parameter α significantly changes the electromotive force of electromagnetic induction E. With the decrease of α such force increases and the time during which it fades grows.

According to formula (6.65), function E is characterized by several relaxation times, which can differ significantly in order of magnitude. The first summand in braces is specified by the finite rate of material magnetization. It determines the change in electromotive force with a relaxation time $\tau_0 = 1/\alpha$. The second and third summands depend on two sequences of characteristic times $\tau_n^+ = 1/p_n^+$ and $\tau_n^- = 1/p_n^-$.

According to formula (6.61), the relaxation times τ_n^+, τ_n^- depend not only on the parameter α, but also on the equilibrium magnetic permeability of the material and the geometry of the body. Since for ferromagnets $\mu_r >> 1$, then $\tau_n^+ << \tau_n^-$ and

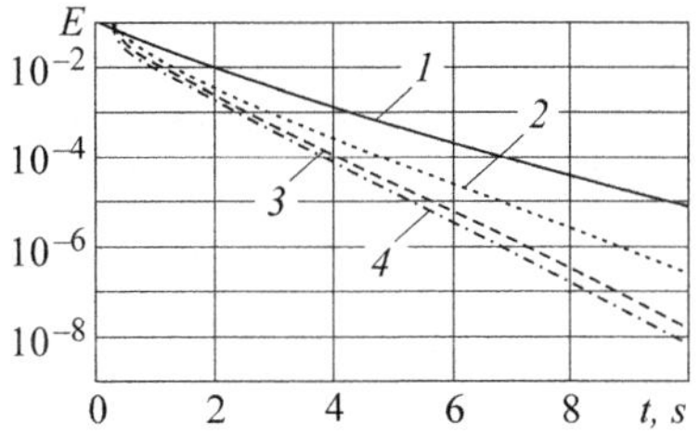

Fig. 6.15 Time dependence of the electromotive force for different α: *1*—1.4; *2*—3.3; *3*—7.5; *4*—10.7

$$\tau_n^+ \approx (\alpha + p_n)\left[\mu_r(\alpha + p_n)^2 - \alpha p_n\right]^{-1},$$
$$\tau_n^- \approx (\alpha + p_n)[\alpha p_n]^{-1}. \tag{6.67}$$

Of two sequences, the relaxation time $\tau_1^- = (\alpha + p_1)/\alpha p_1$ is the largest. If $\alpha << p_1$, then $\tau_1^- \approx \tau_0$, and in the area of large times, the character of the change E depends on the time of relaxation $\tau_0 = 1/\alpha$. Then, from dependences $E(t)$, determined by measurements, it is easy to establish directly the parameter α. If $\alpha << p_1$, then $\tau_1^- \approx 1/p_1$ and $\tau_0 << \tau_1^-$. At the same time the influence of τ_0 on electromotive force will be manifested in the region of small time. Therefore, to determine the parameter α, it is necessary to perform measurement in the region of these times, which is difficult to implement experimentally. Note that the choice of the sample geometry can, according to (6.61), reach p_1 that becomes larger or at least commensurable with α. Then, finding the relaxation time τ_1 from the measurement results, we can establish the value of α by formula (6.67).

Influence of hydrogen concentration on magnetic relaxation. Deformation of the lattice can affect significantly the diffusion processes and thus the kinetics of the reconstruction of the magnetic subsystem of the ferromagnet. As a result, the magnetic relaxation becomes dependent on deformation. Determining parameter α on the basis of measurement results of the magnetic aftereffect and dependence (6.65), we obtain some information about the deformation and, consequently, about the parameters of the other processes that cause it.

In the linear approximation, the dependence of α on the deformation can be represented in the form

$$\alpha = \alpha_0 + A_{ij}\varepsilon_{ij}, \tag{6.68}$$

where A_{ij} are the components of the material second rank tensor, which determines the effect of deformation on the magnetic aftereffect. For materials with cubic symmetry of the crystal lattice, as well as for isotropic bodies, material second rank tensors are spherical. Then the change of the parameter α with deformation is determined by the scalar parameter $A_{ij} = a\delta_{ij}$.

In steels, macroscopic deformation occurs during hydrogen charging, which increases with increasing the concentration of hydrogen atoms (Alefeld and Felkel 1981). In some approximation, the dependence of the deformation ε_{ij} components on hydrogen concentration C_H can be considered as linear (Panasyuk et al. 1984; Fast 1974). The total deformation ε_{ij} is given as the sum of two components: the deformation e_{ij}, caused by external fields (for example, force loads), and the component that takes into account the hydrogen atoms in the lattice

$$\varepsilon_{ij} = e_{ij} + BC_H\delta_{ij}. \tag{6.69}$$

Here B is the material characteristic, which describes the change in the deformation of the lattice with the change in hydrogen concentration.

If deformations e_{ij} are caused by force loading and heating, they are related to stresses σ_{ij} and temperature increase ΔT by thermoelastic relations. Then relation (6.69) will look like

$$\varepsilon_{ij} = S_{ijkl}\sigma_{kl} + \delta_{ij}(\alpha_T \Delta T + BC_{\mathrm{H}}). \tag{6.70}$$

Here S_{ijkl} are components of the elastic compliance tensor; α_T is the coefficient of temperature expansion. In the isotropic case we have

$$S_{ijkl} = \frac{\nu}{E}\delta_{ij}\delta_{kl} + \frac{1+\nu}{2E}\left(\delta_{ik}\delta_{jl} + \delta_{il}\delta_{jk}\right), \tag{6.71}$$

where E and ν are the Young's modulus and Poisson's ratio, respectively.

Taking into account (6.70) the dependence of the parameter α on the stresses and hydrogen concentration in the body will be written as

$$\alpha = \alpha_0 + A\left(\frac{1+4\nu}{E}\delta_{kl}\sigma_{kl} + 3(\alpha_T \Delta T + BC_{\mathrm{H}})\right). \tag{6.72}$$

In the absence of loads and uniform distribution of hydrogen concentration, when $\sigma_{kl} = 0$, we obtain from relation (6.72)

$$\alpha = \alpha_0 + 3ABC_{\mathrm{H}}. \tag{6.73}$$

To find the unknown constants A and B, use the results of measurements of the magnetic aftereffect on material samples under given loading conditions and the degree of the material hydrogen charging.

Thus, by measuring the magnetic aftereffect in a long cylinder and using relation (6.65), parameter α can be determined. Comparing the values of α with formula (6.73), we can estimate the degree of the sample hydrogen charging.

6.5 Express Method for Estimating the Residual Magnetization

The proposed express method is based on the problem of determining the vector value of the residual magnetization of material J. This characteristic is structurally insensitive and therefore best reflects the degree of degradation of ferromagnets.

The method consists in performing such a sequence of actions (Skalsky et al. 2009). First, choose any available indicator tool that is sensitive to changes in the magnetic field intensity, such as a compass. Then select any reference sample with a known magnetic field intensity on its surface or poles. As a reference value of the sample reference H_0 we can, for example, take the magnetic field intensity at the pole of the solenoid. It is calculated by the known formula (Kuhling 1982)

$$H_0 = \frac{NI}{2L}, \tag{6.74}$$

where N is the number of turns; I is current; and L is the length of the solenoid.

After that, the indicator tool is placed on a horizontal surface, and the reference sample is brought to it in the shortest direction at a fixed distance l_0, which is not less than 5...10 of its largest geometric size. This condition allows us to consider the reference sample as a magnetic dipole. In the selected end position, determine the readings of the indicator tool N_0 and measure the value of the distance l_0 to it from the reference sample. The planes of the axes of both these objects must be at the same level. Then, after removing the reference sample, introduce a check test sample with unknown magnetization H_x, on its measuring axis, approaching it so that the readings of the indicator tool N_0 are similar to the previous case, and also measure the distance l_1 between them. Having such experimental data, the magnetic field intensity of the ferromagnetic material H_x is determined by the formula

$$H_x = H_0 l_0^3 \cos^3 \alpha_1 / l_1^3 \cos^3 \alpha_0, \tag{6.75}$$

where α_0 is the angle between the measuring axis and the direction of the arrowhead of the indicator tool at the time of recording the readings N_0 from a distance l_0 for the reference sample and α_1 is the angle between the measuring axis and the direction of the arrowhead of the indicator tool at the time of recording the readings N_0 from a distance l_1 for the check test sample.

Formula (6.75) is obtained as follows. It is known that the magnetic field intensity H_i at a considerable distance from the body with a magnetic moment M in the case of its direction along the longitudinal axis of the magnet will be determined as (Gallagher 1984)

$$H_i = \frac{1}{2\pi \mu_0} \frac{M}{l_i^3}. \tag{6.76}$$

If we take a permanent magnet with a known magnetic field intensity H_0 at a distance l_0 as a reference sample, formula (6.76) will take the form

$$H_0 = \frac{1}{2\pi \mu_0} \frac{M_0}{l_0^3}. \tag{6.77}$$

For the check test sample with unknown magnetization, dependence (6.76) can be written as

$$H_x = \frac{1}{2\pi \mu_0} \frac{M_x}{l_1^3}. \tag{6.78}$$

If we approach the reference sample and the check test sample to the indicator tool, causing the deviation of its arrow by the same value of N_0, it means that for different distances $l_0 \neq l_1$ we obtain the action on the arrow with the same magnetic moments $M_0 = M_x$. Then the relation between the magnetic field intensities, based on the dependences (6.77) and (6.78), will be

$$\frac{H_0}{H_x} = \frac{l_1^3}{l_0^3}. \tag{6.79}$$

From here

$$H_x = H_0 l_0^3 / l_1^3. \tag{6.80}$$

Since the greatest moment acts on the arrowhead, dependence (6.80) must be corrected for the angular distribution, which is taken into account in formula (6.75).

The method of express evaluation of the residual magnetization of ferromagnetic materials is carried out as follows (Nazarchuk et al. 2013). The indicator tool (compass) *2* with a digital limb *3* is placed on the horizontally located base *1* (Fig. 6.16). Its arrow is set in the direction of the earth magnetic field along the *a-a* axis. The preselected reference sample *4* with a known magnetic field intensity H_0 at its poles is applied to the *b-b* measuring axis so that it lies in the plane of the arrow of the indicator tool *2*. The reference sample *4* is approached by the indicator tool along the measuring axis to deflect its arrow from position *a-a* to position N_0, the value of which is chosen arbitrarily. Then the reference sample is fixed, and the distance l_0 from it to the indicator tool is measured. The reference sample is then removed, and a check test sample *5* with an unknown residual magnetization value J_x is added to the measuring axis *b-b*. Repeat the procedure of deviation of the indicator device arrow to the position H_0, and then measure the distance l_1 similarly. Having the measurement data in the above cases, with account of the scheme shown in Fig. 6.17 by dependence (6.74) and taking into account the results of calculations by formula (6.75), determine the residual magnetization of the ferromagnetic material J_x.

Thus, the described method allows us to significantly simplify the method of estimating the residual magnetization of ferromagnetic materials, which increases the manufacturability of diagnostic process and reduces its cost.

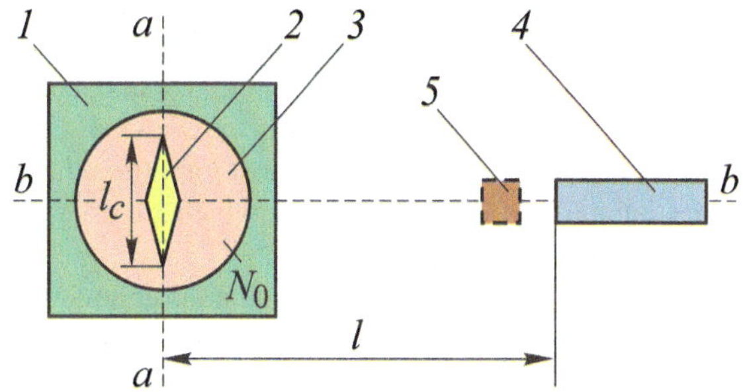

Fig. 6.16 Block diagram of the implementation of the express method

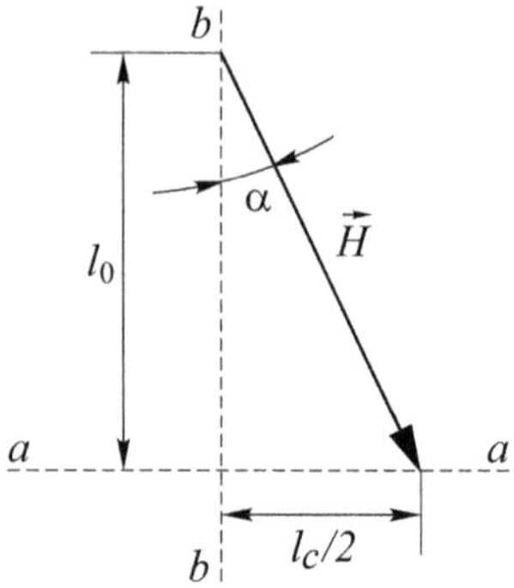

Fig. 6.17 Calculation scheme of the method

6.6 Calculation of the Magnetic Field Distribution Around the Magnetized Bodies of Finite Dimensions

The finite element method (FEE) was used for the calculation. FEE is a numerical technique to find solutions to integral equations and partial differential equations. The solution is based either on the complete elimination of the differential equation for stationary problems or on their expansion into an approximating system of ordinary differential equations, which are then solved using some standard techniques, such as Euler and Runge-Kutt methods (Gallagher 1984).

When solving differential equations with partial derivatives, the main goal is to create an equation that approximates the equation under study and is numerically stable, i.e., errors in input data and intermediate calculations do not accumulate and do not lead to meaningless results. There are many ways to do this, each with its positive and negative features. The finite element method is a good choice for solving partial differential equations that describe complex media (such as machines or oil pipelines) due to their variability, when the desired accuracy is different in different parts of the environment, or when the solution lacks smoothness. For example, when modeling the frontal division of the machine, it is possible to increase the accuracy of modeling in important areas, such as the machine front, and reduce it to calculate what happens to the rear of the machine (thereby reducing modeling resources).

The finite element method arose from the need to solve complex problems of elasticity and structural analysis in civil, marine, and aeronautical engineering. Its development can be traced back to the works by Alexander Khrennikov (1941) and Richard Courant (1942). Although the views of the two scientists were incredibly different, they still agreed on the most important thing: the splitting of a large continuous domain into smaller domains, which are usually called discretization (splitting) elements (Gallagher 1984).

In his work, Khrennikov distributed domains using the lattice principle. At the same time, Courant splitted the region into a finite number of triangular subdomains corresponding to the solutions of the second-order elliptic differential equations (the latter arise from the problem of cylinder rotation). Courant's contribution was evolutionary, that is, it was based on the vast knowledge about such differential equations accumulated by Reilig, Ritz, and Galorkin.

The development of the finite element method began in the mid-1950s for air tube and structural analysis and received its greatest development at the University of Stuttgart in the work of John Argeris and the University of Berkeley, or rather in the work of Ray W. Klog in the 1960s for use in civil engineering. By the end of the 1950s, the key concepts of the stiffness matrix and the collection of elements already existed in almost the same forms in which they are used today. In 1965, at the request of NASA, the NASTRAN program was written as a software built to implement the FEM. The method itself was proved in 1973 in the publication by Strang and Fix—"Finite Element Method Analysis", and since then has been generalized to a separate field of applied mathematics and mathematical modeling of physical systems in many engineering disciplines such as electromagnetism or fluid dynamics (Gallagher 1984).

The finite element method, usually at the design and development stage of software products, uses many disciplines, mostly in the field of mechanical engineering (such as aerospace, marine, biometric, and automatic industries). The structural modeling of FEM package is very helpful in generating rigid and force visualizations in the places of shifts and bends, as well as mapping the distribution of forces and displacements.

The most important advantages of the finite element method are:

- Properties of materials of adjacent elements may be different. This allows us to apply this method to bodies composed of several materials.
- Finite elements are simple areas (straight lines, triangles, rectangles, pyramids, prisms). Thus, this method can be used to approximate bodies with a complex shape of edges.
- Element sizes can be variable. This allows us to increase or decrease the elements of the lattice.
- Using the FEM it is easy to consider boundary conditions with a discontinuous surface load, as well as mixed boundary conditions.
- The finite element algorithm allows us to create general programs for solving problems of different classes.
- The problem is reduced to solving a system of equations of large-scale algebra. However, a good conditionality of the system of algebraic equations allows us to obtain fairly accurate solutions for systems of equations of size 5–10 million and more.

The main disadvantage of this method is that a large amount of computer memory and high computing speed are required. Currently, the development of computers has virtually eliminated this shortcoming.

The distribution of the magnetic field around the cylinder, ellipsoid, parallelepiped, and sphere is calculated for the needs of the research by the finite element method, presented here. To see the effect of shape on the magnetic field distribution, all samples had the same volume. The samples were magnetized in a solenoid, after that the magnetic field was recorded, and at a certain moment modeling was carried out by this numerical method.

Samples were split using three-dimensional finite elements in the form of a tetrahedron with 10 nodes. A schematic view of such a finite element is given in Fig. 6.18.

A similar split into finite elements was used for the solenoid. Figure 6.19 shows an ellipsoid and a solenoid splitted into finite elements.

Figure 6.20 shows the distribution of the magnetic field intensity around the ellipsoid.

Modeling was performed as follows. The parameters and dimensions of the solenoid as well as the sample were the same as in the experiments on residual magnetization. A direct current was supplied to the solenoid that increased from 0 to 0.7 A for 0.5 sec, after that the current was turned off, and the distribution of the magnetic field around the sample was calculated. The ellipsoidal sample of steel 30

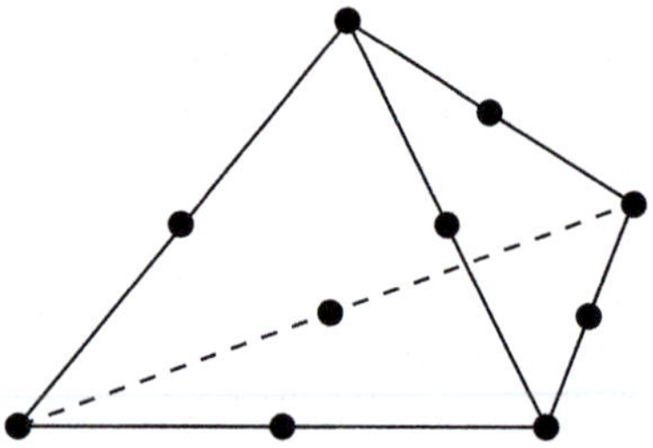

Fig. 6.18 View of a three-dimensional 10-node finite element

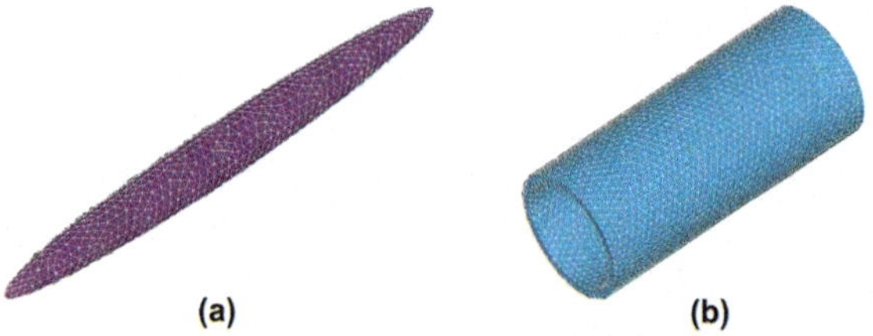

Fig. 6.19 Ellipsoid (**a**) and solenoid (**b**) splitted into finite elements

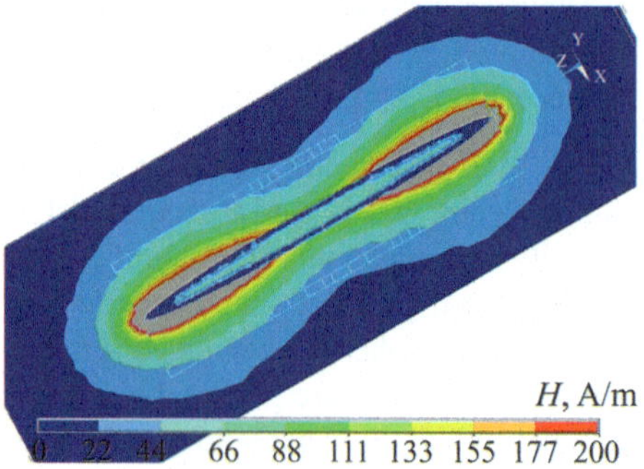

Fig. 6.20 Distribution of the magnetic field around the ellipsoid of size 100×10 mm

of a length of 100 mm and a diameter of 10 mm and the dimensions and parameters of the solenoid are described above.

The distribution of the magnetic field intensity for the cylinder, parallelepiped, and sphere was calculated under the same conditions (Figs. 6.21, 6.22 and 6.23).

A cylindrical specimen was of a diameter of 8 mm and a length of 100 mm, and a parallelepiped had 7 × 7 × 100 mm. The sphere radius was 11 mm. All the samples were of the same volume.

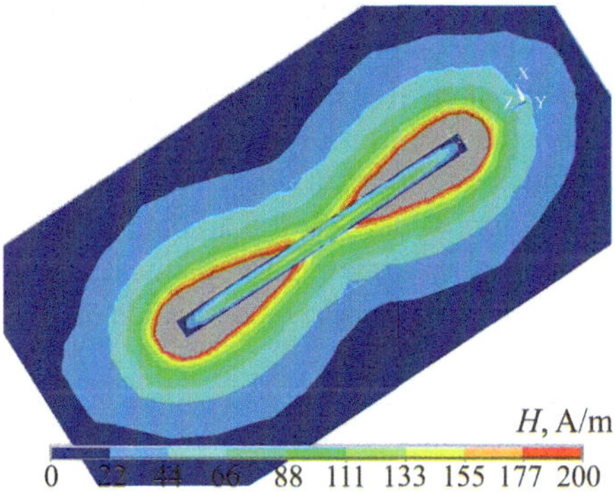

Fig. 6.21 Distribution of the magnetic field around the cylinder

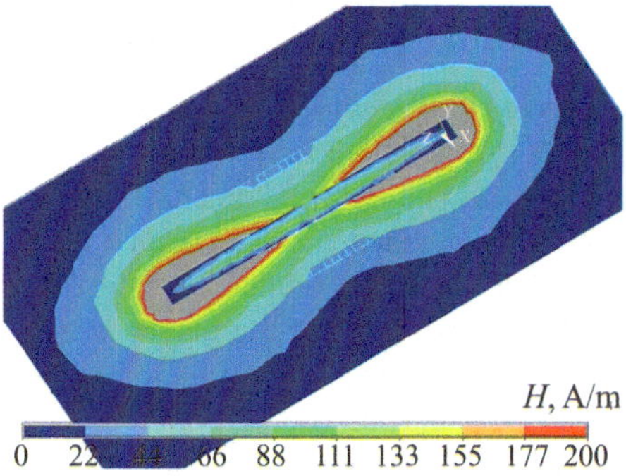

Fig. 6.22 Distribution of the magnetic field around the parallelepiped

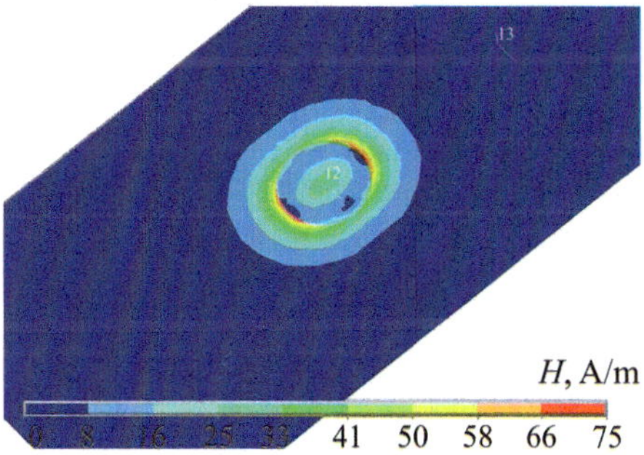

Fig. 6.23 Distribution of the magnetic field around the sphere

The results show that the magnetic field is distributed unevenly around the sample and increases at the poles. Comparing the distribution of the magnetic field around the different shaped samples, it was concluded that for a cylindrical sample and a parallelepiped, the distribution of the magnetic field is almost the same, and its power is greater than around the other samples. Therefore, in the future, the investigations on magnetic ductility and residual magnetization of ferromagnets the cylindrical specimens were used, while for comparison characteristics the ellipsoidal specimens were investigated.

Thus, the calculation of the distribution of magnetic field intensity near bodies of different geometry and sizes by the finite element method shows its heterogeneity, which is typical of each of them. This distribution is the most optimal for the sphere. In long bodies, it is distorted, and the intensity decreases in the case of an ellipsoid, cylinder, and a prism of the same volume. The highest intensity is concentrated in the region of the poles. The obtained data allowed us to implement experimental methods for estimating the change in residual magnetization and magnetic ductility of ferromagnets under the influence of hydrogen-mechanical factor (Nazarchuk et al. 2013).

6.7 Experimental Evaluation of Magnetic Ductility of Structural Steels

Experimental technique. Prior to the magnetic ductility tests, the change in the earth's magnetic field at the test site was measured (Fig. 6.24). Measurements were performed with a magnetometer LEMI-D11B for three days. The figure shows that the earth's magnetic field varies in a very small range, and its absolute value will not affect investigations on the magnetic ductility of ferromagnetic materials.

At the beginning of the magnetic ductility measurements, a magnetometer was adjusted—the sensitivity of its ferroprobe to the magnetic field intensity was calibrated. For this purpose, any reference sample with a known magnetic field intensity on its surface or poles was selected. The magnetic field intensity at the solenoid pole was chosen as the reference value of the reference sample H_0. It was calculated by formula (6.74).

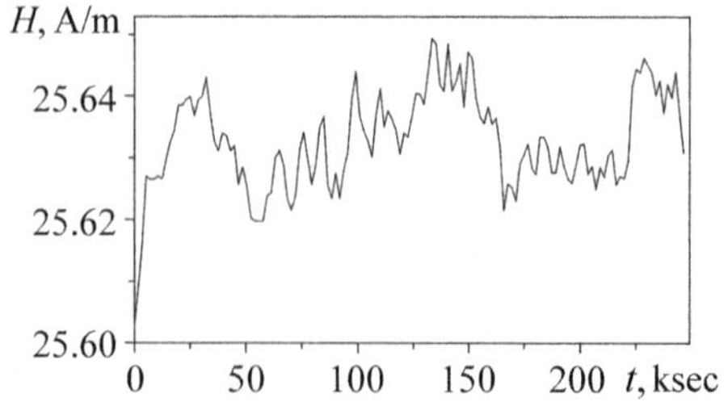

Fig. 6.24 Temporal change of the background intensity of the earth's magnetic field

After that, the magnetometer was placed on a horizontal surface, and it gradually approached the reference sample in the shortest direction at a fixed distance l_0, which was 5–10 values of its largest geometric size (Fig. 6.25). This condition allows us to consider the reference sample as a magnetic dipole.

The measurement scheme is shown in Fig. 6.26.

In the selected end position, the display of the magnetometer N_0 was recorded and the distance l_0 to it from the reference sample was measured. The axial planes of both of these objects were at the same level. Then, removing the reference sample, a sample of ferromagnetic material with unknown magnetization H_x was added to its measuring axis, approaching it so that the readings of the magnetometer N_0 were similar to the previous case. Then the distance l_1 between them was measured. Having such experimental data, the magnetic field intensity of the ferromagnetic material H_x was evaluated by the formula

$$H_x = H_0 l_0^3 / l_1^3. \tag{6.81}$$

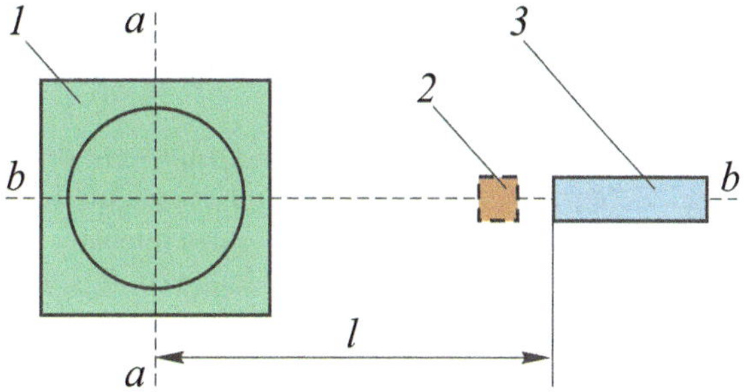

Fig. 6.25 Block diagram of the experiment: *1*—magnetometer LEMI-D11B; *2*—reference sample; *3*—studied ferromagnet

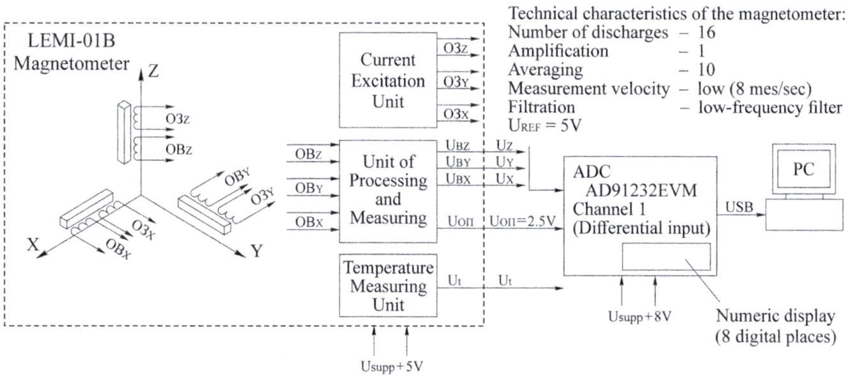

Fig. 6.26 Block diagram of residual magnetization measurements with LEMI-D11B magnetometer

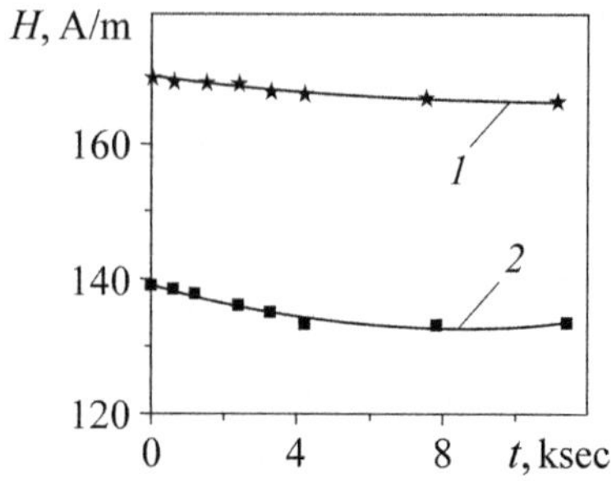

Fig. 6.27 Change of magnetic field intensity at the cylindrical sample pole on the relaxation time: 1—non-hydrogen-charged samples; 2—hydrogen-charged samples

Formula (6.81) follows from the substantiation of the express method for estimating the residual magnetization described in the previous subchapter.

The experimental results and their discussion. Cylindrical samples of structural U8 steel (see Appendix) with a diameter of 8 mm and a length of 100 mm were investigated (Nazarchuk et al. 2013). They were prepared of a bar of material in the as-delivery state. One group of the samples was hydrogenated in gaseous hydrogen at a temperature of 42 °C and a pressure of 0.2 MPa for 7.2 k s. After that they were kept in the air for 31.5×10^6 k s. The second group of the samples was tested without additional processing.

Prior to magnetization, the samples from each group were demagnetized by an alternating magnetic field in the solenoid (each in the same mode). Then they were magnetized in the solenoid in a constant magnetic field of intensity $\vec{H}_0$ (magnetization modes of each sample before the test were also identical). Residual magnetization was measured for 12 k s. Figure 6.27 presents the research results.

We see that the residual hydrogen dissolved in the metal affects the magnetization and changes the residual magnetization and magnetic aftereffect (magnetic ductility). Thus, the stress H at the pole (edge) of the cylindrical sample of hydrogen-charged steel is 16–18% less than that of non-hydrogen-charged steel.

A similar experiment was performed on the 15 steel (see Appendix) samples. One group of rods was plastically deformed by compression with a force of 4.3 tons. After that the samples of cylindrical shape with a diameter of 8 mm and a length of 100 mm, as well as the samples of ellipsoidal shape with a length of 100 mm, diameter 10 mm, and the same volume as cylindrical ones, were prepared. Another group of samples was tested without further processing. All the test conditions were identical to the previous experiment. Residual magnetization was measured for 15 k s. Figure 6.28 shows the research results.

We can conclude from the figure that the plastic deformation of material affects the magnetization changes the residual magnetization and magnetic aftereffect (magnetic ductility). Thus, the intensity $\vec{H}$ at the pole (edge) of cylindrical and ellipsoidal plastically deformed specimens is higher than for specimens from the original material. The shape of the sample also has a significant effect on the residual magnetization. It can be seen that the intensity $\vec{H}$ at the poles of cylindrical specimens is almost twice as high as that of ellipsoidal specimens, although they have the same volume.

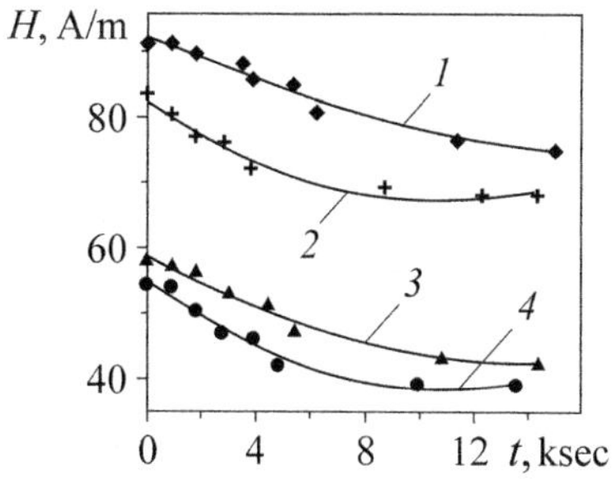

Fig. 6.28 Change in magnetic field intensity at the sample pole on the relaxation time: *1*—cylindrical plastic-deformed sample; *2*—cylindrical original sample; *3*—ellipsoidal plastic-deformed sample; *4*—ellipsoidal original sample

So, the proposed express method for estimating the residual magnetization of ferromagnets can significantly simplify the procedure of determining the residual magnetization of ferromagnetic materials, which increases the manufacturability of diagnostic process and reduces its cost. Studies have shown that residual hydrogen dissolved in the metal affects the magnetization and changes the residual magnetization and magnetic aftereffect (magnetic ductility). Thus, the intensity $\vec{H}$ at the pole (edge) of the cylindrical sample of hydrogen-charged steel is 16–18% less than in the non-hydrogen-charged sample. In addition, plastic deformation of the material also affects the magnetization, causing similar changes. The intensity $\vec{H}$ at the pole (edge) of the cylindrical and ellipsoidal plastically deformed samples is higher than that of the samples of the original material. We can see that the shape of the sample has a significant effect on the residual magnetization. Thus, the intensity of cylindrical specimens is almost twice as high as that of ellipsoidal specimens, although they have the same volume.

References

Andreikiv OE, Chekurin VF, Gembara OV (2002) Mathematical model of magnetic aftereffect in hydrogenated ferromagnets. Mater Sci 38(4):534–543

Alefeld G, Felkel I (eds) (1981) Vodorod v metallakh. T. 1. Osnovnyie svoistva (Hydrogen in metals. Vol. 1. Basic properties). Mir Publishing House

Baryakhtar VG (ed) (1996) Fizika tviordogo tela (Physics of a solid). Naukova dumka Publishing House

Fast JD (1974) Vzaimodeisviie metallov s gazami (Interaction of metals with gases). Metallurgiya Publishing House

Gallagher R (1984) Metod konechnykh elementov. Osnovy (Finite element method. Basics). Mir Publishing House

Goodwin ARH, Trusler JPM (2003) Speed of sound. In: Goodwin ARH, Marsh KN, Wakeham WA (eds) Measurement of the thermodynamic properties of single phases. Elsevier, pp 237–323

Kolsky H (1964) Stress waves in solids. J Sound Vib 1:88–110

Kronmuller H (1981) Vodorodnoie magnitnoie posledeistviie v ferromagnitnykh splavakh i metallakh (Hydrogen magnetic aftereffect in ferromagnetic alloys and metals. In: Alefeld G, Felkel I (eds) Vodorod v Metallakh. T. 1: Ostovnyye svoistva (Hydrogen in metals, Vol. 1: Basic propertiers). Mir Publishing House

Kuhling H (1982) Spravochnik po fizike (Handbook of Physics). Mir Publishing House

Landau LD, Lifshitz EM (1957) Elektodinamika sploshnykh sred (Electrodynamics of continuous media). State Technical Publishing House

Lykov AV (1967) Teoriia teploprovodnosti (Theory of heat condactivity). Vysshaya shkola Publishing House

Lynnworth LC (1975) Industrial applications of ultrasound—a review: II. Measurements, tests, and process control using low-density ultrasound. IEEE Trans Sonics Ultrasonics 22:71–101

Mishin DD (1991) Magnitnyie materialy (Magnetic materials). Vysshaya shkola Publishing House

Mykhalchuk VB, Plakhtii RM (2005) Metodyka otsinky zalyshkovoi namagnechenosti tsylindrychnykh zrazkiv (A method of estimation of residual remagnetization of cylindrical samples). Zbirnyk prats XIX vidkrytoi naukovo-telhnichnoi konferentsii molodylh vchenykh i spetsialistiv FMI im. H. V. Karpenka NANU "KMN-2005" (Proceedings of the XIX conference of young scientists and specialists of H. V. Karpenko PhMI "KMH-2005"). Lviv

Nazarchuk ZT, Andreykiv OYe, Skalskyi VR (2013) Otsinyuvannia vodnevoi degradatsii feromagnetykiv u magnetnomu poli (Evaluation of hydrogen degradation of ferromagnets in a magnetic field). Naukova dumka Publishing House

Panasyuk VV, Andreikiv AE, Obukhivskii OI (1984) A calculation model of crack growth in metals under the action of hydrogen. Mater Sci 20(3):199–202

Skalsky VR, Nazarchuk ZT, Mikhalchuk VB, Klym BP, Pochapsky YeP, Dolishnii PM (2009) Sposib ekspres-otsinky zalyshkovoi namagnichenosti feromagnitnykh materialiv (A method for express-estimation of the residual magnetization of ferromagnetic materials). Pat. No. 91142. IPC (2009): G01N27/72. 25 June, 2010, Bull. No. 12

Skalskyi V, Mykhalchuk V, Selyvonchyk T, Semegenivsky R (2006) Otsenka vliayaniia vodoroda na ostatochnuiu namagnichennost stali (Evaluation of the influence of hydrogen on the residual magnetization of steel). Materialy Mezhdunarodnoi nauchno-technicheskoi konferentsii "Sovremennyie problemy i perspektivy mekhaniki" (Proceedings of the International Scientific and Technical Conference "Modern Problems and Prospects of Mechanics"), Tashkent

Skalskyi VR, Nazarchuk ZT, Mykhalchuk VB, Klym BP, Pochapskyi YeP, Dolishnii PM (2009) Sposib otsinky stupenia navodnennia ferromagnetnykh splaviv (A method of evaluation of the degree of hydrogenation of ferromagnetic alloys). Patent № 88249. IPC G01N29/14; G01N17/00, 25. June.2009, Bull. №12

Solodyak MT (2000) Mathematical model of magnetothermoelasticity of ferromagnetic bodies. Mater Sci 36(1):1–13

Tikadzumi S (1983) Fizika ferromagnetizma. Magnitnyie svoistva veshchestva (Physics of ferromagnetism. Magnetic properties of matter). Mir Publishing House

Tikadzumi S (1987) Fizika ferromagnetizma. Magnitnyie kharakteristiki i prakticheskiie primineniia (Physics of ferromagnetism. Magnetic characteristics and practical applications). Mir Publishing House

Vonsovsky SV (1971) Magnetizm. Magnitnyie svoistva dia-, para-, ferro-, anti-ferro i ferromagnetikov (Magnetism.Magnetic properties of dia-, para-, ferro-, antiferro- and ferromagnets). Nauka Publishing House

Vonsovsky SV (1984) Magnetizm (Magnetism). Nauka Publishing House

Vonsovsky SV, Shur YaS (1948) Ferromagnetizm (Ferromafnetism). OGIZ (Association of state book and magazine publishing houses)

Williams H, Shockley W, Kitel C (1952) Izucheniie skorosti dvizheniia granits mezhdu ferromagnitnymi oblastyami. V: Ferromagnitnyi rezonans (Study of the speed of movement of boundaries between ferromagnetic regions. In: Vonsovsky SV (ed) Ferromagnetic resonance. Publishing House of Foreign Literature

Chapter 7
Methodology of Investigation of the Hydrogen Influence on Ferromagnet

The vast majority of equipment for thermal power plants, pipeline transport, mechanical engineering products, etc. are made of steels that are ferromagnetic materials. In contact with hydrogen-containing environment under long-term operation, ferromagnetic materials degrade and lose their strength properties and other physical characteristics. This is caused by changes in the ferromagnet structure and the emergence of sites of increased local damage. The latter, for the most part, occurs under the influence of hydrogen factor and load, which leads to a sharp deterioration of the ferromagnet properties and thus to the microcracks initiation and propagation. It is in such places that the beginning of the initiation of structural elements macrofracture takes place.

Destructive and non-destructive test (NDT) methods are used to detect hydrogen damage in ferromagnetic structural materials. While the former have found their application in laboratory test conditions, the latter are increasingly used in engineering practice to monitor the technical state of equipment both during its operation and during technological shutdowns.

In particular, NDT methods, such as X-ray, eddy currents, and ultrasonic have gained some recognition. But even today they no longer meet the growing scientific and technical requirements for diagnosing objects of long-term operation. This has stimulated the development of new highly sensitive effective methods of NDT for detecting local damage to ferromagnets (especially hydrogen) and means for their implementation, which are based primarily on new phenomena and effects of magnetism. The investigations described below in this chapter are aimed at a partial solution to this problem.

V. Skalskyi and Z. Nazarchuk, *Magnetoelastic Acoustic Emission*, Springer-AAS Acoustics Series, https://doi.org/10.1007/978-981-99-4032-5_7

7.1 Problem State-of-the-Art

As noted in previous chapters, recently a special interest when solving the problem of technical diagnostics of long-term operating objects has been shown in the NDT methods, constructed using the BE (Barkhausen 1919). It should be noted that the main physical mechanisms, on which the BE is based, have been known for a long time (Becker and Döring 1939). However, rapid practical use of this effect began after an experimentally found explicit relationship between the various parameters of electrical signals of the BE and the elastic waves of acoustic emission. Such waves appear, due to internal, fatigue, and external elastic stresses in ferromagnetic steels, as well as with their heat treatment (Lomaiev 1977; Lomaev et al. 1984; Gorkunov et al. 1999; Kuleev et al. 1986).

As a result, many authors emphasize the fundamental possibility of solving with the help of BE and, especially MAE, one of the most important problems of NDT—to determine the internal and fatigue elastic stresses arising in steel products both in their manufacture and operation and hence the sites of local hydrogen damage.

The widespread and effective use of MAE in NDT is somewhat constrained by the lack of knowledge of some essential details of its physical nature. In fact, a majority of available studies are limited to experimental investigations of the influence of various factors on the MAE with establishing different degrees of correlation with elastic stresses for different steels and different experimental conditions of BE manifestation, without considering the physical causes of such correlations in each separate case.

For example, the authors of Shibata and Ono (1981), Ono and Shibata (1980) discuss their experiments involving a theoretical model (Rudiak 1970), which presents the solution of the equations of the theory of elasticity by Green's method for a known external jump-like change of deformation $\Delta\varepsilon$ in an infinite elastic medium. The movement of dislocations or the crack propagation is considered, for example, to be the source of such change. However, it is unknown whether such an approach to domain wall jumping in ferromagnets can provide new information about MAE, not to mention the experiment recommendations, as it ignores almost all the features of domain wall jumping in ferromagnets. The character of these jumps is determined by both the distribution of potential barriers and the structure of the domain walls, and their energy in the external magnetic field, depending on time; the influence of eddy currents, the inhomogeneity of the internal field due to the shape effect, as well as the geometry of the mutual location of the magnetic moments of the adjacent domains (Williams et al. 1952). This character can be determined only by the joint solution of Maxwell's equations, the equations of domain wall movement, and the equations of the theory of elasticity for different types of domain structure.

Recently the physical aspects of remagnetization of ferromagnets have been described, and the character of changes in their domain structure under the influence of the applied external magnetic field has been studied (Kuleev et al. 1986; Tikadzumi 1987). However, the physical aspects of the hydrogen effect on the magnetic properties of these materials have not yet been sufficiently investigated. In view of the above said, it makes sense to establish the influence of the hydrogen factor based on modern ideas on the domain structure and the character of interactions within ferromagnets and to demonstrate unequivocal regularities, arising in this case as well as their manifestation at the macrolevel.

7.2 Calculation of Exchange Interaction Energy for Iron and Nickel Clusters in the Presence and Absence of Hydrogen

In (Brechko et al. 2004), it is experimentally confirmed that the above-mentioned hydrogen-induced processes lead to a significant rearrangement of the domain structure of the ferromagnet (Fig. 7.1), which significantly affects its magnetic properties.

Magnetically ordered crystals are characterized by the fact that the magnetic moments of atoms are directed parallel to each other within macroscopic regions called domains. At the same time, their spins try to orient themselves parallel to each other due to the presence of a special form of Coulomb interaction—exchange interaction (Krinchik 1976). Violation of spin parallelism is caused by the thermal motion of electrons. In addition, there are domain boundaries—layers with non-parallel spins that separate domains with different magnetization orientations. During the formation of the domain structure and during technical magnetization, this additive to the value of exchange energy (EE) plays a decisive role due to the non-parallelism of

Fig. 7.1 Domain structure of the amorphous Fe78B12Si9Ni1 alloy in the original state (**a**) and after hydrogen saturation (**b**)

spins, because it is proportional to the value of the Zeeman energy of the magnet in the external magnetic field. Consider these issues in more detail.

As shown in the example of the hydrogen molecule, the calculation of the exchange interaction is not very simple. For atomic systems, it is easier to use approximate calculation methods and appropriate software packages for their realization. In particular, the molecular orbital method (MO) is used for this purpose. The theory of MO method gives an idea of the electron density distribution and explains the properties of molecules. In this theory, quantum mechanical dependences for the atom are extended to a more complex system—a molecule. A molecule is considered an all-in-one and not a set of individual atoms. In a molecule (as in an atom) there are discrete energy states of individual electrons (molecular orbitals) with their self-consistent motion in the field of each other and of all nuclei of the molecule. It is assumed that all the electrons of a given molecule (as in an atom) are distributed in the corresponding orbitals. The state of an electron in an atom is described by one-electron wave function ψ, which is a solution of the Schrödinger equation. A wave function that has a specific mathematical form and satisfies the condition of normalization and unambiguity is called a molecular orbital by analogy with an atomic one. Each orbital is characterized by its own set of quantum numbers that represent the properties of electrons in each energy state. Unlike one-center orbitals of atoms, the orbitals of molecules are multicenter, i.e., molecules have common orbitals for two or more atomic nuclei. Each molecular orbital has a certain energy and is approximately characterized by the corresponding ionization potential.

Molecular orbitals are formed by combining atomic orbitals during sufficient approaching. The set of orbitals of the molecule with the indication of its type and the number of electrons on it gives the electronic configuration of the molecule. There are 3 types of MO: those that bind, loosen, and do not bind. Electrons on binding MOs strengthen the bond; on loosening—destabilize (loosen). A molecule is stable only if the number of electrons in the binding orbitals exceeds the number of electrons in the loosening ones. Electrons in non-binding molecular orbitals do not participate in the formation of a chemical bond. From the original atomic orbitals arises n MO. Thus, during the formation of a diatomic H_2 molecule from H atoms, from the s-orbitals of two H atoms two two-center MOs appear: one is more energy advantageous (binding), and the other is less favorable (loosening) than the original atomic orbitals. On the binding MO, the electron most of the time is between nuclei (electron density increases), promoting their chemical bonding. On the loosening MO, the electron is behind the nuclei most of the time, causing their repulsion.

To calculate the quantum mechanical characteristics of molecules by this method as well as self-consistent field approach, MOPAC software is used as one of the known and available software packages. HyperChem is used to prepare the MOPAC input file. It can significantly facilitate the process of creating an input file due to the possibility of the visual construction of the studied structures.

HyperChem program. The HyperChem software package was created and developed in 1995 by HyperCube Inc. and is currently one of the most well-known and widespread tools for molecular calculations and visualization of the results. The program has a user-friendly and well-thought-out interface and a large number of different ways to visualize the results. The program interface complies with the standards adopted for Windows applications.

The MOPAC software is based on the method of molecular orbitals. It was created by James Stewart (J. J. P. Stewart, US Air Force Academy) and allows one to calculate the quantum mechanical properties of molecules containing up to 100 heavy and up to 150 light atoms. The program uses the semi-empirical Hamiltonians MNDO, MINDO/3, AM1, or PM3 and PM6, developed by M. J. S. Dewar, to construct molecular orbitals and establishing the heat of formation of the molecule under study. When minimizing the heat of the molecule formation, its geometry is optimized and the energy of the lower state and some excited states, the spatial distribution of charges at the atoms, and the components of the dipole moment of the molecule are determined. In addition, depending on the problem, the program, on the user's inquiry, can optimize the geometry of the excited or ionized state of the molecule, consider the interconfiguration interaction, determine the order of bonds, calculate the contribution of σ- and π-electrons to the bond, and so on.

Using the HyperChem program, various structures from nickel and iron atoms were built with the addition of one hydrogen atom and without it (Figs. 7.2, 7.3, 7.4 and 7.5). Using the MOPAC program, the EE of these systems was calculated, and the effect of hydrogen on these two metals was compared. During the preparation of the input file for the MOPAC software, the atomic structure was geometrically optimized, considering the distances between atoms and their interaction with hydrogen in different geometric structures. A semi-empirical calculation approach modified for the calculation of transition metals such as iron and nickel was used (Fig. 7.7).

The figures below show the calculated structures of iron and nickel in the presence and absence of hydrogen atoms, as well as tables with the values of the total EE of pure metal and under interaction with one hydrogen atom. The maximum ratio of metal atoms to hydrogen atoms was 35.

Although the density of hydrogen adsorbed in iron is much lower on average, it can accumulate in some areas where its local concentration will be much higher than the average one in the metal, thus causing an embrittlement effect. Experimental confirmation of high hydrogen concentration has been reported in Antonov et al. (1998), Castellote et al. (2007). Such areas are characterized by changes in the elastic properties of metal, the development of microdefects is possible due to reduced lattice cohesion, decreased surface energy, high molecular hydrogen pressure in microvoids, and other mechanisms (Panasyuk et al. 1981; Kolachev 1985a). Local areas of metal

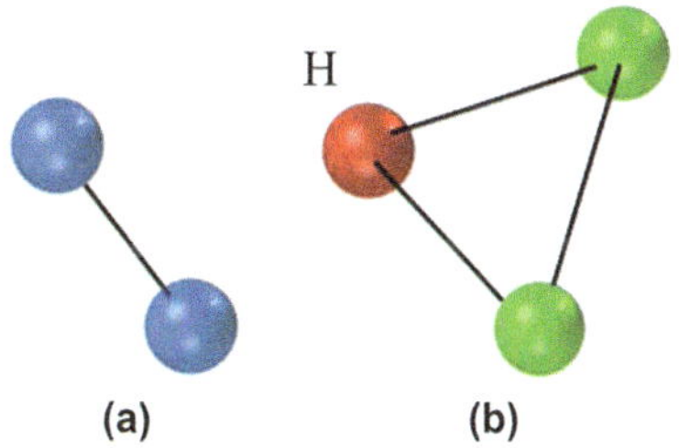

Fig. 7.2 Configuration of two metal atoms: without hydrogen (**a**) and with one hydrogen atom (**b**)

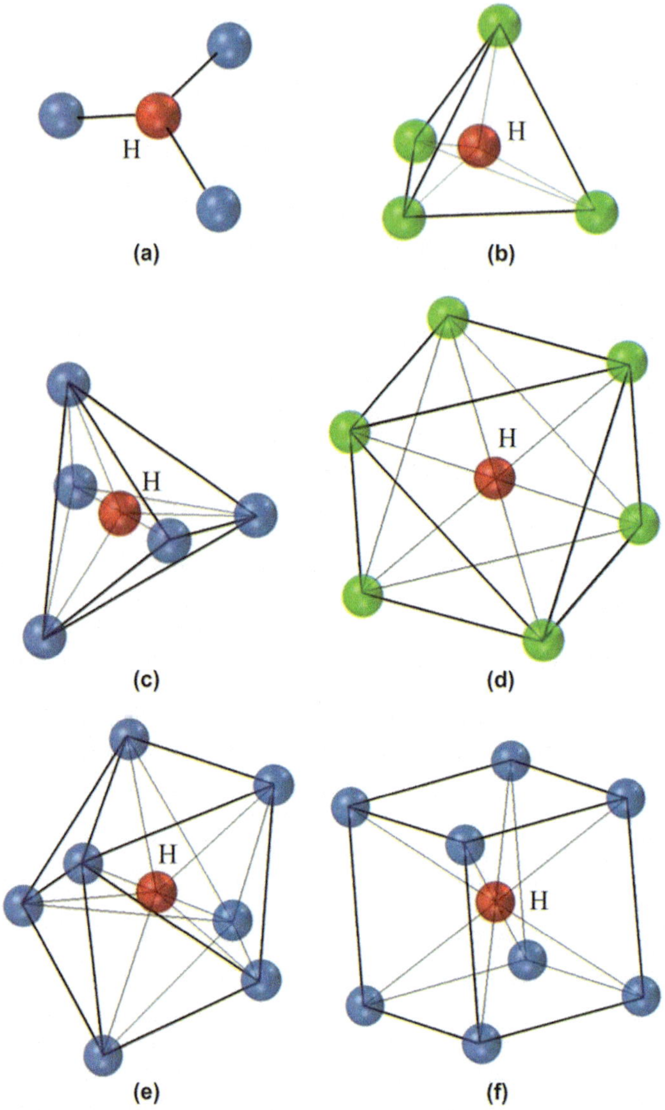

Fig. 7.3 Configuration of 3–8 metal atoms with one hydrogen atom

can be enriched with hydrogen in various ways, for example, by directed diffusion in the region of maximum triaxial tensile stresses or during hydrogen transfer by dislocations, when in the zone of dislocation accumulation, formed on obstacles of various types, local increase in hydrogen concentration becomes much higher than the average concentration in the metal volume. However, even in the case of a perfect lattice, the uniform distribution of hydrogen atoms in the volume is energy unfavorable.

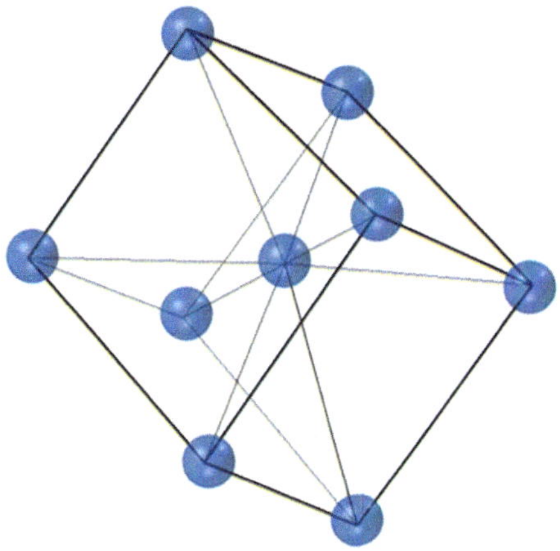

Fig. 7.4 Body-centered iron lattice

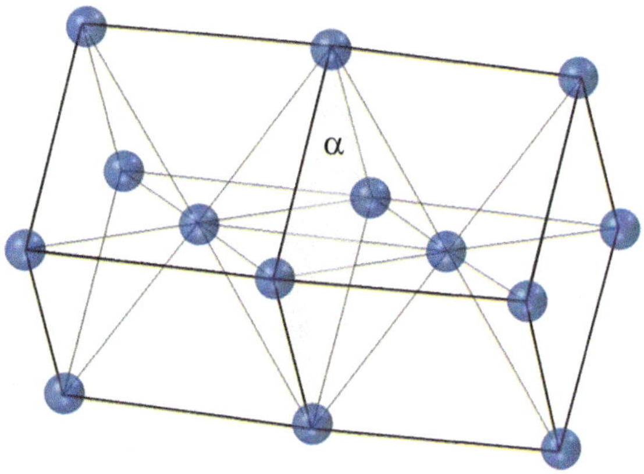

Fig. 7.5 Two body-centered metal cells: α is the plane in which the hydrogen atom is located

Therefore, they can join in clusters, thereby reducing the energy of the metal-hydrogen system. The possibility of the existence of hydrogen clusters is indicated by direct quantum mechanical calculations of the configurational part of the average energy of the ion-proton system interaction (Vavrukh and Slobodian 1985). However, due to the rather large potential barrier, the formation of clusters in metal is complicated. External tensile forces significantly reduce the potential barrier.

As it is known, hydrogen in the body-centered lattice of iron can occupy an octahedral and tetrahedral position (Fig. 7.6) (Sanchez et al. 2008). Therefore, the calculation of the EE system shown in Fig. 7.6 was carried out. The obtained results show that the values of EE for these two cases differed slightly: for the tetrahedral arrangement of hydrogen $E_{\text{ex}} = -27.076$ eV and for the octahedral arrangement $E_{\text{ex}} = -27.096$ eV.

The molecular orbital method was used to calculate EE in iron and nickel clusters in the presence and absence of hydrogen atoms. Calculations were performed in particular, for octa- and tetrapores of the body-centered lattice. These calculations

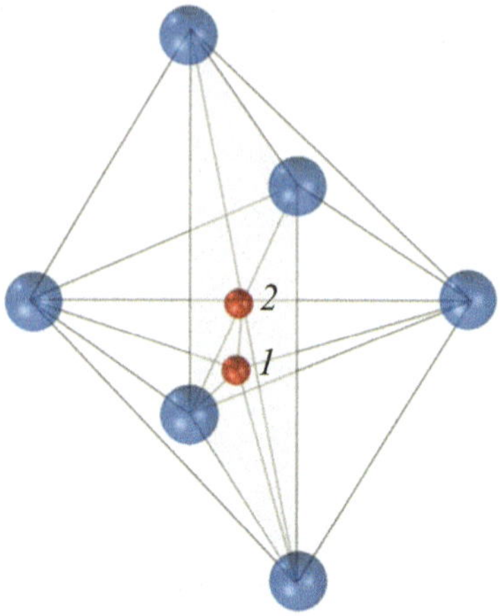

Fig. 7.6 Tetrahedral (*1*) and octahedral (*2*) location of the hydrogen atom on α plane (Fig. 7.5) between α-iron crystals

Fig. 7.7 Location of hydrogen in iron (35 Fe atoms)

show (Table 7.1) that the value of EE in the calculated Fe clusters is smaller in absolute value than for similar Ni systems. The effect of hydrogen is ambiguous. In cases when the ratio of the number of metal atoms to hydrogen is small (does not exceed 7), hydrogen increases the modulus of the exchange interaction energy. For larger values, the ratio of the number of metal atoms to hydrogen EE decreases slightly by modulus (for iron) or does not change.

In general, the EE changes depending on the hydrogen concentration, type of cluster, and the number of metal atoms in it is ambiguous, which requires further research.

Table 7.1 Values of total exchange energy in Me–H structures

Fig. no	Number of Me atoms	E_{ex} without hydrogen, eV		E_{ex} with hydrogen atoms, eV	
		Fe	Ni	Fe–H	Ni–H
5.1**b**	2	−5.24	−5.07	−6.5	−6.21
5.2**a**	3	−8.24	−7.47	−8.53	−9.1
5.2**b**	4	−9.13	−11.42	−9.27	−12.62
5.2**c**	5	−10.35	−14.54	−10.11	−18.9
5.2**d**	6	−10.25	−18.91	−11.38	−18.98
5.2**e**	7	−11.97	−24.77	−12.48	−21.47
5.2**f**	8	−17.45	−25.19	−13.87	−25.29
5.5	14	−31.09	−41.46	−27.08	−44.13
5.6	35	−88.27	−103.16	−65.16	−104.44

7.3 Evaluation of the Sensitivity of the MAE Method to the Plastic Deformation Level

Basic concepts of the finite element method. When designing structures, the engineer is faced with the task of finding the stress distribution or stress fields. Sometimes, to find out whether the specified gaps between the structural parts are violated, the engineer needs to calculate the displacement only at certain points in the system. In some cases, especially if the load and behavior of the structure depend on time, the designer must calculate the total distribution of displacements or displacement field. For the calculated stress field, the equilibrium conditions must be satisfied at each point, and the displacements under these conditions must be continuous (i.e., the compatibility conditions must be satisfied).

In some problems, to find stresses and displacements, it is necessary at first to set the defining equations, which in one form or another ensure the fulfillment of the conditions of equilibrium and compatibility. The main difficulty that arises, in this case, is to solve the problem: of whether these equations can adequately represent the requirements that are set when designing the structure. Moreover, it is necessary to take into account the complexity of the structure geometry, as well as the character of the loads and the properties of the material.

Considering the differences that arise due to the circumstances described above in the behavior of the structure and its model, it is possible to start solving the selected equations. Exact solutions of such equations are seldom available, and it is not often possible to construct adequate approximate solutions with a small number of approximation terms. A large number of these terms are required to obtain a sufficiently accurate solution.

The development of electronic computers has radically changed the situation in solving differential equations. Most practicing engineers now have access to a numerical study of the problem assigned to them, so the number of terms of the series representing the stress field or displacements, that are taken into account, can be large. Finite difference methods are also used in which differential equations are approximated by means of discrete values given at selected points. The advantage of these methods follows from the long history of their development, which resulted in the emergence of convergence theorems. In addition, the algebraic equations that arise in these methods, which need to be solved numerically, often have a particularly simple form.

Today, the most common numerical method for solving the problems of deformable solid mechanics is the finite element method (FEM) (Morozov and Nikishkov 1980).

The naturalness of the FEM mechanical interpretation not only allows us to build calculation schemes, proceeding from the physical formulation of the problem but also helps to identify gross errors at any stage of the computational process. Besides, the variational-difference character of the FEM provides a rigorous mathematical analysis of computational models as well as the construction of the reliable estimates of the accuracy and convergence of the method.

FEM is an analytical procedure that has been intensively developed over a relatively short period of time. The key idea of the method for analyzing the behavior of structures is as follows: continuum (structure as a whole) is modeled by splitting it into areas (finite elements), in each of which the behavior of the continuum is described by a separate set of selected functions, describing stress and displacement in the specified area. These sets of functions are often given in such a way as to satisfy the conditions of continuity in the whole continuum. In other cases, the chosen representations of the fields do not provide continuity, but give the opportunity to obtain a satisfactory solution and, unlike completely continuous models, there is no complete certainty that the solution is correct. If the behavior of the structure is described by a single differential equation, an approximate solution of this equation can be obtained both by the FEM and by the technique of splitting into series or finite difference schemes. If the structure as a whole is heterogeneous and consists of a large number of individual structural elements, the behavior of each of which is described by its differential equation, then in this case, as a rule, only the FEM can be applied directly.

Together with the mentioned alternative methods of numerical solution of applied problems of structural mechanics, it is necessary to build and solve a system of algebraic equations in the FEM. The special advantages of the method are the convenience of forming equations and the ability to represent completely irregular and complex structures and loading conditions.

As noted above, the FEM is developing rapidly. Since 1955, it has become the most promising direction of numerous investigations on the problems of mathematical physics: design calculations, heat transfer, fluid motion, propagation of electromagnetic waves, and so on. The popularity of the method and interest in it are explained by the above ability to represent the real aspects arising in applied design problems.

The distribution of the FEM practical applications is a consequence of the development of technology in the mid-fifties. The main prerequisite for the development of the method above is the ability to effectively build and solve high-order system of algebraic equations. The development of electronic computers in the mid-fifties allowed satisfying these requirements.

General issues concerning the use of FEM are described widely in the literature (Panasyuk et al. 1988; Zenkevich 1975; Segerlind 1979). It is assumed that the values of elastic displacements $\{u\}^{\upsilon}$ at any inner point of the υ-th element are unambiguously determined in terms of the components of the displacements of nodes $\{q\}^{\upsilon}$

$$\{u\}^{\upsilon} = [N]\{q\}^{\upsilon}, \tag{7.1}$$

where $[N]$ is a matrix of polynomial interpolating functions called as shape functions. Depending on the order of the shape function, finite elements are divided into elements with linear, quadratic, etc. approximation of displacements.

By introducing the matrix differential operator $[B]$, it is possible to obtain a relationship between the displacement vector at the nodal points and the values of the deformation components inside the element $\{\varepsilon\}^{\upsilon}$ in the form

$$\{\varepsilon\}^{\upsilon} = [B]\{q\}^{\upsilon}. \tag{7.2}$$

Similarly, through the matrix of elasticity $[D]$, stresses are determined

$$\{\sigma\}^{\upsilon} = [D](\{\varepsilon\}^{\upsilon} - \{\varepsilon_{\text{therm}}\}), \tag{7.3}$$

where $\{\varepsilon_{\text{therm}}\}$ is the vector of thermal or other initial deformations.

Considering the above mentioned, the local equilibrium equation for one element will be written as

$$[K]^{\upsilon}\{q\}^{\upsilon} = \{f\}, \tag{7.4}$$

where $[K] = \int_{V_{\upsilon}} [B]^T[D][B]dV$ is the rigidity matrix of the element (index "T" indicates the transposition operation); $\{f\}$ is the local vector of nodal load, which is the sum of real load and fictitious deformation $\{f\} = \{p\} + \{h\}$.

The account of the existing relationships between elements allows us to pass from the initial matrices and vectors for individual elements to global values that refer to the whole model. Thus, the solution of the problem is reduced to solving a system of linear algebraic equilibrium equations of a set of finite elements

$$[K]\{Q\} = \{F\}, \tag{7.5}$$

where global analogues of local values are indicated by capital letters [see formula (7.4)]. The result of the solution of system (7.5) is the determination of the values of the nodal components of elastic displacements. The components of stresses and strains can be obtained by formulas (7.2), (7.3).

The algorithm for constructing a system of the FEM equations in the case of a nonlinear problem for an elastoplastic body is similar to the above mentioned.

However, it is necessary to take into account the nonlinear relationship between the components of the tensors of mechanical stresses and strains. For this purpose use the equation of the material yield strength in the n-dimensional stress space, the general form of which is as follows:

$$F(\{\sigma\}, \kappa) = 0, \tag{7.6}$$

where κ is the strengthening parameter.

Based on the law of plastic yield (or associative law) (Zenkevich 1975), first proposed by Mises, we obtain relations that relate the increase in stresses with the increase in deformations

$$d\{\sigma\} = [D]^*_{ep} d\{\varepsilon\}. \tag{7.7}$$

Here

$$[D]^*_{ep} = [D] - [D]\left\{\frac{\partial F}{\partial\{\sigma\}}\right\}\left\{\frac{\partial F}{\partial\{\sigma\}}\right\}^T [D]\left[A + \left\{\frac{\partial F}{\partial\{\sigma\}}\right\}^T [D]\left\{\frac{\partial F}{\partial\{\sigma\}}\right\}\right]^{-1}$$

is elastoplastic matrix, which relates the increments of the components of the mechanical stresses tensor with the increments of the strain tensor components.

A detailed description of the presented theory of plasticity was first given in Yamada et al. (1968), Zienkiewicz et al. (1969).

Thus, using the yield theory, Eq. (7.5) is written in the form of increments, and the solution is obtained by sequential summation at each step of the calculation.

The equation of motion in the case of a dynamic problem of an elastic-ductile-plastic material using FEM takes the form

$$[M]\{\ddot{Q}\} + [C]\{\dot{Q}\} + [K]\{Q\} = \{F\}, \tag{7.8}$$

where $[M]$ is the global matrix of inertia (mass matrix); $[C]$ is damping matrix (Goodwin 2003); $\{F\}$ is the discretized function of the problem; and the point indicates the differentiation over time. Time integration, as a rule, is carried out according to the difference scheme, and the main difficulty is the need to avoid significant instability of the calculation process.

The FEM is widely used to solve the problems of fracture mechanics: calculation of fracture parameters for areas of different shapes; modeling the fracture processes or behavior of bodies with a fixed or growing crack in conditions for which adequate mathematical models have not yet been built; analysis of the strength of real structures in terms of brittle fracture resistance; etc.

At the present stage of computer technologies development, there are a number of effective software packages for the implementation of the FEM problems, such as ANSYS, FEMLAB, ABAQUS, and NASTRAN. In our research, we used the NASTRAN software package.

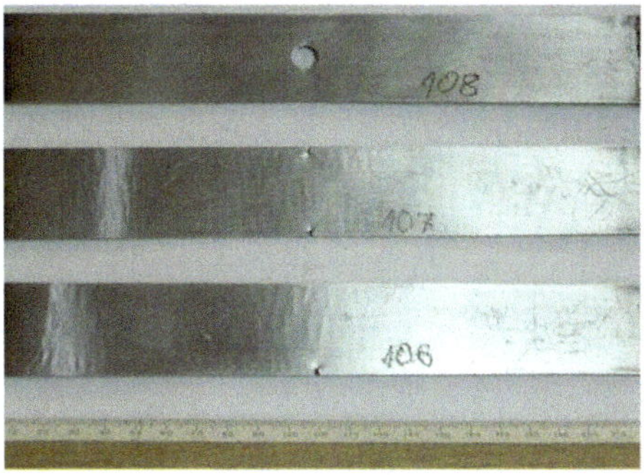

Fig. 7.8 Test specimens

Numerical calculation of the field of plastic deformations in the vicinity of the smooth stress concentrator tip. Plate specimens of 15 steel and 65G steel (see Appendix) with concentrators of two types: edge notches and a circular hole were investigated (Fig. 7.8).

The specimens dimensions are 240 × 30 × 2 mm, and the depth of the edge notch is 1.3 mm with a curvature radius of 0.45 mm.

To calculate the stress concentration at the holes and notches, the mechanical characteristics of the materials were experimentally determined. The obtained data were included in the FEM calculation program.

As mentioned above, to determine the fields of plastic deformations in the vicinity of the smooth stress notch tip of the specimens (Fig. 7.8) under their tensile loading, a numerical calculation was performed by the finite element method based on the NASTRAN software package. Tetragonal spatial finite elements with four nodal points were used to construct the finite element model. The specimen volume modeled on a personal computer was split by a finite element grid into 108,200 finite elements. Near the stress notches, the splitting of the grid was thickened 10 times (Fig. 7.9). The material was modeled as elastic–plastic on the basis of the experimentally constructed quasi-static tensile diagram, which was included in the calculation program. After introducing all the initial data into the calculation program, including the tensile load level and the method of its application, nonlinear static analysis of the elastoplastic field distribution of mechanical displacements, deformations and stresses in the working part of specimens with stress notches, was performed.

As a result of numerical calculations, the volume of the plastic deformation zone at the stress notches and the corresponding values of tensile load and the Mises stress maximum were determined (Fig. 7.10). The results obtained are shown in Table 7.2.

Figure 7.10 presents the experimental results of the dependence of the sum of the MAE signals amplitudes on the applied external magnetic field intensity for specimens of the investigated steels at different degrees of their elastoplastic deformation according to the calculations by the finite element method (see Table 7.2).

According to the obtained results, the sum of the MAE signal amplitudes in plastically deformed material increases with increasing level of applied external magnetic field and depends on the volume of the formed plastic zone, i.e., the larger

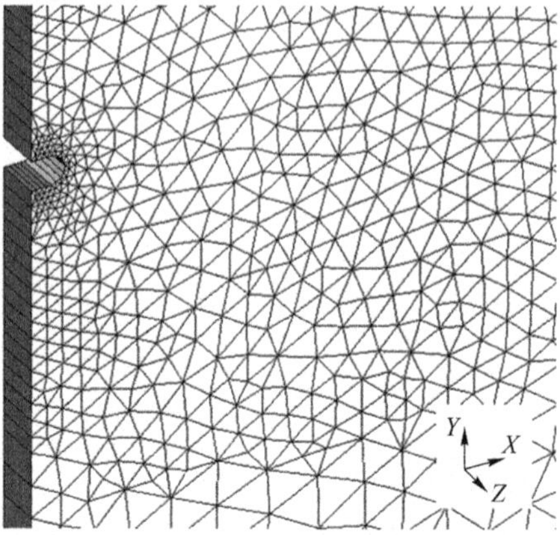

Fig. 7.9 Fragment of a finite element grid in the vicinity of a stress notch

the zone of plastic deformations, the higher the sum of amplitudes of the recorded MAE signals (Fig. 7.11).

Thus, as shown by the results of computational and experimental studies, the MAE method is sensitive to relatively small changes in the volume of the plastic zone in ferromagnetic materials. This fact allows us to involve the MAE phenomenon to create a new non-destructive test method for estimating the size plastic zones formed in the pre-fracture areas of structural steels and thus to diagnose bulk damage of structural elements made of such ferromagnetic materials.

7.4 Influence of Absorption of Electrochemical Hydrogen on the Young's Modulus of Structural Steel

Usually, the Young's modulus E at the macrolevel is determined from the slope of the rectilinear section of the tensile curve of materials according to the appropriate standard. However, it can be calculated using the acoustic approach as well, because the propagation velocity of elastic waves significantly depends on the elastic properties of the material (Kolsky 1964; Lynnworth 1975). This velocity is measured by various methods, which can be divided into two categories: resonant method using the passage time or by pulse reflection (Lynnworth 1975; Goodwin 2003). The method of estimating the wave passage time is based on the Newton–Laplace equation

$$c = (P/\rho)^{0.5},$$

where P is the bulk modulus of elasticity and ρ is the density of the material.

For thin rods, this equation is written as

$$c_E = (E/\rho)^{0.5}, \tag{7.9}$$

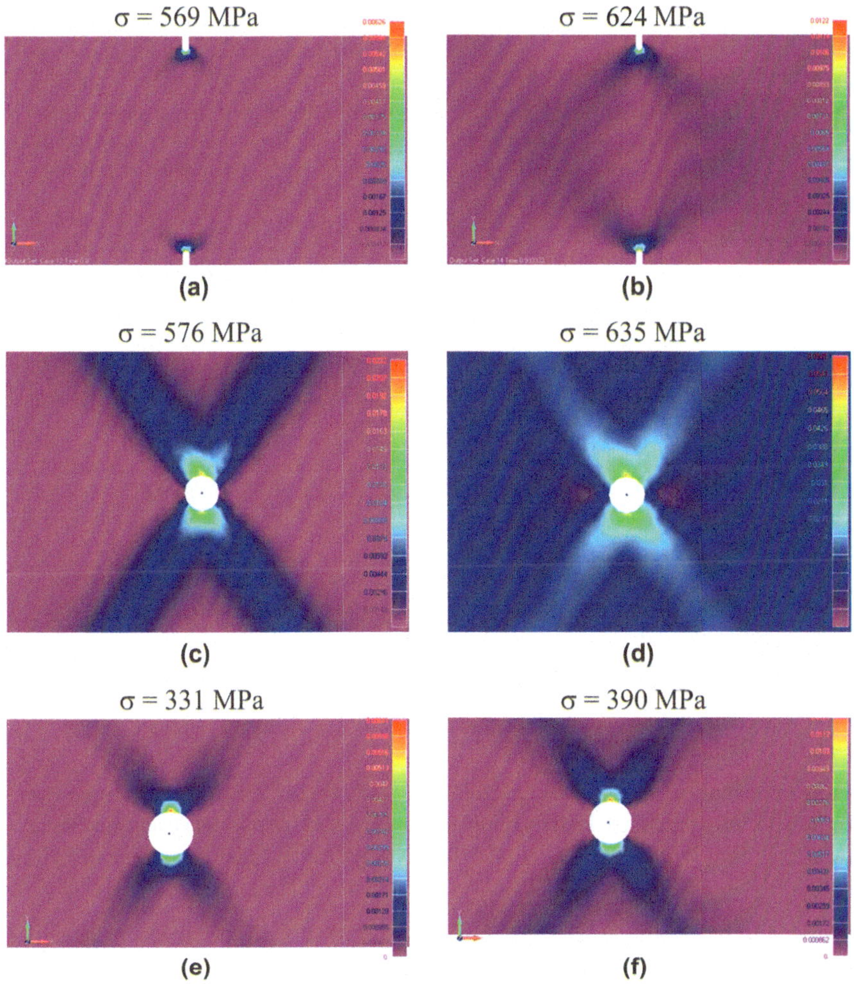

Fig. 7.10 Results of the FEM calculation of the field of elastoplastic deformations in 65G steel (**a**–**d**) and 15 steel specimens (**e**, **f**)

where c_E is the velocity of the longitudinal wave, provided that the diameter of the rod is significantly smaller than the wavelength. Note that this velocity is slightly less than the velocity of longitudinal waves in the volume of isotropic material.

Relation (7.9) is quite convenient for determining the elastic properties of materials subjected to various influences, such as thermal, magnetic, or chemical. Particularly it is interesting for studying the effects of hydrogen, which is known by its embrittlement action in iron-based structural materials even in mass concentrations of the order of 10^{-6} ppm (Goodwin and Trusler 2003; Tsuchida et al. 2010). Despite the extremely low solubility in α-Fe, even such a concentration of absorbed hydrogen (which is usually much lower than the concentration of many non-metallic inclusions:

Table 7.2 Results of the FEM calculations of the volumes of plastic zones formed at the stress notches in specimens of materials under study in their tension, as well as the corresponding maximum stresses according to Mises

Notch type	Maximum stress in stress notch (according to Mises) σ, МПа	Volume of plastic zone V, mm^3	% of plastic zone to sample volume/sample volume in solenoid
65G steel			
Hole ($d = 6$ mm)	506	–	–
	576	5.4	0.04/0.06
	599	14.32	0.1/0.15
	632	32.72	0.23/0.35
	671	117.92	0.82/1.27
Two notches	514	–	–
	569	10.72	0.7/0.11
	590	18.56	0.13/0.2
	624	40	0.28
	640	53.72	0.37/0.58
15 steel			
Hole ($d = 4$ mm)	331	79.4	0.55/0.85
	374	107.92	0.75/1.16
	390	1440	100

sulfur, phosphorus, arsenic, etc.), combined with thermomechanical fatigue over a certain period of time can lead to aging and unexpected fracture of structural elements (Maier et al. 1987; Watanabe et al. 2012; Nykyforchyn 2002; Kryzhanivs'kyi and Nykyforchyn 2011).

Problem state-of-the-art. The bibliography related to the study of the hydrogen effect on the modulus of elasticity of iron-based materials is quite short (Gabetta et al. 2008; Lunarska et al. 1977; Carpenter and Fawks 1981; Zhang et al. 1985, 1986; Armstrong and Carpenter 1988; Ortiz and Ovejero-Garcia 1992; Enos and Scully 2002; Ricker and Pitchure 2009). Most studies have been published on high levels of hydrogen charging, which is accompanied by irreversible damage to the microstructure of the material and is represented by the formation of numerous internal pores and micro-cracks, especially near the surface. Such studies should take into account a large number of effects: hydrogen-induced cracking in the form of surface and internal cracks and cavities, hydrostatic pressure of molecular hydrogen H_2 in internal cavities; complete redistribution of the deformation field due to partial damage to the material and the action of hydrogen atom dissolved in the crystal structure or trapped by defects in the material, for example, at intergranular or interfacial boundaries or at any 0D, 1D, 2D, or 3D defects of the crystal matrix. For such a complex case, it

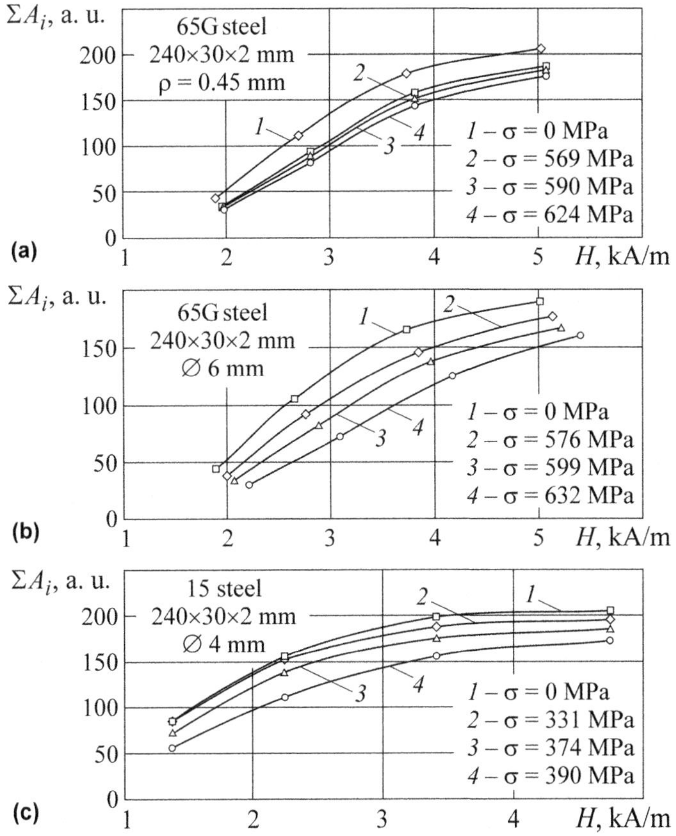

Fig. 7.11 Graphs of the dependence of the sum of the MAE signal amplitudes on the magnetic field intensity for specimens with edge notches (**a**) and circular holes (**b, c**) under different elastoplastic deformation

is very difficult to determine the component of the change in the mechanical properties of the material, which would concern exclusively the influence of hydrogen, especially in its various manifestations.

For example, in Lunarska et al. (1977) the properties of armco-iron of two different degrees of purity under strong cathodic hydrogen charging were studied by measuring the change in the resonant frequency of a standing longitudinal elastic wave in a rod. The gradual decrease in frequency (and hence the calculated value of the Young's modulus) under cathodic hydrogen charging was mainly caused by formation of a significant number of voids and internal microcracks. Similar investigations were described in Carpenter and Fawks (1981), Zhang et al. (1985), which studied the effect of hydrogen on the modulus of elasticity under conditions of even more intense hydrogen charging by measuring the change in resonant frequency during the bending of thin rectangular specimens made of iron of both high and commercial purity. The authors concluded that the changes in the Young's modulus recorded during the experiments were the result of structural changes and stress relaxation, rather than the effect of hydrogen in a solid metal solution.

These conclusions are cautious because they evaluate the effect of hydrogen on mechanical properties by comparing the hydrogen-undamaged material with that in which the microstructure is irreversibly damaged. In order to get an idea of the mechanisms of hydrogen-induced processes, they should be studied under reversible hydrogen charging conditions, i.e., when hydrogen is absorbed/desorbed as an interstitial impurity in a solid solution before irreversible damage to the microstructure occurs. However, such studies encounter a serious problem of the extremely low solubility of hydrogen in the α-Fe crystal lattice, which is of the order of 10^{-7}, i.e., 10^{-3} ppm. Besides, since industrial materials contain very high concentrations of impurities (compared to the concentration of hydrogen) and a high concentration of structural defects, the impact of hydrogen in terms of H–Fe interaction can be claimed only with a high degree of uncertainty. That is why any attempt to study the H–Fe system, in which the action of other elements is minimized, is of great importance. This concerns in particular the theory of hydrogen-enhanced local plasticity (HELP) proposed by Byrnbaum (Birnbaum and Sofronis 1994) and the theory of hydrogen strength reduction outlined by Oriani and Josephic 1977. If hydrogen has an influence on the reduction of the interatomic forces, changes should be expected in the elastic constants, such as the Young's modulus. When hydrogen acts only to facilitate the dynamics of dislocations, according to HELP, anelastic behavior should be detected (Enos and Scully 2002; Ricker and Pitchure 2009; Birnbaum and Sofronis 1994; Oriani and Josephic 1977).

Among the studies of the effect of hydrogen on the modulus of elasticity, only a few sources can be noted. In particular, using the internal friction technique and measuring the frequencies of free oscillations, Lunarska and co-authors (Zhang et al. 1986) noticed changes in the shear modulus of ultrapure polycrystalline α-Fe (containing only 5 ppm non-metallic and 23 ppm metallic inclusions) under the influence of about 3 ppm absorbed hydrogen. It can be concluded that the decrease in the shear modulus G of the investigated material is about 0.05% per 1 ppm of absorbed hydrogen, although its concentration in iron specimens is not accurately determined.

Recent papers in this field (Enos and Scully 2002) have provided a continuation of this discussion. The effect of hydrogen on the elastic and inelastic reactions of iron, in which the total concentration of impurities exceeded 1000 ppm, was investigated using a dynamic mechanical analyzer, and it was found that both effects were present. Absorbed diffusion-mobile hydrogen (although it is unknown how low its concentration is) caused a slight decrease in the modulus of elasticity (0.5–1%, as estimated from the stress–strain curve), and also caused increased inelastic relaxation, which could reflect, for example, the movement of dislocations (although other relaxation mechanisms were also possible). In this case, no structural damage to the material occurred.

Apart from the above-mentioned investigations on pure iron specimens, no other sources have been found that would establish a relationship between absorbed hydrogen and changes in the elastic properties of iron. Although the mentioned studies show a slight decrease in the modulus of elasticity of iron under the action of absorbed hydrogen, it should be noted that they use only the resonance method, which is sensitive not only to the physical properties of the material but also represents the influence of the environment around the specimens and the state of its surface, especially when the environment is liquid. Therefore, it is necessary to check the effect of hydrogen on the Young's modulus using another method that is insensitive to the environment, namely by estimating the time of the wave passage (echo pulse) under simultaneous (or almost simultaneous) cathodic hydrogen charging of the metal.

Therefore, the effect of electrolytic hydrogen charging on the change in the modulus of elasticity of low-carbon steel was evaluated by the method of measuring the time of the elastic wave passage in a metal rod.

Experimental technique. Cylindrical specimens 2120 mm long and 12 mm in diameter were made of cold-rolled low-carbon 15 steel (see Appendix). Such specimen was placed in the chamber of electrochemical cell—a plastic tube with a diameter of 50 mm (Fig. 7.12). It was filled with a 0.1 N NaOH solution, which protected the metal surface against excessive corrosion. No catalysts to intensify the process of hydrogen absorption into steel were used in the investigations. Two steel rods with a diameter of 8 mm served as anodes. The anodes and the specimen (cathode) were connected to a stabilized DC source to obtain atomic hydrogen on the metal surface. Hydrogen charging took place in two stages, with measuring the time of the wave passage before and after each of them.

During the experiment, the amount of charge transferred through the specimen surface was 14 kK in the first stage and 3 kK in the second stage, which took place immediately after the first one. Based on the assumptions:

1. that all the electric charge was spent exclusively for hydrogen reduction from the electrolyte according to the reaction

$$H^+ + e^- \rightarrow H$$

2. that about 0.1–1% of reduced hydrogen atoms were absorbed by the metal, while the other 99–99.9% were converted into molecular form

$$2H \rightarrow H_2$$

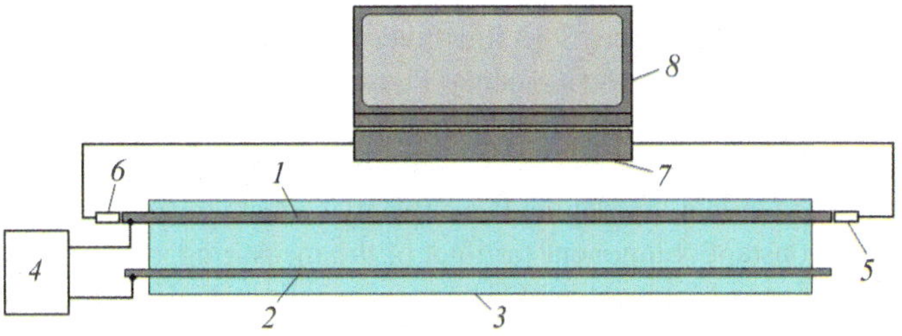

Fig. 7.12 Block diagram of the experiment: *1* is the specimen cathode; *2* is the anode; *3* is the electrochemical cell tube; *4* is DC source; *5* is the source of excitation of elastic waves; *6* is the receiver of elastic waves; *7* is the acoustic emission measuring system SKOP-8; and *8* is computer

and were removed from the metal surface in the form of H_2 gas bubbles, an average concentration in the range of 0.1–1 ppm of hydrogen in the metal specimen was obtained. This value is commensurable with the concentrations proposed by Hirth and Carnahan—0.2 ppm during electrolytic hydrogen charging (Hirth and Carnahan 1978). Obviously, this concentration does not include metallurgical hydrogen absorbed by the steel during production and storage, i.e., that was present in the specimen before testing. It is considered that this "proper" hydrogen is immovable, located in traps such as dislocations, grain boundaries, etc. and is not released during sorption–desorption. According to this simplified model, electrolytic hydrogen is considered to be diffusion-movable and irreversly absorbed, and the chemical potentials of hydrogen on the surface of the metal and in its core determine the driving force of the sorption process.

The electrolyte temperature at the measuring point ($\sim$ 500 mm from the specimen edge) according to the readings of the electronic digital thermometer varied in the range of 27.1–28.1 °C: before hydrogen charging it was 27.6 °C; after the first stage of hydrogen charging 27.2 °C; and after the second 27.1 °C. The test chamber was protected against possible heating effects by the current passing through the electrolyte by gradual renewal of the electrolyte. Note that the increase in temperature under adiabatic conditions, according to our estimates, should be about 2 °C, if the electrolyte would not be renewed.

Piezoceramic transducers were attached to both ends of the rod under study, one of which served as a source of elastic waves and the other as a receiver. The first transducer was connected to a rectangular pulse generator and the receiver to the acoustic emission measuring system. All of these units were placed in a single equipment and synchronized with a single timer. Each response from the action of the elastic pulse was digitized at a frequency of 4 MHz. It contained 8200 reference points, covering 2.05 ms of signal duration. In our case, we managed to record not only the first arrival of the wave through the entire length of the rod but also its second arrival after passing the length of the rod three times. The recorded acoustic signal contained well-defined oscillations of longitudinal waves, which allowed us to calculate the sound propagation velocity. The diameter of the rod (12 mm) was much smaller than the wavelength of the carrier frequency (40–100 mm), providing conditions for direct determination of the change in the isothermal Young's modulus based on the time of longitudinal waves passage.

Investigation results. Figure 7.13 illustrates a typical AE signal recorded by the measuring AE system SKOP-8 (Karpenko Physico-Mechanical Institute of the NAS of Ukraine).

Twenty such signals were recorded at intervals of about two seconds, and they turned out to be identical. Averaging these twenty acoustograms and digital filtering to eliminate a constant component (artifact of the measuring system) allowed us to remove most noise and false echoes. The wave velocity, calculated by the delay time between the first and second arrival of the longitudinal wave to the receiver in the time intervals 410–420 μs and 1240–1250 μs, is $c_E = 5087$ m/s, which corresponds to the modulus of elasticity $E = 203.7\,\text{GPa}$, if the density is $\rho = 7870\,\text{kg/m}^3$. This

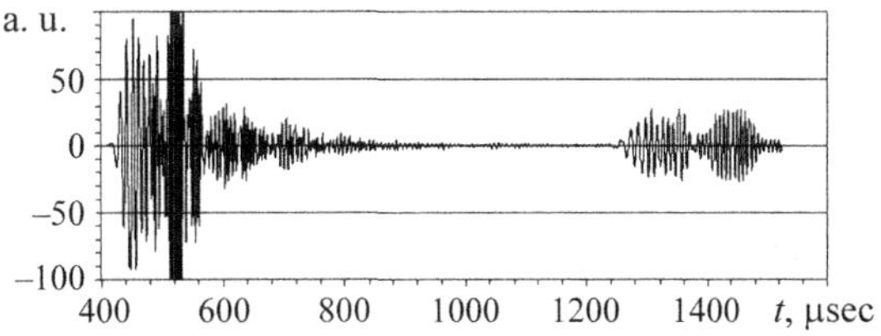

Fig. 7.13 Typical average acoustic signal recorded by the acoustic emission system and obtained by analysis of 20 signals after their digital filtering

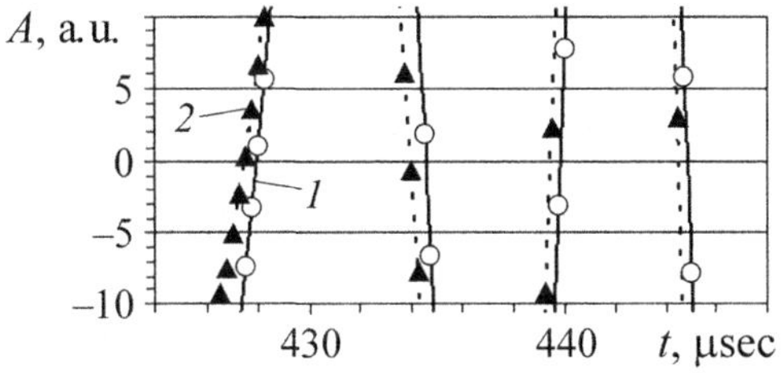

Fig. 7.14 Fragments of superimposed signals recorded before (*1*) and after (*2*) hydrogen charging of the metal and presented in the time window of the longitudinal wave first arrival

value of the Young's modulus agrees with the known reference data for iron and low-carbon steel.

The effect of absorbed hydrogen on the time of the first arrival of the longitudinal wave to the receiver is shown in Fig. 7.14 with tenfold scaling along the *y*-axis in a narrow period of time. It can be seen that the sound velocity has increased, contrary to the data obtained from the literary sources. To evaluate these changes, we linearly interpolate the obtained results to receive the moments of time at which the recorded signal changes its polarity (the phase velocity is compared), starting from the first half-wave at moment $\sim 419\ \mu s$ to the fifth half-wave at moment $\sim 445\ \mu s$.

Similar calculations were performed for the second arrival of the longitudinal wave. Averaging the differential times, i.e., the shift of the signal curve during its intersection with the *x*-axis, for both stages of hydrogen charging an increase in the velocity of 3.3–4.0 m/s after the first stage and 3.9–4.6 m/s after two stages of hydrogen charging was obtained. These figures obviously correspond to an increase in the Young's modulus by 0.13–0.16% after 14 kK of charge used and by 0.15–0.18% for 17 kK, considering that the phase velocity of an isothermal longitudinal wave must be directly proportional to $E^{0.5}$, and the change in the modulus of elasticity ΔE is determined by the dependence

$$\Delta E \approx 2E\Delta c/c.$$

Note that the values of the differential times are significantly scattered. For example, the time shifts for the phase of the *x*-axis intersection (zero amplitude) after 14 kK of hydrogen charging change in the sequence of 0.17–0.46–0.52–0.29–0.21 μs (all negative). A similar trend is observed after the second stage (17 kK). This suggests that there is a systematic distortion of the acoustic signal shape due to the presence of hydrogen in the metal. The signal from the second arrival of the wave, unfortunately, is not as smooth and clear, both due to its attenuation and the interference of the elastic oscillations reflected from the faces.

Discussion of results. There are several circumstances that should be taken into account when studying the change in the elastic properties of the metal: (a) change in the specimen length; (b) unequal density of the material; (c) variation in the physical properties of the material.

Let us briefly analyze the possible influence of some physical factors.

Linear expansion due to hydrogen absorption. Assuming a uniform distribution of 0.1–1 ppm of hydrogen in the rod, its extension can be calculated according to

$$\Delta L/L_0 = 1/3C(\mathrm{H})[V_m(\mathrm{H})/V_m(\mathrm{Fe})],$$

where ΔL is the absolute extension of the rod; L_0 is its initial length; $C(\mathrm{H})$ is hydrogen concentration in the metal; and $V_m(\mathrm{H})$ and $V_m(\mathrm{Fe})$ are the partial molar volumes of hydrogen in iron and of iron, respectively.

Taking $V_m(\mathrm{H}) = 2$ cm^3/mol, we obtain $\Delta L = 1 - 11$ μm, which corresponds to a 0.5–5 ppm increase in the rod length. Increase of the specimen length is not caused by the reduced time of the wave arrival, but rather its growth. Therefore, this factor is not decisive, but competing, and even very insignificant.

Influence of temperature variation. Temperature is important factor that affects the sound velocity in material. In fact, temperature causes a change in both the material density and its elastic constants. For the body-centered cubic α-Fe and most structural steels, the temperature coefficient of sound velocity is about 0.6 m s^{-1} K^{-1}. For example, for 1008 steel and 1020 steel this coefficient is equal to 0.68 m s^{-1} K^{-1}, and for 1035 steel 0.63 m s^{-1} K^{-1} (Skal's'kyi et al. 2009). Since the liquid in the electrochemical cell was constantly renewed during the experiment, a slight decrease in temperature of about 0.5 °C, was recorded. This would correspond to an increase in the sound velocity by $\sim$ 0.3 m/s, which would be 7–8% of the total velocity increase. To get the velocity change caused exclusively by temperature, it would have to be reduced by 6–7 °C during the experiment, which we consider impossible. Thus, the detected positive changes in the sound velocity cannot be attributed to the influence of temperature. We must acknowledge, however, that the stability of this parameter is important, and this fact needs special attention.

Density change. Assuming 0.1–1 ppm to be the concentration of dissolved hydrogen and 2 cm^3 to be its partial molar volume in iron, we obtain 2–16 ppm reduction of metal density due to hydrogen absorption. Since $E = \rho c^2$, the decrease in density causes a decrease in the modulus of elasticity, not its increase, although in any case these changes are insignificant.

Redistribution of stresses in the rod cross-section. Due to the significant partial volume of hydrogen in iron, the "swelling" of the outer layers of the metal should occur, where the chemical potential of hydrogen is high compared to the rod core, if time is insufficient to achieve equilibrium, corresponding to our research. This "swelling" of the volume is restrained by the rod core, which is subjected to tensile stresses, while the surface layer is compressed. It is likely that even before hydrogen charging, the subsurface layer is already in a compressed state due to the rod cold rolling during its manufacture. This distribution of residual stresses accelerates hydrogen diffusion into the material (the Gorsky effect) with increasing filling of the rod core with hydrogen after overcoming the compressed outer layer of metal.

The effect of redistribution of deformations in the rod can be one of the most important factors influencing the change in the modulus of elasticity under the action of absorbed hydrogen. If this effect exists, it will represent the dependence of the elastic waves propagation rate on stress. In a simplified form, in relation to the rod, it can be presented in the form (Kruger and Damm 2006)

$$c = c_0(1 - \beta\sigma),$$

where c_0 is the initial velocity in the absence of stresses and $\beta = g_{222}/f_{22}$ (g_{222} and f_{22} are nonlinear and linear elastic constants, respectively).

Experimental studies show that for industrial M16 bolts made of carbon steel, the change in the sound velocity is about 0.1% per 100 MPa of tensile stresses, and this relation is linear, at least up to 450 MPa. Extrapolating this dependence to the change in the sound velocity of 0.08–0.09% (as in our experiments), we obtain compressive stresses of the order of 80–90 MPa (Jhang et al. 2006).

Similar values are obtained from previous studies, which show an increase in velocity by 0.1% for every 20 ksi of compressive stresses in low-carbon 1018 steel (Lynnworth 1975; Skalskyi et al. 2011). By converting these data into a metric system, we obtain values of the corresponding compressive stresses of about 105–125 MPa for the longitudinal wave in the steel rod ($\sim$ 5100 m/s).

Therefore, the change in the field of internal stresses under the influence of absorbed hydrogen is one of the possible factors that cause the change in the velocity of the elastic wave and the modulus of elasticity of the material.

Magnetoelastic effect. Continuing the discussion on the possible influence of the stress field in the material due to its hydrogen charging, we should note that α-iron, which makes up most of the studied material, is a ferromagnet with magnetoelastic properties (Hsu 1974; Nazarchuk et al. 2009). This means that the stress (external or internal) caused by the absorbed hydrogen has an influence on the magnetization vectors of the magnetic domains contained in separate grains of the polycrystalline material. Although this effect is probably small, there is still a noticeable deviation from the perfect cubic crystal structure with certain orientation dependences of the moduli of elasticity, including the Young's modulus. The calculated value of the Young's modulus of the ferromagnetic material E_m can be obtained by formula (Morales et al. 2010)

$$E^{-1} = E_0^{-1} + 0.4\lambda_s/\sigma_i,$$

where E_0 is the Young's modulus of the demagnetized material, λ_s is the saturation magnetostriction, and σ_i is the internal stress.

Taking into account the obtained results concerning the change of the Young's modulus, as well as taking λ_s of the order of 10^{-6}, we obtain $\sigma_i \approx 10^2$ MPa, i.e., the quite possible order of magnitude, although this parameter requires further study and more accurate estimation. It is also not excluded that the distortion of the first oscillating pulse shape of the longitudinal wave is caused by the stress redistribution due to hydrogen absorption by the metal.

Considering the obtained results, it can be stated that the effect of absorption of electrochemical hydrogen on the Young's modulus of a low-carbon steel rod with a diameter of 12 mm can be recorded by estimating the time of the elastic wave passage into material. It is established that after specimen hydrogen charging, an increase in the elastic wave velocity is observed, which is caused by an increase in the modulus of elasticity of material. As the analysis of the obtained results shows, this factor is the most significant in this effect.

7.5 Method of Estimating the Concentration of Hydrogen Absorbed from the Hydrogen Gaseous Phase

After remagnetization of the ferromagnetic specimens with a frequency of 9 Hz with recording MAE, they were hydrogenated from the gas phase, theoretically calculating at first the concentration and distribution of hydrogen in the plate specimens.

The field of hydrogen concentration $C_{\mathrm{H}}(x, y, t)$ was obtained from the Fick's laws. The hydrogen diffusion equation is expressed by the dependence

$$\frac{\partial C_{\mathrm{H}}}{\partial t} = D_{\mathrm{H}}(T(t))\left[\frac{\partial C_{\mathrm{H}}^{2}}{\partial x^{2}} + \frac{\partial C_{\mathrm{H}}^{2}}{\partial y^{2}} + \frac{\partial C_{\mathrm{H}}^{2}}{\partial z^{2}}\right], \tag{7.10}$$

where $D_{\mathrm{H}}(T)$ is the diffusion coefficient of hydrogen in material and T is the absolute temperature of the specimen.

Dependence $D_{\mathrm{H}}(T)$ was presented in the Arrhenius form with the corresponding coefficients taken from the literature for α-iron (Kolachev 1985b)

$$D_{\mathrm{H}}(T) = 1.03 \times 10^{-7} \cdot \exp\left[-\frac{11{,}300}{RT}\right] \mathrm{m^2/s},\ (298\,\mathrm{K} < T < 1873\,\mathrm{K}); \tag{7.11}$$

for nickel (Ovchinnikov and Rassada 1990)

$$D_{\mathrm{H}}(T) = 7.85 \times 10^{-7} \cdot \exp\left[-\frac{40{,}800}{RT}\right] \mathrm{m^2/s}\ (400\,\mathrm{K} < T < 1600\,\mathrm{K}), \tag{7.12}$$

where R is a universal gas constant.

The boundary condition was set on the basis of the Sievert's law

$$C_{\mathrm{H}}^{s} = K_{\mathrm{H}}(T(t))\sqrt{p_{\mathrm{H}_2}}, \tag{7.13}$$

where C_{H}^{s} is hydrogen concentration on the specimen surface, K_{H} is the solubility coefficient of hydrogen on the plate surface, and p_{H_2} is the partial pressure of hydrogen in the environment. The initial condition is assumed to be zero $C_{\mathrm{H}}(x, y, 0) = 0$.

Temperature dependence of the solubility coefficient of hydrogen on the plate surface on temperature $K_{\mathrm{H}}(T)$ is taken from experimental data described in the literature and approximated by linear dependences for α-iron (Lomayev 1977)

$$K_{\mathrm{H}}(T) = 7.055 \times 10^{-3} \cdot T - 3.382\,(\mathrm{cm}^3/100\,\mathrm{g}),\quad (570\,\mathrm{K} < T < 770\,\mathrm{K}); \quad (7.14)$$

for nickel (Tkachev et al. 1999).

$$K_{\mathrm{H}}(T) = 9.170 \times 10^{-3} \cdot T - 2.967\,(\mathrm{cm}^3/100\,\mathrm{g}),\quad (587\,\mathrm{K} < T < 793\,\mathrm{K}). \quad (7.15)$$

To solve this problem, the numerical finite elements method was used in the form of FEMLAB 3 software package. The 15 steel (see Appendix) and nickel plates were splitted as follows: steel—3567 elements; nickel—1234 elements. Based on the performed calculations, the distribution of hydrogen concentration in the plates was obtained after 2 h hydrogen charging (Fig. 7.15).

The analysis of the obtained calculations demonstrates that the maximum and minimum values of hydrogen concentration over the plate volume differ slightly, so when calculating thin plates it is enough to find the hydrogen concentration on their surfaces.

Figure 7.16 presents graphs of changes in time of maximum and minimum hydrogen concentration in the plate volume.

After calculations, the specimens were hydrogenated in a specially designed chamber in a gaseous hydrogen environment at the temperature and pressure in the chamber shown in Fig. 7.17. Based on theoretical data, the hydrogen charging regimes were chosen such to bring the concentration of hydrogen in metal as close as possible to the real one in ferromagnets (0.5–5.0 ppm), which have a long-term service life.

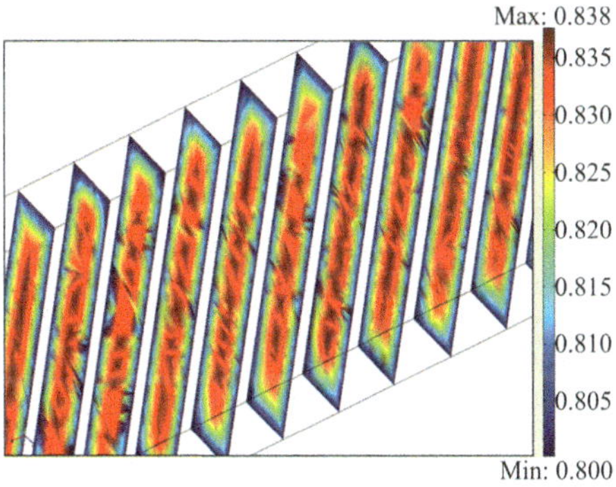

Fig. 7.15 Distribution of hydrogen concentration in a plate of 15 steel (thickness 4 mm) after two hours of hydrogen charging

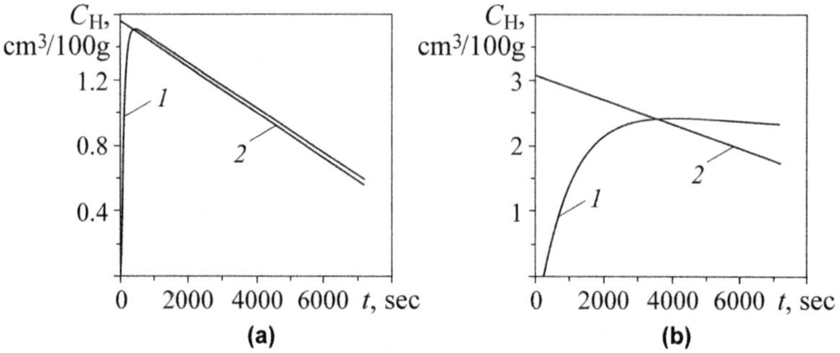

Fig. 7.16 Time change of maximum (curve *1*) and minimum (curve *2*) hydrogen concentration in the volume of plates of 15 steel (**a**) and nickel (**b**)

Fig. 7.17 Heating and hydrogen charging modes of specimens: *1* is time dependence of temperature change; *2* is time dependence of pressure change

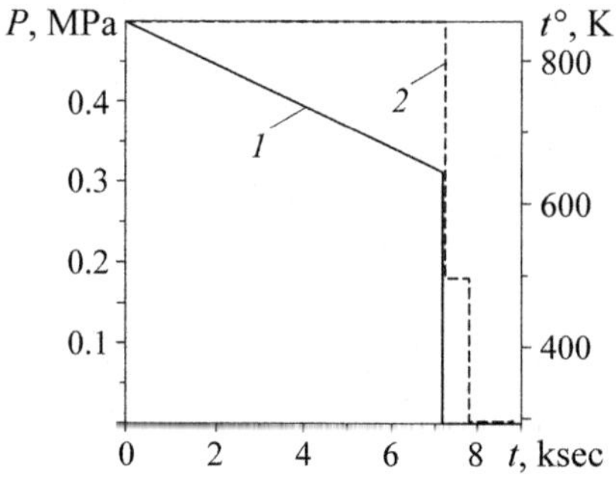

After hydrogen charging, the specimens were retested in the solenoid. Figure 7.18 presents the obtained results for specimens of 15 steel.

Similar experiments were performed on nickel specimens. Figure 7.19 presents the dependence of the total signal amplitudes on the change of the magnetic field intensity for a non-hydrogen-charged and hydrogen-charged nickel plate at a frequency of remagnetization of 9 Hz.

The results show that the amplitudes of the MAE signals increase with increasing external magnetic field intensity H in both non-hydrogen-charged and hydrogen-charged specimens. However, in the second case, they are significantly larger for

Fig. 7.18 Dependence of the sum of signal amplitudes on the change of the magnetic field intensity for non-hydrogen-charged (*1*) and hydrogen-charged (*2*, *3*) steel plates at a frequency of remagnetization of 9 Hz

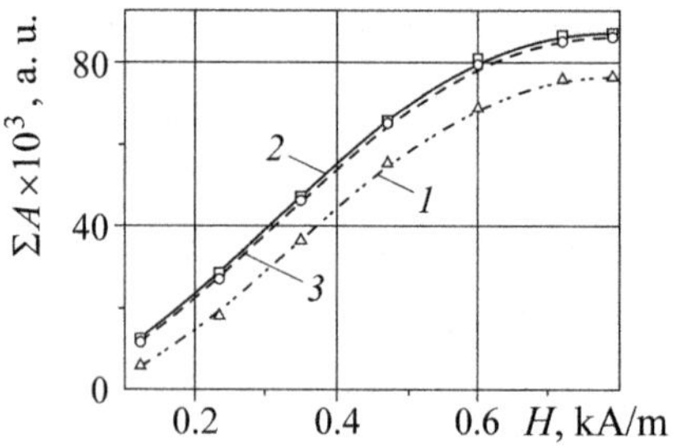

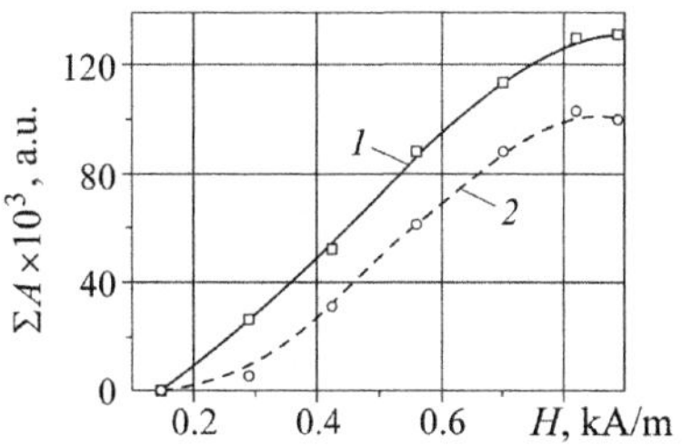

Fig. 7.19 Dependence of the sum of signal amplitudes on the change of the magnetic field intensity for non-hydrogen-charged (*1*) and hydrogen-charged (*2*) nickel plates at a frequency of remagnetization of 9 Hz

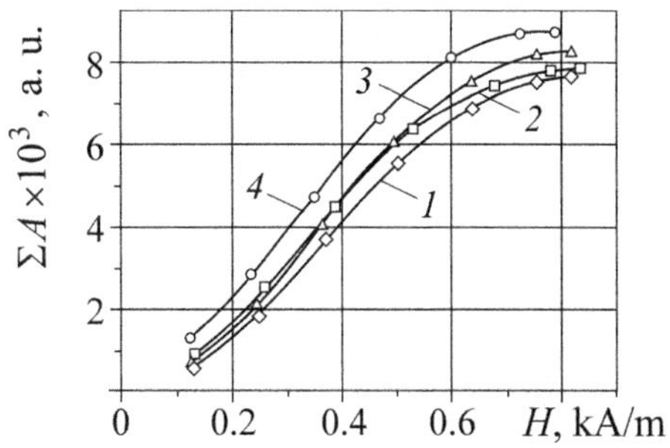

Fig. 7.20 Dependence of the sum of amplitudes on the magnetic field intensity for non-hydrogen-charged—*1*, hydrogen-charged, saturated with argon *2* and helium *3* steel plates with a thickness of 2 mm at a frequency of remagnetization of 9 Hz

steel specimens (Fig. 7.18) and smaller for nickel specimens (Fig. 7.19), which is due to the impact of hydrogen on the material domain structure.

In order to verify whether other gases affect the domain structure of the material, similar study was performed on the specimens saturated with argon and helium. Figure 7.20 shows the effect of different gases on the amplitudes of MAE signals from the magnetic field intensity for a steel plate.

Thus, under the influence of hydrogen there is a change in the exchange energy of the magnetic interaction of the ferromagnet. This causes the change in the value of the domain walls jumps under the action of an external magnetic field, which creates magnetostriction in the ferromagnet, accompanied by the MAE emission.

Based on the obtained experimental data, it is found that the amplitudes of MAE signals increase with increasing external magnetic field intensity H in both non-hydrogen-charged and hydrogen-charged specimens. As the amount of hydrogen increases, the amplitude of the MAE signals grows. The BJ are best observed in the low frequency range of the ferromagnet remagnetization.

7.6 Modeling of Domain Wall Movement

A detailed study of the magnetization curve in the area of the domain walls displacement demonstrates that with a gradual change in the intensity of the external remagnetization field, the change in the ferromagnet magnetization occurs jump-like. As shown above, such jump-like changes in magnetization are called BJ or remagnetization jumps (Fig. 7.21) (Skal's'kyi et al. 2016). The equilibrium value of the magnetic field H corresponds to the coordinate x_O. If it is increased, the reverse displacement of the wall will occur until it, leaving its equilibrium state, reaches a point with a coordinate x_A. It has a maximum of a certain internal pressure, the role of which is played by the surface energy density gradient of the domain wall $\partial y/\partial x$, and the external field reaches a critical value $H = H_{\text{cr}}$.

Further movement of the domain wall can occur without increasing the external magnetic field to a point with a coordinate x_B, because for a range of values $x_A < x < x_B$ the internal pressure is less than the external pressure. As a result, the domain wall moves by a jump from point x_A to point x_B, and the magnetization changes jump-like by a certain value. This process is irreversible, because with a further magnetic field decrease, the wall does not return back from the position x_B to the point x with the coordinate x_A, but will irreversibly pass to the position x_C and move jump-like to x_H.

The reverse movement of domain walls occurs in the area of weak fields and does not cause significant hysteresis with changes in the magnitude and direction of the magnetic field. Because of the inhomogeneous internal stresses, changes in the gradient of the surface energy density of the boundary layer and other inhomogeneities at the stage of displacement of the boundary between neighboring domains, in addition to inverse ones, there are also irreversible displacements of these boundaries. Irreversible shifts of boundaries are Barkhausen jumps (BJ) and take place in strong fields. In the magnetic field absence, the wall does not return to its original state, but is fixed in a new position, which is the cause of BJ. In the fields close to saturation, the magnetization is due to the rotation of vectors I_S.

The investigation of the action of external remagnetization fields on ferromagnetic materials is of interest to scientists from both scientific and practical points of view, in particular, the analysis of approaches to mathematical description of the processes of movement of 180° and non-180° domain walls (Gallaktionova 1967).

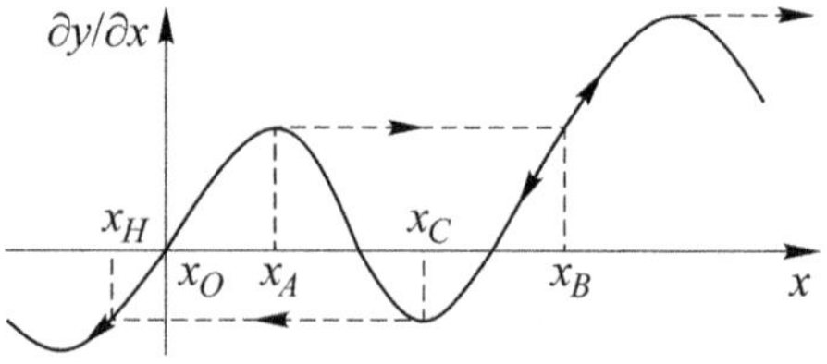

Fig. 7.21 Mechanism of a jump-like movement of domain walls

The Becker potential-energy model. In (Skal's'kyi et al. 2016) Becker formulated a model of a separate BJ, with which he described the basic regularities of occurrence of such jumps due to the irreversible movement of the domain wall. Since the initial location of domain walls in the absence of an external magnetic field in the conditions of thermodynamic equilibrium determines the minimum free energy of the ferromagnetic specimen (Durin and Zapperi 2006), it is expressed for statistical analysis of Barkhausen jumps by random and deterministic functions of the domain wall position.

If to study the domain wall movement in a crystal with plane-parallel 180° domain walls, it should be required to consider the energy of the domain wall and the energy of the demagnetization fields only. If such a crystal is placed in a magnetic field with intensity H that is parallel to the direction of magnetization of the domains, the thermodynamic equilibrium will be disturbed due to the appearance of force F_H in the crystal, since this will cause the displacement of all domain walls. The new position of the walls will be determined from the condition of forces equilibrium (Skal's'kyi et al. 2016)

$$F_H - F_\gamma - F_N = 0, \tag{7.16}$$

where $F_H = 2S(x)J_S H$ is the force caused by the action of the external field H; $F_N = \frac{2N(x)J_S^2 S(x)}{x_0} x$ is caused by the presence of demagnetization fields; $F_\gamma = S(x)\frac{\partial \gamma(x)}{\partial x}$ is caused by the displacement of the domain wall by the value of $\mathrm{d}x$; J_S is saturation magnetization; $S(x)$ is the area of the domain wall, which depends on the coordinate of the domain wall x; $N(x)$ is demagnetizing factor; and $\frac{\partial \gamma(x)}{\partial x}$ is surface energy density gradient.

Provided that $N(x) = \text{const}$ the internal forces can be divided into deterministic F_d and random components F_r. If the structure of the crystal is homogeneous, the component F_γ plays the main role and the internal stresses σ are expressed through $\gamma = (K_{\text{eff}} A/a)^{1/2}$; $K_{\text{eff}} = \alpha k_1 + \beta \lambda \sigma$; α, β are constant coefficients for this crystal; k_1 is crystallographic anisotropy constant; and λ is constant of magnetostriction. If the crystal's structure is heterogeneous in the presence of defects the contribution of the component, which can be expressed as $F_\gamma = \gamma \frac{\partial S(x)}{\partial x}$, increases (Skal's'kyi et al. 2016).

During the domain wall movement, the deterministic component of the force changes linearly $F_d(x) = bx$, where $b = 2NJ_S^2 S_0/x_0$; S_0 is the area of the moved domain wall. The random component $F_r(x) = K(\omega, x)$ has the form of a stationary random function of the argument x, and ω determines the law of distribution of this function. It is proved that for the Poisson distribution of defects and stress peaks in a crystal, $K(\omega, x)$ is a normal function (Vonsovsky 1971).

In (Skal's'kyi et al. 2016) the equation of the magnetic moment of the ferromagnetic specimen during BJ is derived as well

$$m(t) = 2S_0 J_S x(t), \tag{7.17}$$

where $x(t)$ is the movement of the domain wall, and the equation of movement of the domain wall looks like

$$m_{\text{eff}}\frac{\mathrm{d}^2x}{\mathrm{d}t^2}+\beta\frac{\mathrm{d}x}{\mathrm{d}t}+bx+K(\omega,x)=2S_0J_SH(t), \tag{7.18}$$

where m_{eff} is its effective mass.

Random potential model. Micromagnetic models describe spin rotations, and magnetic anisotropy well, but do not take into account three-dimensionality, range of interaction, demagnetization effect, and so on. To avoid these complications, it is necessary to formulate such approaches that would represent the most significant aspects of the Barkhausen effect, namely the disorder (random character) of the magnetic system. This can be done, in particular, leaving only a homogeneous magnetostatic energy E_m, and replacing the rest of the terms by the random magnetization function m

$$E-F(m)-mH_{\text{eff}} \tag{7.19}$$

where $H_{\text{eff}}=H+H_{\text{dem}}$.

To understand the origin of Eq. (7.19), consider the movement of one rigid domain wall that divides the specimen into two domains. In this case, the magnetization is proportional to the position of the wall x: $m=M_s(2x/L-1)$, where M_s is the saturation magnetization and L is the width of the specimen. The specimen is magnetized when the applied magnetic field pushes the domain wall in a random field with potential $F(m)$. To formulate the problem, specify the statistical properties of the function $F(m)$, such as its distribution and correlation parameters. Neel was the first who propose such a model with a random potential to study the origin of the hysteresis loop in the Rayleigh section (Lomayev 1977; N'eel 1942). In his model, the random function has the form of the sum of parabolas with a random curvature. Barkhausen noise based on this class of models can be explained as follows. When the effective field increases, the domain wall remains stationary as long as the condition $H_{\text{eff}}<W(m)$ is fulfilled, where $W(m)\equiv-\frac{\mathrm{d}F}{\mathrm{d}m}$. When H_{eff} reaches the value of the local maximum of function $W(m)$, the domain wall performs a series of successive jumps until the condition $H_{\text{eff}}<W(m)$ is met again.

The dimensions of the Barkhausen jumps S are the changes in the magnetization Δm, that occurrs as a result of these jumps in the random field $W(m)$. This model with a random field in the form proposed by Neel does not allow us to correctly describe Barkhausen noise, because the random field is significantly uncorrelated, as a result of which the distribution of jump sizes is exponential, while experiments indicate the power form of distribution (N'eel 1943; McMichael et al. 1993; Durin et al. 1995; Spasojevic et al. 1996; Durin and Zapperi 2000). To obtain exactly the power form of the size distribution of jumps, assume that the random field is far correlated. This approach was first proposed in Durin and Zapperi (2002), Bertotti (1986).

ABBM model of the equation of motion. In (Bertotti 1986; Bertotti et al. 1987; Alessandro et al. 1990), the authors for the first time formulated an approach related to the application of the Brownian field distribution. This theoretical model of domain wall movement in ferromagnetic materials describes Barkhausen noise statistics well and is known as the ABBM model (Durin and Zapperi 2006; Alessandro et al. 1990). The authors model the change in magnetization under the action of an external magnetic field on the material, considering the movement of one 180° domain wall, which divides the specimen into two domains with opposite directions of magnetization. The following stochastic differential equation is proposed to describe the position of the domain wall $u(t)$:

$$\Gamma \partial_t u(t) = H(t) - ku(t) + F(u(t)), \tag{7.20}$$

where Γ is the effective ductility; H is the applied external magnetic field; k is demagnetizing coefficient; and $F(u(t))$ is a random field that depends only on the position of the domain wall. As a result of numerical solution of Eq. (7.20), a trajectory resembling the results of Barkhausen noise measurements is obtained.

According to the proposed approach, it is possible to analytically calculate the stationary distribution of domain wall velocities and the distributions of sizes and duration of avalanche jumps. To describe the stationary movement of the domain wall, the external magnetic filed $sH(t) = ct$ is set linearly, c is the growth rate of the field H, and t is the time. One of the ways to find the result is to solve the Fokker–Planck equation, as it was done in Bertotti et al. (1987).

In (Fisher 1985), a phenomenological approach was proposed that took into account the energy of inhomogeneous exchange and the energy of magnetic anisotropy in the vicinity of the dislocation nucleus. The influence of such energies can cause a significant change in internal stresses during the remagnetization of the ferromagnetic material. Expressions for magnetic forces acting on dislocations in a magnetic field are also obtained.

In (Fisher 1985), the model of domain wall movement taking into account internal mechanical stresses in ferroelectric ceramics is described. Ferromagnetic materials by their properties are similar to ferroelectric ones, because in both cases the phenomenon of hysteresis is observed when the material is remagnetized, which is explained by the presence of a domain structure.

Under the mechanical loads action, additional local mechanical stresses appear between grains, which delay the movement of non-180° domain walls, and it moves jump-like. According to the developed model, it is possible to determine the contribution of the movement of 180° and non-180° domain walls to the polarization currents in ferroelectric ceramics.

Therefore, the actual problem now is diagnosing of metal products and structural elements, as well as equipment that has exhausted its scheduled (normal) life and needs to be replaced. The vast majority of them are made of low-alloy steels, which are ferromagnetic materials. In contact with hydrogen-containing or hydrogen sulfide environemnts during long-term operation, undergoing changes in temperature, external environment and natural factors, stress, ferromagnetic materials degrade, losing their strength, and crack growth resistance characteristics.

7.7 Stochastic Mathematical Model of 90° Domain Walls Movement in Ferromagnetic Materials

The magnetization of ferromagnetic material during formation of the domain structure with closing 180° and 90° domain walls, which covers the entire volume of the body, occurs due to the movement of the domain walls. Consider the movement in the integral magnetic field of a flexible single 90° domain wall of ferromagnetic material that separates two domains with perpendicular directions of magnetization $\vec{M}_1$ and $\vec{M}_2$, so that $|M_1| = |M_2| = |M|$, and its position can be described by function $h(x, y, t)$ (Fig. 7.20), assuming that the movement of the domain walls is independent. The total energy E of the ferromagnetic material in the external magnetic field can be described as

$$E = E_m + E_{DW} + E_{dis} = E_H + E_{dem} + E_{DW} + E_{dis}, \tag{7.21}$$

where $E_m = E_H + E_{dem}$ is the magnetostatic energy of the material; E_H is the energy in the external field of remagnetization; E_{dem} is demagnetizing energy caused by free magnetic poles at the specimen ends; E_{DW} is the energy of the formed domain walls; and E_{dis} is the energy of disorder.

The energy of the magnetized material in an external magnetic field with intensity $\vec{H}$ can be written as

$$E_H = -\vec{H}\vec{M}_S = -HM_S \cos\varphi, \tag{7.22}$$

where φ is the angle between $\vec{M}_S$ and $\vec{H}$. The energy will reach a minimum value provided that the directions of magnetization and field intensity coincide ($\varphi = 0$).

A magnetized body of finite dimensions possesses energy even in the absence of an external field, as a result of the interaction of the ferromagnet magnetization with its proper field.

On the specimen ends, the magnetization changes jump-like from M in the material to zero outside it, what is considered to be due to the appearance of free magnetic poles on the specimen. The notion about the presence of free magnetic poles is similar to the concept of surface charge in electrostatics. In fact, magnetic poles do not really exist, and the concept of magnetic poles is only a convenient way to mathematically describe the state of magnetization of a body of finite dimensions.

The proper field of a ferromagnetic body formed by its magnetic poles is proportional to the magnetization M, and inside the sample its direction is opposite to the magnetization. This field is called demagnetizing and is generated by the presence of free magnetic poles at the specimen boundaries, resists the applied magnetic field, and is proportional to the magnetization of the material, and its energy can be described as

$$E_{dem} = -NM, \tag{7.23}$$

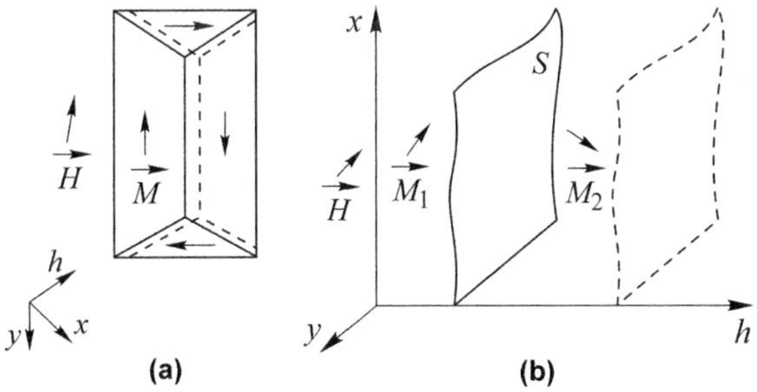

Fig. 7.22 Domain structure of the ferromagnet (**a**) and the scheme of movement of the 90° domain wall (**b**) in the external field of remagnetization

where N is the demagnetizing coefficient.

Magnetic poles inside and on the surface of the body create stray fields. Inside the body, the stray field is directed against magnetization (demagnetization field) and is the cause of domain formation. In the external field applied along the Oy axis so that the motion of the 90° domain wall in it occurs quasi-parallel to itself (Fig. 7.22).

Magnetostatic energy is determined by the relation

$$E_m = E_H + E_{\text{dem}} = -\mu_0(\vec{H} + \vec{H}_{\text{dem}})\vec{M}_S \iint\limits_S h(x, y, t)dxdy, \tag{7.24}$$

where μ_0 is the magnetic permeability of the vacuum; $\vec{H}$ is the vector of external magnetic field intensity; $\vec{H}_{\text{dem}}$ is the demagnetizing field vector; and S is the projection of the domain wall $h\,(x, y, t)$ on the xOy plane.

In the simplest case, the term $\vec{H}_{\text{dem}}$ is proportional to the total magnetization. For the magnetization from a constant field we have

$$\vec{H}_{\text{dem}} = -\frac{k\vec{M}_s}{V} \iint\limits_S h\,(x, y, t)\,\mathrm{d}x\mathrm{d}y, \tag{7.25}$$

where k is the coefficient that takes into account the domain structure and shape of the ferromagnetic specimen and V is its volume.

In real ferromagnetics, the demagnetization field is homogeneous only if they are of an ellipsoidal shape and are uniformly magnetized. In the bodies of other shapes, such as cylinders and rods of rectangular cross-section, the demagnetization field is inhomogeneous. To estimate the internal demagnetization field in them the average values of their demagnetizing coefficient are usually used, averaging it by the body volume (magnetometric demagnetizing coefficient) or by the average cross-section of the body perpendicular to the direction of the demagnetization field (ballistic demagnetizing coefficient).

The balance between the energy of magnetocrystalline anisotropy and exchange energy determines the width of the domain wall and its surface energy. To describe the energy of the 90° domain wall caused by balance, obtain

$$E_{DW} = \gamma_{DW} \iint\limits_{S} \sqrt{1 + \left|\vec{\nabla} h(x, y, t)\right|^2} \mathrm{d}x\mathrm{d}y. \tag{7.26}$$

Here $\gamma_{DW} \approx \sqrt{AK_0}$ is the surface energy of the 90° domain wall; A is a constant of exchange interaction; and K_0 is a constant of anisotropy.

For small gradients, expression (7.26) takes the form

$$E_{DW} = \gamma_{DW} S_{DW} + \frac{\gamma_{DW}}{2} \iint\limits_{S} \left|\vec{\nabla} h(x, y, t)\right|^2 \mathrm{d}x\mathrm{d}y, \tag{7.27}$$

where S_{DW} is the area of the undeformed 90° domain wall.

Ferromagnetic materials have different sources of inhomogeneities, which are crucial for explaining the fluctuations of the BJ; therefore in well-ordered systems, such noise is much smaller. There are several types of inhomogeneities that contribute significantly to the energy of disorder. In particular, in crystals such inhomogeneities are caused by vacancies, dislocations, non-magnetic impurities, which are the centers of attachement of the 90° domain wall during its movement in the material.

In most cases, the movement of the 90° domain wall occurs under strong damping, and the equation of motion will look like (Fisher 1985)

$$\Gamma \frac{\partial h(x, y, t)}{\partial t} = -\frac{\delta E(h(x, y, t))}{\delta h(x, y, t)}, \tag{7.28}$$

where Γ is the ductility coefficient, determined by the dissipation rate, neglecting the inertial effects $\Gamma = 1$; $E(h(\vec{r}, t))$ is the total energy functional of the studied material.

Let us introduce the notation $\rho(x, y) = h(x, y, t)$, $f = -\mu_0 \vec{H} \vec{M}_s h(x, y, t)$. We obtain the expression for the functional

$$F[\rho] = \iint\limits_{S} f(x, y, \rho(x, y), \vec{\nabla}\rho(x, y)) \mathrm{d}x\mathrm{d}y, \tag{7.29}$$

for the functional derivative

$$\frac{\delta F[\rho]}{\delta(\rho(x, y))} = \frac{\partial f}{\partial \rho(x, y)} - \vec{\nabla} \frac{\partial f}{\partial \vec{\nabla}\rho(x, y)}, \tag{7.30}$$

and for the functional derivative of the product

$$\frac{\delta(FG)}{\delta(\rho(x, y))} = \frac{\delta(F)}{\delta\rho(x, y)} G + F \frac{\delta(G)}{\delta\rho(x, y)}. \tag{7.31}$$

Apply (7.29–7.31), considering (7.28), to the functional derivative of the magnetostatic energy of the 90° domain wall (7.24)

$$
\begin{aligned}
\frac{\delta E_m}{\delta h(x,y,t)} &= \frac{\delta\left\{-\mu_0(\vec{H}+\vec{H}_{\text{dem}})\vec{M}_s \iint_S h(x,y,t)\mathrm{d}x\mathrm{d}y\right\}}{\delta h(x,y,t)} \\
&= \frac{\delta\left\{-\mu_0\vec{H}\vec{M}_s \iint_S h(x,y,t)\mathrm{d}x\mathrm{d}y - \mu_0\vec{H}_{\text{dem}}\vec{M}_s \iint_S h(x,y,t)\mathrm{d}x\mathrm{d}y\right\}}{\delta h(x,y,t)} \\
&= \frac{\delta\left\{-\mu_0\vec{H}\vec{M}_s \iint_S h(x,y,t)\mathrm{d}x\mathrm{d}y\right\}}{\delta h(x,y,t)} \\
&\quad + \frac{\delta\left\{-\mu_0((-k\vec{M}_s/V)\iint_S h(x,y,t)\mathrm{d}x\mathrm{d}y)\vec{M}_s \iint_S h(x,y,t)\mathrm{d}x\mathrm{d}y\right\}}{\delta h(x,y,t)} \\
&= -\mu_0\vec{H}\vec{M}_s\frac{\partial h(x,y,t)}{\partial h(x,y,t)} + \frac{2\mu_0 k\vec{M}_s^2}{V}\iint\limits_S h(x,y,t)\mathrm{d}x\mathrm{d}y \\
&= -\mu_0\vec{H}\vec{M}_s + \overline{k}\tilde{h},
\end{aligned}
\tag{7.32}
$$

where $\overline{k} = \frac{2\mu_0 k\vec{M}_s^2}{V}$; $\tilde{h} = \iint_S h(x,y,t)\mathrm{d}x\mathrm{d}y$.

For the functional derivative of energy of the 90° domain wall (7.24), considering that $\vec{\nabla}\rho(\mathrm{x,y}) = \vec{\nabla}h(x,y,t)$, we obtain the expression

$$
\begin{aligned}
\frac{\delta E_{DW}}{\delta h(x,y,t)} &= \frac{\delta\left\{\gamma_{DW}S_{DW} + \frac{\gamma_{DW}}{2}\iint_S \left|\vec{\nabla}h(x,y,t)\right|^2 \mathrm{d}x\mathrm{d}y\right\}}{\delta h(x,y,t)} \\
&= \frac{\delta(\gamma_{DW}S_{DW})}{\delta h(x,y,t)} + \frac{\gamma_{DW}}{2}(-1)\vec{\nabla}\frac{\partial\left|\vec{\nabla}h(x,y,t)\right|^2}{\partial(\vec{\nabla}h(x,y,t))} \\
&= -\frac{2\gamma_{DW}}{2}\vec{\nabla}\vec{\nabla}h(x,y,t) = -\gamma_{DW}\vec{\nabla}^2 h(x,y,t).
\end{aligned}
\tag{7.33}
$$

Violation of the order, as mentioned earlier, can take the form of non-magnetic inclusions, dislocations, or residual stresses that cause deformation and attachement of the domain walls. It can be modeled by introducing a random potential $V(x,y,h)$, the derivative of which determines the force field $\eta(x,y,h)$, acting on the 90° domain wall (Fisher 1985). In the partial case of point non-magnetic inclusions, the random force will be

$$
\eta(x,y,h) = -\sum_i f_p(x-x_i, y-y_i, h-h_i), \tag{7.34}
$$

where (x_i, y_i, h_i) are the coordinates of the center of attachement; $f_p(x,y,h)$ is the force of attachement of the domain wall, which application area is proportional to the width of the domain wall (7.19).

For the case when the distance between the centers of the domain walls attachement is less than the averaging scale, the distribution will take the form of δ-correlated Gaussian noise

$$\langle \eta(x,y,h)\eta(x',y',h') \rangle = \delta^2(x - x', y - y')R(h - h'), \tag{7.35}$$

where $R(x)$ fades sharply for large x. The specific form of the function in the case of both random relations and a random field does not significantly affect the laws of similarity of distributions of avalanche-like processes that characterize BN and are related with the movement of domain walls (Gorobets 2009).

Therefore, taking the derivative of the total energy functional, Eq. (7.28) of the domain wall's motion can be rewritten as

$$\frac{\partial h(x, y, t)}{\partial t} = \mu_0 \vec{H} \vec{M}_s - \overline{k} \tilde{h} + \gamma_{dw} \vec{\nabla}^2 h(x, y, t) + \eta(x, y, h). \tag{7.36}$$

For the purpose of numerical solution (7.36) we discretize the surface of the 90° domain wall, splitting it into N identical elements, and assume the motion of one averaged point in a random field of attachement as the domain wall movement. Assume that all N elements h_i are at the same distance from each other, then

$$\frac{\mathrm{d}h_i}{\mathrm{d}t} = \mu_0 H M_s - \overline{kh} S + \eta_i(h), \tag{7.37}$$

where $\overline{h} = \sum_{i=1}^{N} \frac{h_i(t)}{N}$ describes the average position of the 90° domain wall (Fisher 1985).

Summing up both parts of expression (7.37) by all the centers i we obtain the dependence for the total magnetization $m \equiv \overline{h}N$

$$\frac{\mathrm{d}m}{\mathrm{d}t} = \mu_0 H M_s N - \tilde{k} m + \sum_{i=1}^{N} \eta_i(h), \tag{7.38}$$

where $\sum_{i=1}^{N} \eta_i(h)$ is the random field of fixation of the 90° domain wall, $\tilde{k} = \frac{2\mu_0 k \vec{M}_s^2}{b}$, and b is the length of the specimen.

Under remagnetization of the specimen of finite dimensions introduced into the external field, the magnetic charges appear on its end surfaces, thus promoting the appearance of a demagnetization field $\vec{H}_{\mathrm{dem}}$ directed in the opposite direction. For the elements made of ferromagnetic materials $\vec{H}_{\mathrm{dem}}$ and magnetization $\vec{M}$ are proportional to the external field intensity of the remagnetization $\vec{H}$ according to (7.23) and

$$\vec{M} = \frac{\vec{B}}{\mu_0} - \vec{H}, \tag{7.39}$$

where $\vec{B}$ is the vector of magnetic induction.

Let us calculate the demagnetizing coefficient on the example of a steel plate of length b, used in subsequent experiments, which is demagnetized by the field $\vec{H}$, whose direction is parallel to its longest side. The field intensity created by the face is perpendicular to the direction $\vec{H}$, at a distance r ($r \gg l$, l is the plate thickness)

$$H_{\text{dem}} = \frac{q}{2\pi \mu_0 r}, \tag{7.40}$$

where q is charge per unit length of the face.

Considering the density of the surface charges $\sigma = \mu_0 M_n$, where M_n is the normal component of the magnetization

$$q = \sigma l = \mu_0 M_n l. \tag{7.41}$$

For the central cross-section of the plate the width is the same as the distance from the edge $r = \frac{b}{2}$. For the field created by two faces

$$H_{\text{dem}} = -2\frac{\mu_0 M_n l}{2\pi \mu_0 (b/2)} = \frac{2l}{\pi b} M_n. \tag{7.42}$$

Taking into account (7.25), the demagnetizing coefficient was calculated by

$$k = \frac{2l}{\pi b}. \tag{7.43}$$

For a steel plate of sizes 31 × 152 × 2 mm in the external magnetic field $\vec{H}$, the demagnetizing coefficient $k = 83 \times 10^{-4}$.

We have replaced the component $\sum_{i=1}^{N} \eta_i$ in (7.37) with an effective random field of attachement $W(m)$, which describes the disorder caused by the collective movement of 90° domain walls and blocks their movement. In this case, the domain wall jumps between two configurations, leading to W change

$$W(m') - W(m) = \sum_{i=1}^{n} \Delta\eta_i, \tag{7.44}$$

where summation is limited by n-centers along which the wall moved, the total number of which is proportional to the average size of the jump, assuming that it is uncorrelated (the distribution of the jumps sizes is exponential).

The field $W(m)$ is a random function of the position of the 90° domain wall. It is described by the stochastic Wiener process (Koreneva et al. 2013; Narayan and Fisher 1993), like the description of the field of fixation of 180° walls. This field has the following properties:

(1) independent increments;
(2) $\langle W(0)\rangle = 0$;
(3) $\langle |W(m) - W(m')|^2 \rangle = D|m - m'|$, D—quantifies fluctuations W;
(4) one-dimensional normal distribution law.

So Eq. (7.38) will look like

$$\frac{dm}{dt} = \mu_0 H M_s N - \tilde{k}m + W(m). \quad (7.45)$$

Numerical calculations (7.45) were performed for realization of the random field of attachement of a single 90° domain wall (Fig. 7.23), and the dependence of the temporal change in magnetization was obtained (Fig. 7.24). The calculations were carried out under the initial condition $m(0) = 0$ and taking into account that $\frac{dm}{dt} \geq 0$.

As a result of numerical calculations, a sequence of jumps in the magnetization of the ferromagnet Δm, due to the movement of 90° domain walls in a random attachement field $W(m)$, with increasing external magnetic field intensity is obtained (Fig. 7.25).

The main parameters that characterize the BJ are the jump duration Δt of the 90° domain wall, the change in the volume of remagnetization ΔV, that occurred as a result of a single jump and its size $S = \Delta m = M_S \Delta V$. Using the results of numerical calculations, according to the mentioned mathematical model of the 90° domain walls movement in the ferromagnetic material, namely in a steel plate of specified dimensions with calculated demagnetizing coefficient, a change in the volume ΔV of the remagnetization areas, their duration, and size was obtained for maximum (max) and minimum (min) magnetization jumps (Table 7.3).

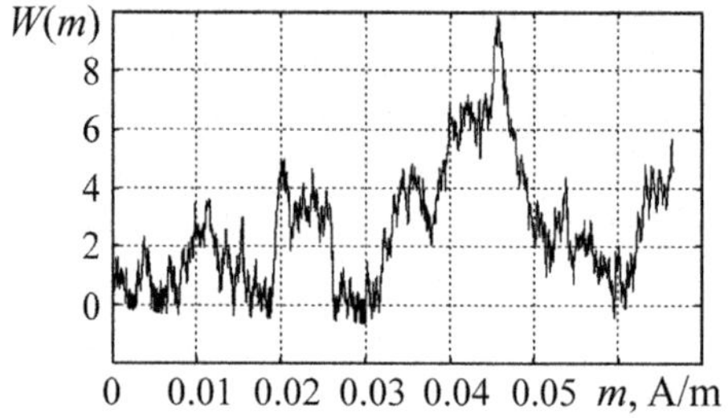

Fig. 7.23 Realization of a random attachement field $W(m)$ of a single 90° domain wall

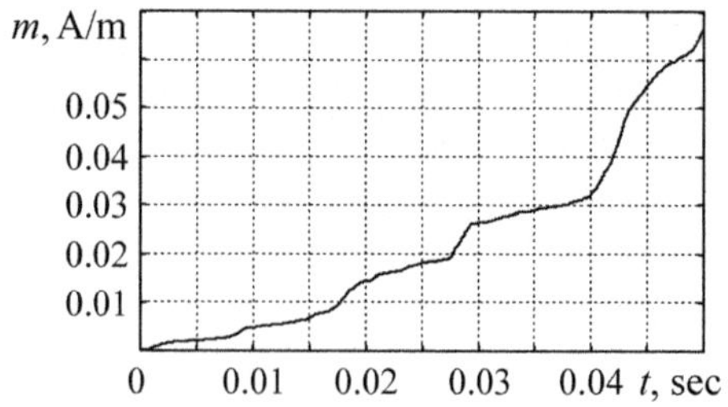

Fig. 7.24 Realization of time dependence of ferromagnet magnetization caused by 90° domain walls jumps

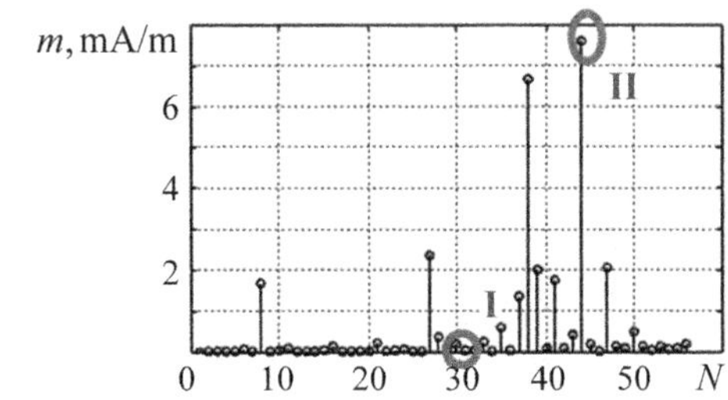

Fig. 7.25 Sequence of magnetization jumps *m* caused by the movement of 90° domain walls in ferromagnetic material

Table 7.3 Results of calculations of the main BJ parameters

$\Delta V \cdot 10^{13}$, m^3		$\Delta t \cdot 10^4$, s		$\Delta m \cdot 10^3$, A/m	
min	max	min	max	min	max
0.059	1.5	0.16	6.2	0.3	7.7

Analyzing the mechanism of domain wall propagation under the action of the external magnetization field and taking into account the influence of the random fields due to inclusions and structural inhomogeneities that delay the movement of 90° domain walls in the material, we can distinguish its characteristic features:

- randomness of occurrence of individual events in time (domain wall jumps);
- limited time;
- randomness of the jump size.

References

Alessandro B, Beatrice C, Bertotti G, Montorsi A (1990) Domain wall dynamics and Barkhausen effect in metallic ferromagnetic materials. II experiments. J Appl Phys 68(11):2908–2915

Antonov VE, Cornell K, Fedotov VK, Kolesnikov AI, Wipf H (1998) Neutron diffraction investigation of the dhcp and hcp iron hydrides and deuterides. J Alloy Compd 264(1/2):214–222

Armstrong JH, Carpenter SH (1988) Detection of the initiation and growth of hydrogen-induced cracks in Armco iron using continuous modulus measurements. Metall Trans A 19:473–478

Barkhausen H (1919) Zwei mit Hilfe der neuen Verstärker entdeckte Erseheinugen. Physik Zeitschrift 20:401–403

Becker R, Döring W (1939) Ferromagnetismus. Springer

Bertotti G (1986) Statistical interpretation of magnetization processes and eddy current losses in ferromagnetic materials. In: Gorzkowski W, Lachowicz H, Szymczak H (eds) Proceedings of the 3rd international conference on physics of magnetic materials. Word Scientific, Singapore, pp 489–508

Bertotti G, Balucani U, Lovesey SW, Rasetti MG, Tognetti V (eds) (1987) Dynamics of magnetic domain walls and Barkhausen noise in metallic ferromagnetic systems. In: Magnetic excitations and fluctuations II. Springer, Heidelberg

Birnbaum HK, Sofronis P (1994) Hydrogen-enhanced localized plasticity—a mechanism for hydrogen-related fracture. Mater Sci Eng, A 176:191–202

Brechko TM, Scriabina NE, Spivak LV, Bramovich MY (2004) Domennaya struktura i effekt Barkhausena v amorfnom splave $Fe_{78}B_{12}Si_9Ni_1$ (Domain structure and Barkhausen effect in the $Fe_{78}B_{12}Si_9Ni_1$ amorphous alloy). Pisma v Zurnal Tekhnicheskoi Fiziki (Lett Phys J) 30(9):68–72

Carpenter SH, Fawks JE (1981) An investigation of the cathodic charging of pure iron with hydrogen by modulus measurements. Scr Metall 15:699–704

Castellote M, Fuella J, de Viedma PG, Andeade C (2007) Hydrogen embrittlement of high-strength steel submitted to slow strain rate testing studied by nuclear resonance reaction analysis and neutron diffraction. Nuclear Instruments Methods Phys Res B 259:975–983

Durin G, Zapperi S (2000) Scaling exponents for Barkhausen avalanches in polycrystalline and amorphous ferromagnets. Phys Rev Lett 84:4075–4078

Durin G, Zapperi S (2002) On the power spectrum of magnetization noise. J Magn Magn Mater 242–245:1085–1088

Durin G, Zapperi S (2006) The Barkhausen effect. Elsevier

Durin G, Magni A, Bertotti G (1995) Fractal properties of the Barkhausen effect. J Magn Magn Mater 140–144:1835–1836

Enos DG, Scully JR (2002) A critical-strain criterion for hydrogen embrittlement of cold-drawn, ultrafine pearlitic steel. Metall Mater Trans 33:1151–1166

Fisher DS (1985) Sliding charge-density waves as a dynamic critical phenomenon. Phys Rev B 31(8):1396–1427

Gabetta G, Nykyforchyn HM, Lunarska E (2008) In-service degradation of gas trunk pipeline X52 steel. Mater Sci 44:104–119

Gallaktionova NA (1967) Vodorod v metallakh (Hydrogen in metals). Metallurgy Publishing House

Goodwin ARH (2003) Measurement of the speed of sound. In: Goodwin ARH, Marsh KN, Wakeham WA (eds) Measurement of the thermodynamic properties of single phases. International Union of Pure and Applied Chemistry, (IUPAC), pp 237–323

Goodwin ARH, Trusler JPM (2003) Speed of sound. In: Goodwin ARH, Marsh KN, Wakeham WA (eds) Measurement of the thermodynamic properties of single phases. Elsevier

Gorkunov ES, Drogoshansky YuN, Mikhovski M (1999) Effekt Barkhausena i ego ispolzovaniie v strukturoskopii ferromagnitnykh materialov (Obzory 1–3) (Barkhausen effect and its use in structuroscopy of ferromagnetic materials. Reviews 1–3). Defektoskopiya (Defectoskopy) 6:3–23; 7:3–32; 8:3–25

Gorobets YuI (2009) Dyslokatsii u feromagnetykakh u magnitnomun poli (Dislocations in ferromagnets in magnetic field). Naukovi visti TEUU "KPI" (Scientific News of Kyiv Polytechnic Institute) 5:136–138

Hirth JP, Carnahan B (1978) Hydrogen adsorption at dislocations and cracks in Fe. Acta Metall 26:1795–1803

Hsu NN (1974) Acoustic birefringence and the use of ultrasonic waves for experimental stress analysis. Exp Mech 14:169–176

Jhang K-Y, Quan H-H, Ha J, Kim N-Y (2006) Estimation of clamping force in high-tension bolts through ultrasonic velocity measurement. Ultrasonics 44:e1339–e1342

Kolachev BA (1985) Vodorodnaya khrupkost metallov (Hydrogen brittleness of metals). Metallurgy Publishing House

Kolachev BA (1985) Vodorodnaia khrupkost metallov (Hydrogen brittleness of metals). Metallurgiya Publishing House

Kolsky H (1964) Stress waves in solids. J Sound Vib 64(1):88–110

Koreneva VV, Zhoga LV, Nesterov VN, Piunov EM (2013) Model dvizheniia domennykh granits (Model of domain walls movement). Izvestiya Volgogradskogo Gosudarstvennogo Tekhnicheskogo Universiteta (News of the Volgogard State Technical University) 8(23):33–37

Krinchik GS (1976) Fizika magnitnykh yavlenii (Physics of magnetic phenomena). Publishing House of Moscow University

Kruger SE, Damm EB (2006) Monitoring austenite decomposition by ultrasonic velocity. Mater Sci Eng A 425:238–243

Kryzhanivs'kyi EI, Nykyforchyn HM (2011) Specific features of hydrogen-induced corrosion degradation of steels of gas and oil pipelines and oil storage reservoirs. Mater Sci 47:127–136

Kuleev VG, Shcherbinin VE, Zhakov SV, Subbotin YuS, Menshikov NM (1986) Vliyaniie fizicheskikh razlichii mezhdu effektom Barkhausena i akustchekoi emissiyei Barkhausena na ikh primeneniie v nerazrushayushchem kontrole (Effect of physical differences between the Barkhausen effect and Barkhausen acoustic emission on their application in non-destructive testing). Defektoskopiya (defectoskopy) 9:3–17

Lomaev GV, Malyshev VS, Degtiarev AP (1984) Obzor primeneniia effekta Barkhausena v nerazrushayushchem kontrole (Overview of the application of the Barkhausen effect in non-destructive testing). Defektoskopiya (defectoskopy) 3:54–71

Lomaiev GV (1977) Metod magnitnykh shumov v nerazrushayushchem kontrole ferromagnetikov (The method of magnetic noise in non-destructive testing of ferromagnets). Defektoskopiya (defectoskopy) 4:75–94

Lomayev GV (1977) Metod magnitnykh shumov v nerazrushayushchedm kontrole ferromagnetikov (Magnetic noise method in non-destructive testing of ferromagnets). Defektoskopiya (defectoscopy) 4:75–94

Lunarska E, Zielinski A, Smialowski M (1977) Effect of hydrogen on shear modulus of polycrystalline α iron. Acta Metall 25:305–308

Lynnworth LC (1975) Industrial applications of ultrasound—a review: II. Measurements, tests, and process control using low-density ultrasound. IEEE Trans Sonics Ultrason 22:71–101

Maier HJ, Popp W, Kaesche H (1987) A method to evaluate the critical hydrogen concentration for hydrogen-induced crack propagation. Acta Metall 35:875–880

McMichael RD, Swartzendruber LJ, Bennet LH (1993) Langevin approach to hysteresis and Barkhausen noise jump modelling in steel. J Appl Phys 73(12):5848–5850

Morales AL, Nieto AJ, Chicharro JM (2010) Influence of internal stresses on field-dependent elastic modulus and damping in pure nickel. J Magn Magn Mater 322:3584–3594

Morozov EM, Nikishkov GP (1980) Metod konechnykh elementov v mekhanike razrusheniia (Finite elements method in facture mechanics). Nauka Publishing House

N'eel L (1942) Theorie des lois d'aimantation de Lord Rayleigh. I: Le deplacements dun paroi isolee. Can J Phys 12:1–20

N'eel L (1943) Theorie des lois d'aimantation de Lord Rayleigh. II: Multiples domaines et champ coercitif. Can J Phys 13:18–30

Narayan O, Fisher DS (1993) Threshold critical dynamics of driven interfaces in random media. Phys Rev B 48:7030–7042

Nazarchuk ZT, Skal's'kyi VR, Klym BP (2009) Influence of hydrogen on the changes in the power of Barkhausen jumps in ferromagnets. Mater Sci 45:663–669

Nykyforchyn HM (2002) Manifestation of hydrogen and low-temperature brittleness in near-threshold cyclic crack resistance of materials. Mater Sci 38:471–483

Ono K, Shibata M (1980) Magnetomechanical acoustic emission of iron and steels. Mater Eval 38:55–61

Oriani RA, Josephic PH (1977) Equilibrium and kinetic studies of the hydrogen-assisted cracking of steel. Acta Metall 25:979–988

Ortiz M, Ovejero-Garcia J (1992) Effect of hydrogen on Young's modulus of AISI 1005 and 1070 steels. J Mater Sci 27:6777–6781

Ovchinnikov IG, Rassada AE (1990) Analiz eksperimentalnykh dannykh po vliayaniiu vodoroda pri normalnykh temperaturakh na mekhanicheskiie svoistva metallov i splavov i postroyeniie modeli vzaimodeistviia konstruktivnykh elementov s vodorodom. Ch. 1: Problemy vzaimodeistviia vodoroda na metally i puti ee resheniya. Zakonomernosti proniknoveniia vodoroda v konstruktivnyie elementy (Analysis of experimental data on the effect of hydrogen at normal temperatures on the mechanical properties of metals and alloys and construction of a model for the interaction of structural elements with hydrogen. Part I: The problem of the effect of hydrogen on metals and ways to solve it. Patterns of hydrogen penetration into structural elements). Saratov, Deposited in VINITI, № 551, B90

Panasyuk VV, Andreikiv AY, Kharin VS (1981) Teoreticheskii analiz rosta treshchin v metallakh pri vozdeistvii vodoroda (Theoretical analysis of crack growth in metals under action of hydrogen). Fizikochimicheskaya Mechanika Materialov (Physicochemical Mech Mater) 4:61–75
Panasyuk VV, Andreykiv Aye, Parton VZ (1988) Mekhanika razrusheniia i prochnost materialov. T. 1: Osnovy mekhaniki razrusheniia materialov (Fracture mechanics and strength of materials. Vol. 1: Fundamentals of the mechanics of materials fracture). Naukova Dumka Publishing House
Ricker RE, Pitchure DJ (2009) The influence of hydrogen on the elastic modulus and anelastic response of cold worked pure iron. In: Somerday B, Sofronis P, Jones R (eds) Effects of hydrogen on materials. ASM international, pp 219–226
Rudiak VM (1970) Effekt Barkhausena (Barkhausen effect). Uspekhi Fizicheskikh Nauk (Success of Physical Sciences) 111(3):429–462
Sanchez J, Fullea J, Andrade C, de Andres PL (2008) Hydrogen in α-iron: stress and diffusion. Phys Rev B 78(1):014113-1–014113-7
Segerlind L (1979) Primeneniie metoda konechnykh elementov (Application of the finite element method). Mir Publishing House
Shibata M, Ono K (1981) Magnetomechanical acoustic emission—a new method of nondestructive stress measurement. NDT Int 227–234
Skal's'kyi VR, Pochaps'kyi EP, Mel'nyk NP (2016) Modeling of motion of the domain walls in ferromagnetic materials (a survey). Mater Sci 51(6):753–764
Skal's'kyi VR, Serhienko OM, Mykhal'chuk VB, Semehenivs'kyi RI (2009) Quantitative evaluation of Barkhausen jumps according to the signals of magnetoacoustic emission. Mater Sci 45:399–408
Skalskyi V, Hirnyj S, Pochapskyi Ye (2011) The effect of deformation on the parameters of magnetoacoustic signals. Visnyk Ternopilskogo Natsionalnogo Universytetu (Bulletin of Ternopil State Technical University), Special Issue 1:155–161
Spasojevic D, Bukvic S, Milosevic S, Stanley HE (1996) Barkhausen noise: elementary signals, power laws, and scaling relations. Phys Rev E 54(3):2531
Tikadzumi S (1987) Fizika ferromagnertizma. Magnitnyie kharakteristiki i prakticheskiie primeneniia (Physics of ferromagnetism. Magnetic characteristics and practical applications). Mir Publishing House
Tkachev VI, Kholodnyi VI, Levina IN (1999) Rabotosposobnost stalei i splavov v srede vodoroda (Serviceability of steels and alloys in hydrogen environment. Vertical Publishing House
Tsuchida Y, Watanabe T, Kato T, Seto T (2010) Effect of hydrogen absorption on strain-induced low-cycle fatigue of low carbon steel. Proc Eng 2:555–561
Vavrukh MV, Slobodian BB (1985) Lokalizatsiia primesei vodoroda v metalle (Localization of hydrogen admixtures in metal). Fizikochimicheskaya Mechanika Materialov (Physicochemical Mech Mater) 4:26–29
Vonsovsky SV (1971) Magnetizm. Magnitnye svoistva dia-, para-, ferro-, antiferro- I ferromagnetikov (Magnetism. Magnetic properties of dia-, para-, ferro-, antiferro- and ferrimagnets). Nauka Publishing House
Watanabe T, Yamashita T, Tsuchida Y, Tokuno K (2012) Effect of hydrogen charging on low cycle fatigue life and its dependence on cementite morphology. ISIJ Int 52:222–227
Williams Kh, Shokely W, Kittel Ch (1952) Izuchenie skorosti dvizheniia granits mezhdu ferromagnitnymi oblastyami (Studying the speed of movement of boundaries between ferromagnetic regions). In: Vonsovski SV (ed) Ferromagnitnyi rezonans (Ferromagnetic resonance), Moscow
Yamada Y, Yoshimura N, Sakurai T (1968) Plastic stress-strain matrix and its application for the solution of elastic-plastic problems by the finite element method. Int J Mech Sci 10:343–353
Zenkevich O (1975) Metod konechnykh elementov v tekhnike (Finite element method in engineering). Mir Publishing House
Zhang T-Y, Jiang FX, Chu WY, Hsiao CM (1985) Effect of hydrogen on the Young's modulus of iron. Metall Trans A 15:1655–1662
Zhang Y, Jiang FX, Chu FX, Hsiao CM (1986) Effect of hydrogen on the Young's modulus of commercial purity iron. Scr Metall 20:503–507

Zienkiewicz OC, Valliappan S, King IP (1969) Elastic-plastic solution of engineering problems. Initial-stress, finite element approach. Int J Numer Meth Eng 1:75–100

Chapter 8
Mathematical Models of the MAE Signal and Its Informative Parameters

Two main approaches to creating mathematical models of physical signals are used to describe and analyze them. The first one concerns non-random, i.e., deterministic signals, whose values at any time or at any point in space are a priori known, or can be fairly accurately predicted, determined, or calculated. The deterministic signal in this case is given, as a rule, by a mathematical function or a computational algorithm.

The second approach implies the stochastic character of signals, the law of change of which in time (space) is random and which takes specific values with some probability. The model of such a signal is a description of the statistical characteristics of a random process by specifying the laws of the probability distribution, correlation function, spectral energy density, and others.

Randomness can be determined both by the proper physical nature of the signals and by the probabilistic character of the registered signals both in the time and place of their occurrence and in content. From these positions, a random signal can be considered as a representation of a random process or physical properties of the object, determined by random parameters or complex structure of the object, material, environment, and measurement results which it is difficult to predict.

In this chapter the mathematical model of the MAE signal is proposed and substantiated; its informative parameters are considered, and algorithms of their estimation are developed. The influence of excitation conditions on the change of MAE signal parameters has been studied. Based on this, the algorithms for stabilizing the amplitude of the remagnetization current and induction of the remagnetization field have been proposed and developed.

V. Skalskyi and Z. Nazarchuk, *Magnetoelastic Acoustic Emission*, Springer-AAS Acoustics Series, https://doi.org/10.1007/978-981-99-4032-5_8

8.1 Mathematical Model of the MAE Signal

Analysis of the mechanism of generating the MAE signals makes it possible to identify their typical features: random occurrence in time of individual events (random flux of pulses); limited time (finiteness); weakness in energy; random amplitude. These factors allow us to consider the signals within the framework of the random pulse flow model (Buttle et al. 1987).

In this case, the signal can be represented by the superposition of pulses, the shape of which is described by a deterministic function $F(t)$, normalized to one in the maximum, and the pulses may differ in amplitude. The corresponding pulse flow is written as:

$$X(t) = \sum_{k,i} A_{ki} F_{ki}(t - t_{ki}), \, (k-1)T < t_{ki} \leq kT, \quad k, i = 1, 2, \ldots, \tag{8.1}$$

where A_{ki} is the random amplitude; T is the period of remagnetization; k is the period number, t_{ki} is the random moment of the i-th pulse in the k-th period; t_{ki} is called the moment of appearance conditionally.

We assume that it is not necessary that $F_{ki}(t - t_{ki}) = 0$, if $t < t_{ki}$. Then let function $F(t)$ tends to zero fast enough if $|t| \to \infty$, therefore $\int_{-\infty}^{\infty} |F_{ki}(t - t_{ki})| < \infty$. Moments of time t_{ki} can be related to any characteristic point—to any of the extremes $F(t)$ or to any of the points of transition through zero if any, and so on.

A more general interpretation of the random pulse flow is that, in addition to the amplitude, also the shape of the pulses themselves is assumed to be random, which depends on the set of some finite number m of parameters $\mu = \{\mu_1, \ldots, \mu_m\}$:

$$X(t) = \sum_{k,i} A_{ki}(\mu) F_{ki}(t - t_{ki}, \mu). \tag{8.2}$$

In particular, the MAE signal depends on the parameters of the remagnetization field (values of the amplitude intensity, frequency and signal shape) and is sensitive to the structural changes in a ferromagnetic material, heat treatment, plastic deformation, residual stresses, hydrogen charging, and so on.

Degraded material changes its domain structure, which affects the change of MAE parameters. It is founded experimentally that the plastic deformation of the material changes the MAE power the most, and the hydrogen factor, which, in turn, depends on its concentration in the ferromagnet, changes it to a lesser extent (Durin and Zapperi 2002; Nazarchuk et al. 2012).

8.2 Informative Characteristics of the MAE Signal and Algorithms for Their Evaluation

Based on the model (8.1) it is possible to distinguish the pulse information characteristics of the magnetoelastic emission signal. These include the probability density of the pulse amplitude distribution $p(A)$, the sum of the pulse amplitudes for the k-th remagnetization period $\sum_i A_{ki}$, the MAE envelope generated during the remagnetization period, and a set of parameters of the pulse shape $\mu = \{\mu_1, \ldots, \mu_m\}$. The temporal information characteristics include the intensity of the pulse flow $n(t)$ and $q(\vartheta)$ as the density of probabilities of the distribution of intervals $\vartheta_{ki} = t_{k(i+1)} - t_{ki}$ between adjacent pulses of the flow. The final count of pulses for the k-th period of remagnetization is $N_k = \sum_i H(t - t_{ki})$, where $H(\cdot)$ is the Heaviside function.

In the general case, pulse and continuous noise are added to the informative component of the MAE signal, so the model (8.1) can be represented as:

$$X(t) = \sum_{k,i} A_{ki} F(t - t_{ki}) + \sum_j B_j G_j(t - t_j) + S(t) + \gamma(t). \tag{8.3}$$

Here B_j is amplitude, $G_j(t - t_j)$ is function of the shape of the pulse noise generated by the operation of mechanisms, electrical pulses in the power supply network, electromagnetic pulses, etc.; $S(t) = S_0 \sin \omega t$ is a continuous harmonic noise, which occurs due to the influence of the frequency of the industrial power supply (S_0 is its amplitude, ω is frequency); $\gamma(t)$ is a stationary continuous random disturbance generated by radio frequency interference and transmission channel noise. Function $\gamma(t)$ can be described by the correlation theory with mathematical expectation $\overline{\gamma}$, dispersion D_γ, one-dimensional distribution density $f(\gamma)$, and normalized autocorrelation function $r_\gamma(\tau)$ (τ is time shift). Noise can also significantly distort the results of evaluation of the informative characteristics of the signal. Therefore, to obtain a reliable information for correct technical diagnostics, they should be filtered.

We should note the important circumstances related to the emission of signals that occur due to the action of mechanical, physical, chemical, or other factors on the object under study. These signals, as a rule, in many cases do not meet the conditions of "good statistics": accidental error in estimating the statistical parameters of the signal may exceed the allowable error, which guarantees with a certain probability the detection of defect or evaluation of crack formation characteristics. This is due to the fundamental limitation of the signal duration for a given product or structure. The generation of emission signals is uncontrolled and stops after the investigated object fails. In the case of the MAE signal, the random error can be reduced as desired by increasing the magnetization time or the magnetizing field intensity.

The evaluation of such a traditional characteristic of the MAE signal as the sum of the amplitudes for the j sampling (Fig. 8.1) will be as follows:

$$\hat{A}_{\Sigma j}(A_{th}) = \sum_{i=1|(T(j-1)\le t_{ji} < Tj)} A_{ji|(A_{ji} > A_{th})}. \tag{8.4}$$

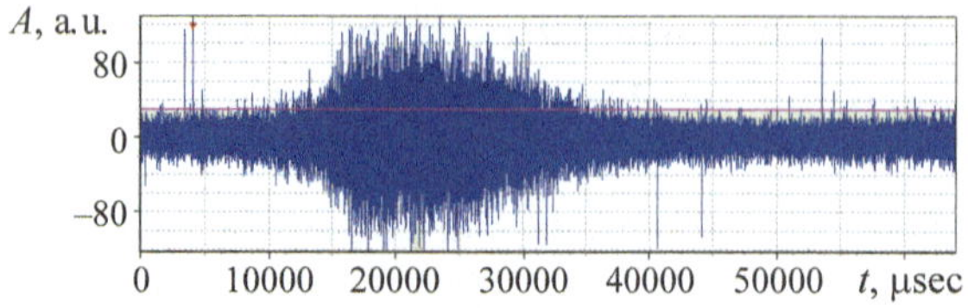

Fig. 8.1 Waveform of the MAE signal recorded during remagnetization of a plane ferromagnetic sample: remagnetization frequency is 9 Hz, amplitude of magnetic field induction is 0.8 T

Expression (8.4) summarizes the amplitudes of the pulses greater than the threshold value $A_{ij} > A_{th}$, which exceeds the noise level. In order to reduce the random error, using M realizations of the signal, to estimate the sum of the amplitudes we obtain the expression:

$$\hat{A}_{\Sigma a}(A_{th}) = \frac{1}{M}\sum_{j=1}^{M}\sum_{i=1|(T(j-1)\le t_{ji}<Tj)} A_{ji|(A_{ji}>A_{th})}, \tag{8.5}$$

and for its variance:

$$D_{\hat{A}_{\Sigma a}(A_{th})} = \frac{1}{M} D_{\hat{A}_{\Sigma j}(A_{th})}. \tag{8.6}$$

Variance is reduced in M times compared to the variance of the estimate for one realization $D_{\hat{A}_{\Sigma j}(A_{th})}$ and as a consequence, the estimated error can be done as small as possible.

Similarly, to evaluate the final count for j realization, we obtain the following expression:

$$\hat{N}_{\Sigma j}(A_{th}) = \sum_{i=1|((A_{ji}>A_{th}),\, T(j-1)\le t_{ji}<Tj)} H(t - t_{ji}), \tag{8.7}$$

for M realizations:

$$\hat{N}_{\Sigma a}(A_{th}) = \frac{1}{M}\sum_{j=1}^{M}\sum_{i=1|((A_{ji}>A_{th}),\, T(j-1)\le t_{ji}<Tj)} H(t - t_{ji}), \tag{8.8}$$

which also reduces the variance in M times compared to the variance of the estimate for one realization $D_{\hat{N}_{\Sigma j}(A_{th})}$:

$$D_{\hat{N}_{\Sigma a}(A_{th})} = \frac{1}{M} D_{\hat{N}_{\Sigma j}(A_{th})}. \tag{8.9}$$

Estimation of the average amplitude in the time interval for the j-th magnetization period T looks like:

$$\hat{A}_j(l,\ A_{th}) = \frac{1}{N_j(l)} \sum_{i=1}^{N_j(l)} A_{ji|(A_{ji}>A_{th})}(l), \tag{8.10}$$

where $N_j(l)$ is the number of pulses recorded in the time window $[(l-1)T_d, lT_d]$, T_d is the signal sampling interval, $l = 1, 2, \ldots$ is a window number.

Accordingly, for M realizations we have

$$\hat{A}_a(l,\ A_{th}) = \frac{1}{M} \sum_{j}^{M} \frac{1}{N_j(l)} \sum_{i=1}^{N_j(l)} A_{ji|(A_{ji}>A_{th})}(l), \tag{8.11}$$

which also reduces the variance in M times.

Estimation of the number of pulses during the time window in the j-th realization is expressed by the formula

$$\hat{N}_j(l,\ A_{th}) = \sum_{i=1|((A_{ji}>A_{th}),\ T_d(l-1)\le t_{ji}<T_d l)}^{N_j(l)} H(t - t_{ji}), \tag{8.12}$$

and for M realizations it looks like

$$\hat{N}_a(l,\ A_{th}) = \frac{1}{M} \sum_{j=1}^{M} \sum_{i=1|((A_{ji}>A_{th}),\ T_d(l-1)\le t_{ji}<T_d l)}^{N_j(l)} H(t - t_{ji}). \tag{8.13}$$

In addition to the traditional MAE signal parameters, the envelope of the MAE signal, in particular its shape and duration, is an important characteristic.

Estimation of the envelope by the moving average algorithm is found by the formula:

$$\hat{s}_j(k) = \frac{1}{N} \sum_{i=0}^{N-1} \left| s_j(k+i) \right|. \tag{8.14}$$

Here $|s(i)|$ is the modulus of the i-th signal reading, N is the number of signal readings that are averaged (the width of the averaging window), $\hat{s}(k)$ is the k-th value of the envelope estimate.

In order to reduce the variance of the envelope estimate, it is also averaged by the number of sampling M registered as a result of the experiment:

$$\hat{s}_a(k) = \frac{1}{M} \sum_{j=1}^{M} \frac{1}{N} \sum_{i=0}^{N-1} \left| s_j(k+i) \right|. \tag{8.15}$$

The envelope estimate by the algorithm of the least squares is found by the formula

$$\hat{s}_j(k) = \sqrt{\frac{1}{N}\sum_{i=0}^{N-1} s_j^2(k+i)}; \tag{8.16}$$

for M sampling we have

$$\hat{s}_a(k) = \frac{1}{M}\sum_{j=1}^{M}\sqrt{\frac{1}{N}\sum_{i=0}^{N-1} s_j^2(k+i)}. \tag{8.17}$$

Here $s^2(i)$ is the square of the i-th signal reading, N is the number of signal readings that are averaged (the width of the averaging window), $\hat{s}(k)$ is the k-th value of the envelope estimate.

The variance of the estimates (8.15) and (8.17) averaged over the number of sampling M decreases and is equal to

$$D_{\hat{s}_a(k)} = \frac{1}{M} D_{\hat{s}_j(k)}.$$

When interpreting the MAE signal as a pulsed random process (8.1), statistical characteristics, namely the shape, parameters, moments and entropy of the amplitude and time distribution of the random moment flux can be additionally used and evaluated.

The amplitude distribution of the MAE signal can be estimated using a multichannel amplitude analyzer with l_0 adjacent amplitude "windows" of width ΔA each or using the program based on the results of digitized data from the signal sampling device. Then the histogram of the amplitude distribution of the MAE signal for j realization will look like:

$$\hat{h}_j^A(l, A_{th}) = \begin{cases} 0, \ A \le A_{th}; \\ N_{jl}^A / \sum_{l=1}^{l_0} N_{jl}^A, \ A \in (A_{th} + (l-1)\Delta A, \ l\Delta A]; \\ 0, \ A > A_{th} + l_0 \Delta A. \end{cases} \tag{8.18}$$

Here, the lower limit of the dynamic range of the pulse amplitudes is determined by the threshold value A_{th}, which depends on the noise level. The upper limit $(A_{th} + l_0 \Delta A)$ is determined by the capacity of the analyzer; N_l is the number of readings in the channel with number l.

Histogram averaged for M realizations is described by formula

$$\hat{h}_a^A(l, A_{th}) = \frac{1}{M}\sum_{j=1}^{M} \hat{h}_j^A(l, A_{th}) \tag{8.19}$$

with estimation variance

$$D_{\hat{h}_a^A(l,A_{th})} = \frac{1}{M} D_{\hat{h}_j^A(l,A_{th})}. \tag{8.20}$$

The estimate of the first and second moments of the amplitude distribution is equal to, respectively:

$$\hat{A} = \Delta A \sum_{l=1}^{l_0} (l \cdot \hat{h}^A(l,\ A_{th})) \tag{8.21}$$

and

$$\hat{A}^2 = (\Delta A)^2 \sum_{l=1}^{l_0} (l^2 \cdot \hat{h}^A(l,\ A_{th})). \tag{8.22}$$

The histogram of the distribution of intervals between adjacent pulses of the MAE signal flux for j realization looks like

$$\hat{h}_j^{\vartheta}(l,\ A_{th}) = \begin{cases} N_{jl}^{\vartheta}(l) / \sum\limits_{l=1}^{l_0} N_{jl}^{\vartheta}(l),\ (A > A_{th}) \wedge (\vartheta \in (\Delta\vartheta(l-1),\ \Delta\vartheta l]); \\ \qquad 0,\ \vartheta > \Delta\vartheta l_0, \end{cases} \tag{8.23}$$

and for the histogram averaged for M realizations we have

$$\hat{h}_a^{\vartheta}(l,\ A_{th}) = \frac{1}{M} \sum_{j=1}^{M} \hat{h}_j^{\vartheta}(l,\ A_{th}) \tag{8.24}$$

and, accordingly, the variance:

$$D_{\hat{h}_a^{\vartheta}(l,\ A_{th})} = \frac{1}{M} D_{\hat{h}_j^{\vartheta}(l,\ A_{th})}. \tag{8.25}$$

Estimation of the first and second moments of intervals distribution are expressed by equalities

$$\hat{\vartheta} = \Delta\vartheta \sum_{l=1}^{l_0} (l \cdot \hat{h}^{\vartheta}(l,\ A_{th})) \tag{8.26}$$

and

$$\hat{\vartheta}^2 = (\Delta\vartheta)^2 \sum_{l=1}^{l_0} (l^2 \cdot \hat{h}^{\vartheta}(l,\ A_{th})). \tag{8.27}$$

Testing the hypothesis on the law of distribution of the MAE signal pulse amplitudes. To test the hypothesis on a specific law of distribution, we can use the so-called fitting criteria known from the literature, which are conventionally divided into two classes—general and special. The general criteria can be divided into three main groups:

1. those based on the study of the difference between the theoretical density of distribution and the empirical histogram;
2. those built on estimating the distance between theoretical and empirical probability distribution functions;
3. the correlation-regression criteria, which are based on the study of correlation and regression relationships between empirical and theoretical order statistics.

Comparison of the empirical histogram of the distribution of a random variable with its theoretical density forms the basis of criteria χ^2, vacuous intervals, Barnett–Eisen intervals and others.

However, it is known that the estimation of the distribution density by the histogram gives an offset error (Skalskyi et al. 2014). You can reduce the offset error by appropriate narrowing of the interval ΔA or $\Delta\vartheta$. This leads to an increase in the histogram estimate variance, which can be reduced by increasing the number of averaging realizations.

Experimental studies were performed using the MAE method on 3 steel (see Appendix) and nickel specimens. According to algorithm (8.18) for one sampling the histograms of the pulse amplitude distributions of the MAE signals were found (Fig. 8.2a, b). A general view of histograms allows us to choose the exponential law of distribution as a zero hypothesis.

If to test the hypothesis of the exponential law of distribution of amplitudes the fitting criterion χ^2 and evaluation $\hat{h}(l)$ are used. It is necessary to calculate the statistics

$$\chi^2 = N_{\Sigma} \sum_{l=1}^{l_0} (\hat{h}(l) - p_l)^2 / p_l, \tag{8.28}$$

where $p_l = \int_{(l-1)\Delta A}^{l\Delta A} \lambda \cdot exp(-\lambda A) \mathrm{d}A$ is theoretical probability of fitting the amplitude in the l window, λ is the distribution parameter, N_{Σ} is the specimen size.

Statistics (8.28) has distribution χ^2

$$p(x) = \frac{(1/2)^{k/2}}{\Gamma(k/2)} x^{k/2-1} e^{-x/2}, \tag{8.29}$$

where k is the number of degrees of freedom.

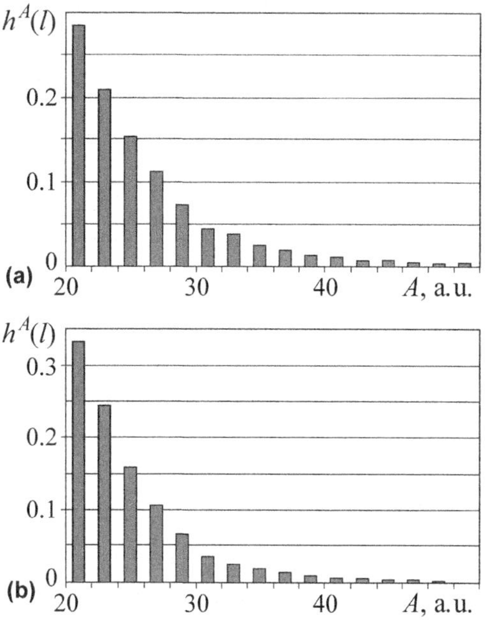

Fig. 8.2 Histograms of the distribution of the amplitudes of the MAE signal pulse for steel (**a**) and nickel (**b**) specimens: the amplitude of the magnetic field induction is 0.8 and 0.4 T, respectively

If the distribution of a random variable is known with an accuracy to its parameters, then $k = l_0 - 1$. If the parameters of the hypothetical distribution law are evaluated by the sampling itself, the number of degrees of freedom is reduced by the number of parameters m, that are evaluated. The exponential distribution is one-parametrical, therefore $m = 1$, and accordingly $k = l_0 - 2$.

The power χ^2 of the criterion is greatly influenced by the number of partitioning intervals (window width ΔA) of the histogram l_0. In practice, it is believed that statistics can be used if $N_\Sigma p_l \geq 5$.

The rule of hypothesis testing is simple (Skalskyi et al. 2014): if

$$N_\Sigma \sum_{l=1}^{l_0} (\hat{h}(l) - p_l)^2 / p_l > \chi_\alpha^2, \tag{8.30}$$

then at the level of significance α, i.e., with $(1 - \alpha)$ reliability, the hypothesis of the exponential law of distribution is rejected. Otherwise, the hypothesis is accepted at a given significance level. Here χ_α^2 is the α-quantile of the distribution (8.29).

To construct the histogram of the amplitudes distribution for the steel specimen, take the number of intervals $l_0 = 16$. Set the confidence probability $\alpha = 0.95$, the sample size $N_\Sigma = 3371$. Intermediate calculation results are used to calculate the criterion value $\chi^2 = 15.8$. For the selected confidence probability $\alpha = 0.95$ and the number of degrees of freedom $k = 16 - 2$ the critical value of the statistics of the criterion $\chi_{0,05}^2 = 23.7$ is found in the table. Since the critical value proves to be higher (23.7 > 15.8) than that calculated from statistical data, the hypothesis that the sample belongs to the exponential distribution must be accepted.

For the histogram averaged for $M = 10$ realizations, the calculated criterion value is $\chi^2 = 6.1$, i.e., that is, the condition that the sampling belongs to the exponential distribution is strengthened.

The performed calculations allow us to state that the amplitude of the MAE signal pulses is distributed according to the exponential law, and its parameter λ can be used as informative one when diagnosing ferromagnetic objects by the MAE method.

8.3 Investigation of Changes in Informative Parameters of MAE Signals Under Different Remagnetization

Informative parameters of the MAE signal are sensitive to the structural changes in the ferromagnetic material, heat treatment mode, plastic deformation, residual stresses, hydrogen charging, etc. (Skalskyi et al. 2014; Shibata and Ono 1981; Skalsky et al. 2011; Perevertov and Stupakov 2015; Skal'skii et al. 2012; Nazarchuk et al. 2013).

On the other hand, the process of MAE signal generation is determined by the parameters of the remagnetizing field (amplitude, frequency and shape of the signal). It depends on the shape and size of remagnetization objects (solenoid, overhead electromagnet) and location of the field source (Skalskyi et al. 2014). Therefore, an important problem when using the MAE method for diagnosing ferromagnetic structural elements is to consider the influence of experimental factors on the signal, in addition to the structure of the ferromagnetic material.

The peculiarities of the influence of experimental factors on the MAE signal parameters are the amplitude of induction of the remagnetizing field; the magnitude of the non-magnetic layer between the overhead electromagnet and the specimen surface, and the thickness of the latter have been investigated (Pochaps'kyi et al. 2018).

Dependence of the parameters of magnetoelastic acoustic emission signals on the amplitude of induction of the remagnetizing field. Regularities of the process of ferromagnetic material magnetization are graphically illustrated in the form of a magnetization curve, which illustrates the dependence of the induction amplitude $\vec{B}$ on the magnetic field intensity $\vec{H}$. Under cyclic variation of the external magnetic field a symmetrical hysteresis loop is obtained. The loop for which the state of technical saturation is reached is called boundary, all the other loops that are inside are called partial (Nazarchuk et al. 2013).

Plane specimens made of nickel and 3 steel with a thickness of 1, 2, and 5 mm were prepared for investigations. Remagnetization was performed applying a sinusoidal signal with a frequency of 9 Hz, using an overhead electromagnet (Pochaps'kyi et al. 2018) for several values of the amplitude of the magnetic field induction, which correspond to several partial hysteresis loops.

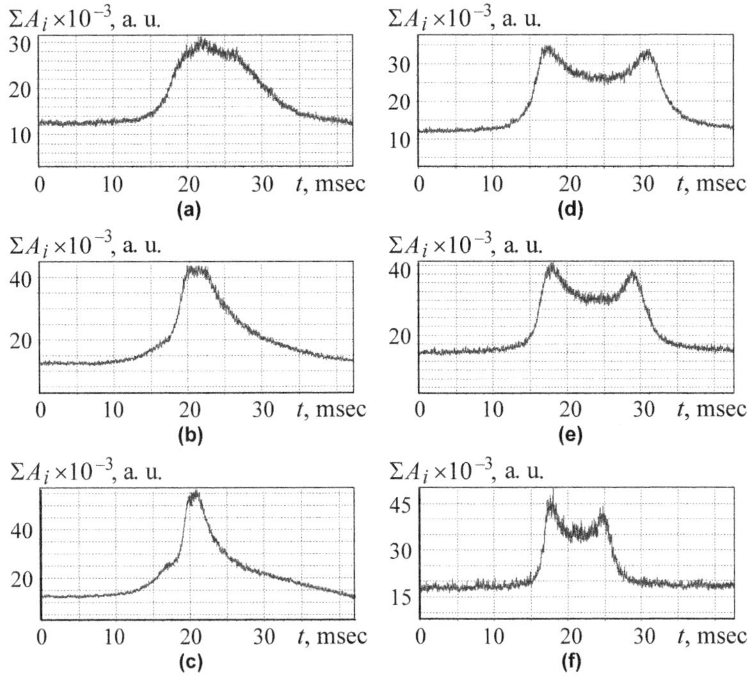

Fig. 8.3 MAE signals envelope: **a–c** nickel plate (B = 0.29 T; 0.48 T; 0.60 T); **d–f** steel plate (B = 0.72 T; 1.22 T; 1.84 T)

The influence of the amplitude of induction of the remagnetization field $\vec{B}$ on the parameters of the generated MAE signals was established. Figure 8.3 presents the envelope of the MAE signals for nickel (a–c) and 3 steel (d–f) specimens, recorded for the ascending branch of the remagnetization loop.

The MAE envelopes for the nickel specimen have one peak, the shape of which changes with increasing B. In particular, it becomes narrower and increases in amplitude. The MAE envelopes for the steel specimen have two peaks, the amplitude of which increases with increasing B, and they are located closer to each other due to the decrease in the MAE signals duration.

For nickel and steel specimens, it is experimentally confirmed that the duration of magnetoelastic acoustic emission signals decreases with the increasing amplitude of magnetization field induction (Fig. 8.4a, b).

It also follows from the obtained experimental results that the sum of the MAE signal amplitudes increases almost linearly with increasing the amplitude of the remagnetization field induction (Fig. 8.5a, b).

The observed behavior of the MAE signal parameters is explained by the fact that with increasing the amplitude of field induction the rate of its change during remagnetization of ferromagnetic material increases, and thus increases the intensity of domain wall jumps.

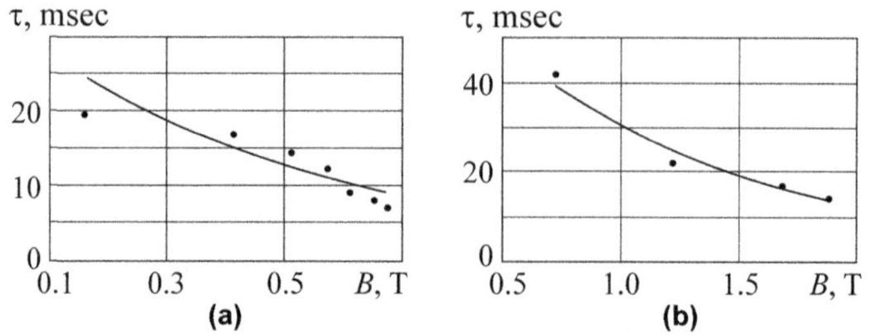

Fig. 8.4 Dependences of the MAE signals duration τ on the amplitude of the remagnetization field induction B: **a** nickel plate; **b** steel plate

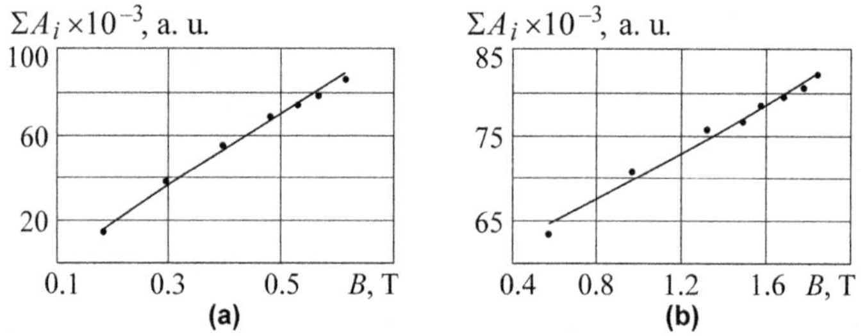

Fig. 8.5 Dependences of the sum of the amplitudes of MAE signals on the amplitude of the magnetization field induction B: **a** nickel plate; **b** steel plate

Dependence of the parameters of magnetoelastic acoustic emission signals on the non-magnetic layer thickness. The influence of the thickness of the non-magnetic layer between the surface of the test specimen and the legs of the magnetic circuit of the overhead electromagnet (observed during large-scale objects diagnostics) on the shape of the MAE envelope signals and their duration was studied (Pochaps'kyi et al. 2018). The presence of a non-magnetic layer causes the appearance of scattering and demagnetization fields (Tikadzumi 1987), as a result of which the magnetizing loop occupies a more inclined position in comparison with the true magnetization curve of the material.

This occurs because in reality the internal field in the material (effective field H_i) will be smaller than the external one H_e due to the appearance of demagnetization field H_r caused by the formation of magnetic poles

$$H_i = H_e - H_r = H_e - NJ. \tag{8.31}$$

Here N is the demagnetization coefficient, J is magnetization of the specimen.

For nickel specimens, when the demagnetizing field is absent, the peaks of MAE activity in the vicinity of the “knees” of the hysteresis and coercive field loops are located very close, and as a result, one continuous peak is formed on the envelope. As the thickness of the non-magnetic layer increases, the demagnetizing field causes an

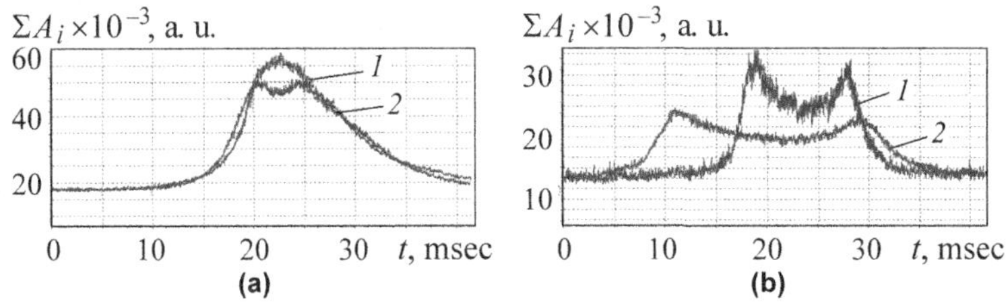

Fig. 8.6 Envelope of MAE signals: **a** nickel plate (curve *1*—$l = 0.05$ mm; curve *2*—$l = 4.7$ mm), $B = 0.48$ T; **b** steel plate (curve *1*—$l = 0.05$ mm; curve 2—$l = 4.7$ mm), $B = 1.68$ T

increase in the duration of MAE signals and a decrease in their amplitude at a constant value of the magnetic induction of the external remagnetization field (Fig. 8.6a).

The MAE envelopes for steel have two peaks, the amplitude of which decreases with increasing layer thickness, and they move away, i.e., the duration of magnetoelastic acoustic emission signals increases (Fig. 8.6b). The appearance of two peaks may be associated with the instability of the change rate of the external demagnetizing field dH/dt and requires additional studies in a wide range of remagnetization frequencies (Perevertov and Stupakov 2015).

When diagnosing with the MAE method, it is advisable to control the rate of the remagnetizing field, as well as the shape of the field, as the peaks of the MAE envelopes can be caused by the formation of different magnetic phases in the material and specific form of dH/dt. The best way is to perform magnetization at a constant field rate, but technically this is quite difficult to implement.

It is experimentally confirmed that for a constant value of induction of the magnetizing field and with increasing magnitude of the demagnetizing fields (layer thickness) the duration of magnetoelastic acoustic emission signals increases for both nickel and steel (Fig. 8.7a).

From the obtained experimental results we also see that the sum of the amplitudes of the MAE signals decreases almost linearly with increasing thickness of the non-magnetic layer (Fig. 8.7b).

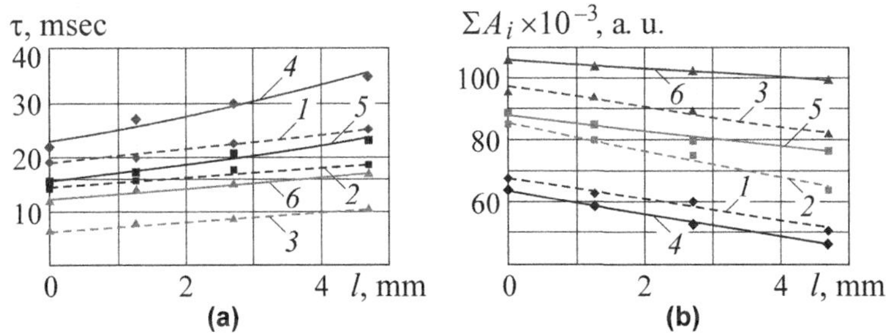

Fig. 8.7 Dependences: **a** duration τ and **b** sum of amplitudes of MAE signals on the thickness *l* of the non-magnetic layer (nickel plate: curve *1*—$B = 0.29$ T; curve *2*—$B = 0.48$ T; curve *3*—$B = 0.60$ T; steel plate: curve: *4*—$B = 0.72$ T; curve *5*—$B = 1.22$ T; curve *6*—$B = 1.84$ T)

Dependence of the parameters of magnetoelastic acoustic emission signal on the thickness of the investigated ferromagnetic specimen. The peculiarities of the magnetoelastic acoustic emission envelope for nickel and 3 steel specimens of thickness $h = 1$; 2 and 5 mm have been studied at constant values of the remagnetizing field and magnetic induction (for nickel—$B = 0.48$ T; for 3 steel—$B = 1.84$ T) (Pochaps'kyi et al. 2018).

The amplitude of the signals of magnetoelastic acoustic emission increases with increasing thickness of the test specimens, both for those made of nickel and 3 steel (Fig. 8.8a, b). There is also a noticeable separation of the MAE signals at the envelope of signal peaks and a significant increase in signal duration. The observed behavior of changes in the intensity and increase in the MAE signals duration can be explained by the increasing influence of eddy currents that occur in conductors during the temporal variation in the magnetic flux and cause phase shifts. As the thickness of the specimens decreases, the separation of the peaks becomes clear and distinct. This is due to the fact that almost all areas of the specimen are in the same remagnetization phase. Thicker specimens are characterized by a larger phase shift of the magnetizing field for deeper layers of ferromagnetic material (Augustyniak et al. 2006).

It has been experimentally confirmed that the duration of magnetoelastic acoustic emission signals increases with increasing thickness of the investigated nickel and steel specimens at a constant value of the remagnetizing field induction (Fig. 8.9).

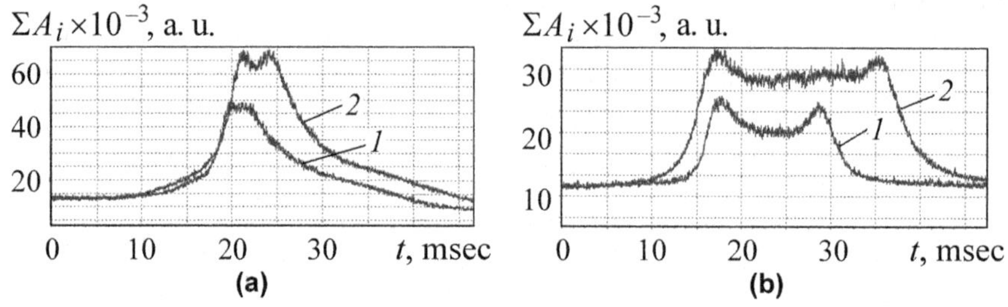

Fig. 8.8 Envelopes of MAE signals: **a** nickel plate (curve *1*—$h = 1$ mm; curve *2*—$h = 2$ mm), $B = 0.48$ T; **b** steel plate (curve *1*—$h = 1$ mm; curve *2*—$h = 2$ mm), $B = 1.84$ T

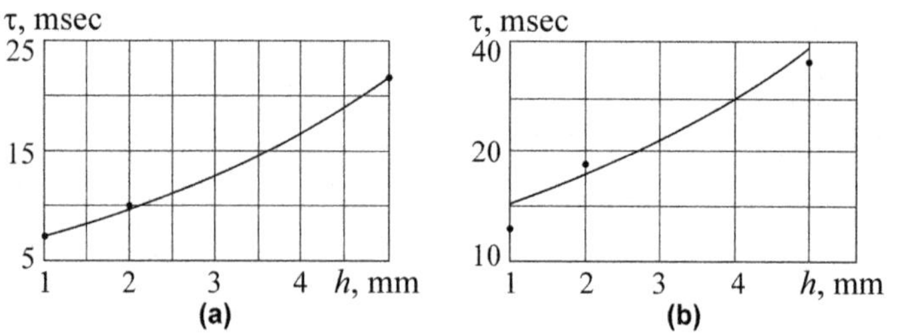

Fig. 8.9 Dependences of the MAE signals duration on the thickness h of the investigated ferromagnetic specimen: **a** nickel plate, $B = 0.48$ T; **b** steel plate, $B = 1.84$ T

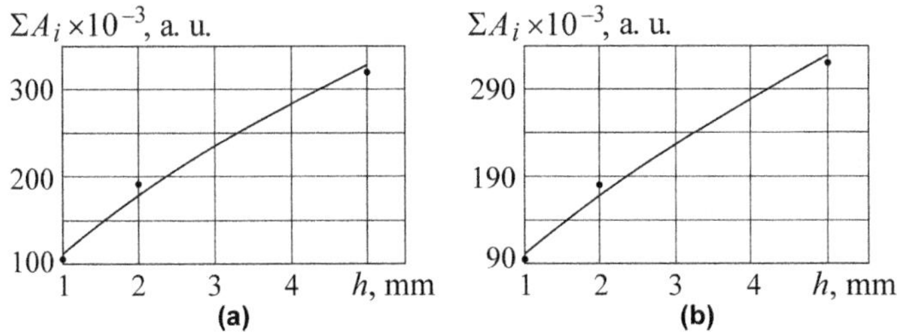

Fig. 8.10 Dependences of the sum of the MAE signals amplitudes on the thickness h of the investigated ferromagnetic specimen: **a** nickel plate, $B = 0.480$ T; **b** steel plate, $B = 1.840$ T

From the obtained experimental results we also see that the sum of the amplitudes of the MAE signals increases almost linearly with increasing thickness of the studied specimens at a constant amplitude of induction of the remagnetizing field (Fig. 8.10). This is explained by the total increase of the signal generation sources (increase in the number of non-180° domain walls) due to the increase in the volume of the remagnetization area.

Based on the research results, we can draw a conclusion on the influence of experimental factors on the MAE signal generation. To ensure comparability of the results of diagnosing objects made of ferromagnetic materials by the MAE method, it is necessary to create the same conditions for each individual diagnostic experiment.

8.4 Excitation of MAE Signals

The process of MAE signals excitation is an important aspect of the method of magnetoelastic acoustic emission (Augustyniak et al. 2008; Gaunkar et al. 2014).

Figure 8.11 shows a corresponding chart, the main components of which are the unit for generating the remagnetization signal *1*, an overhead electromagnet, which includes an U-shaped magnetic circuit 2 and an excitation coil *3*, a test specimen *4*, a non-magnetic layer between magnetic conductor and sample *5*, a measuring coil *6*, a resistor for measuring excitation current *7*.

Fig. 8.11 MAE signal excitation scheme

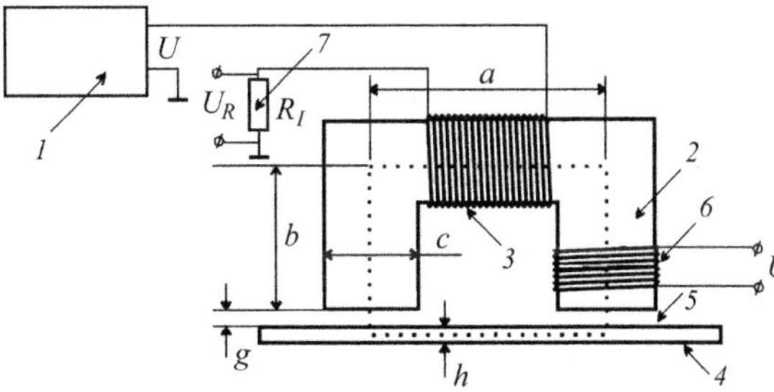

Known dependences were used to calculate the magnetic field induction of the magnetic circuit and the overhead electromagnet. Thus, the Kirchhoff's second law for a magnetic circuit formed by a magnetic conductor, a non-magnetic layer and a specimen can be written as follows

$$F_m = \sum_i U_i^m. \tag{8.32}$$

Here the magnetomotive force is

$$F_m = NI, \tag{8.33}$$

where N and I are the number of turns and current in the winding of the overhead electromagnet.

Magnetic voltage, respectively, is

$$U_i^m = H_i l_i, \tag{8.34}$$

where H_i and l_i are the magnetic field intensity and the length of the i-th section of the magnetic circuit, respectively.

Taking into account the field intensity

$$H_i = \frac{B_i}{\mu_i \mu_0}, \tag{8.35}$$

where B_i is the magnetic field induction, μ_i is the relative magnetic permeability of the ferromagnetic material, μ_0 is the magnetic constant.

The magnetic resistance of the i-th section of the circuit looks like

$$R_i^m = \frac{l_i}{\mu_i \mu_0 A_i}, \tag{8.36}$$

where A_i is the cross-sectional area of the circuit section.

Then the expression for the magnetic voltage is obtained

$$U_i^m = \Phi R_i^m, \tag{8.37}$$

where Φ is the magnetic flux in the circuit.

Using (8.32)–(8.37) for the magnetic flux in the circuit we obtain the expression

$$\Phi = \frac{NI}{\sum_i R_i^m}, \tag{8.38}$$

that is, the flux is inversely proportional to the magnetic resistance of the circuit.

The magnetic resistance of the circuit is found from the expression

$$R^m = R_c^m + R_g^m + R_s^m, \tag{8.39}$$

where the magnetic resistance of the magnetic circuit

$$R_c^m = \frac{l_c}{A_c \mu_0 \mu_c}, \tag{8.40}$$

the magnetic resistance of the layer

$$R_g^m = \frac{l_g}{A_g \mu_0} \tag{8.41}$$

and the magnetic resistance of the specimen

$$R_s^m = \frac{l_s}{A_s \mu_0 \mu_s}. \tag{8.42}$$

Here μ_c, μ_s is the relative magnetic permeability of the magnetic circuit of material and the specimen, A_c, A_g, A_s is the cross-section of the magnetic circuit, the layer, the specimen, respectively, and $l_c = a + 2b$, $l_g = 2g$, l_s are the lengths of the respective sections of the magnetic circuit.

Figure 8.12 shows the dependence of the magnetic field induction for a U-shaped overhead electromagnet in the specimen on the width of the non-magnetic layer obtained by numerical methods. One can see a decrease in the field induction with increasing layer width, which means, according to (8.41), an increase in the magnetic resistance of the circuit.

When diagnosing by the MAE method to improve the measuring accuracy, it is very important to ensure the stability of the value of the magnetizing field amplitude (or constancy of the remagnetization current) and its independence from changes in the value of the non-magnetic layer (which is typical of the practice of diagnosing large-scale objects) (Ng et al. 1994a, 1994b; White et al. 2007).

For the instantaneous values of current I and voltage U in the excitation circuit of the scheme in Fig. 8.11 we can write

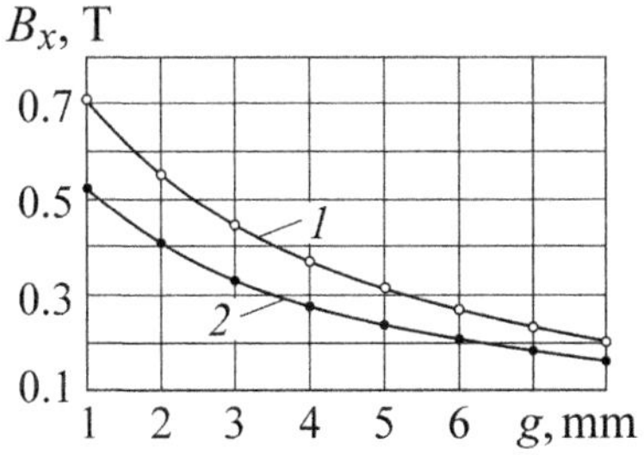

Fig. 8.12 Dependence of magnetic field induction for U-shaped overhead electromagnet in the specimen on the interval width: at a depth of 1 mm—curve *1*; 4 mm—curve *2*

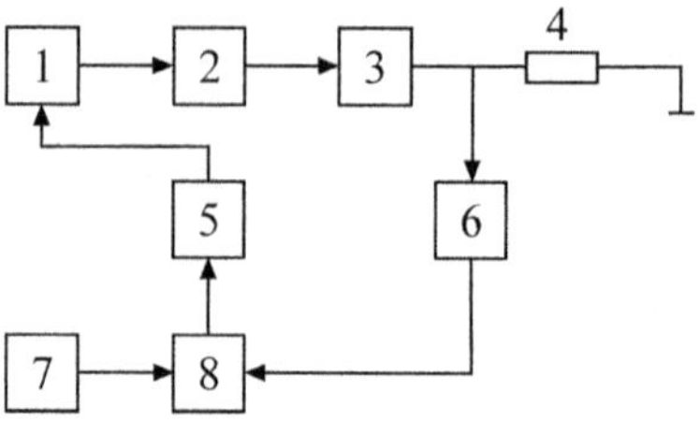

Fig. 8.13 Block diagram of the system of automatic stabilization of the amplitude of the remagnetization current

$$U = L\frac{dI}{dt} + (R_c + R_I)I, \tag{8.43}$$

where L is the circuit inductance, R_c is the active resistance of the excitation coil, R_I is the reference resistance used to measure the current.

The amplitude values of the magnetizing current I^0 and the voltage U^0 are related by the relationship

$$I^0 = U^0/Z = U^0/\sqrt{(R_c + R_I)^2 + \omega^2 L^2}, \tag{8.44}$$

where Z is the total resistance of the circuit, ω is the cyclic frequency of the current.

Figure 8.13 illustrates the proposed scheme for automatic stabilization of the amplitude of the remagnetization current. It consists of a remagnetization signal generator *1*, a signal amplifier *2* with a voltage gain k_0, an overhead electromagnet *3*, a reference resistance *4*, a control signal generator *5*, a current amplitude meter *6*, a remagnetization current amplitude setting unit *7*, a difference measuring unit *8*.

An iterative approach was used to achieve the target value of the remagnetization current amplitude I_t^0, which is set by program in block *7*. During the n-th remagnetization period, the current amplitude I_n^0 is measured in block *6* using the reference resistor *4*. In blocks *8* and *5*, respectively, a difference $(I_t^0 - I_n^0)$ is found and a control signal $(I_t^0 - I_n^0)k$ is formed for the generator of the magnetizing signal *1*. Here, the constant k determines the rate at which the system reaches the target value of the current. In the n-th step of the iterative process, the voltage at the output of the amplifier *2* will be equal to

$$U_n^0 = U_{n-1}^0 + (I_t^0 - I_{n-1}^0)kk_0. \tag{8.45}$$

Because $I_{n-1}^0 = U_{n-1}^0/Z$, then

$$U_n^0 = I_t^0 kk_0 + U_{n-1}^0\left(1 - \frac{kk_0}{Z}\right). \tag{8.46}$$

From the last expression the sum of n terms of a geometric progression with a denominator $q = \left(1 - \frac{kk_0}{Z}\right)$ is obtained:

$$U_n^0 = I_t^0 k k_0 \left[1 + \left(1 - \frac{k k_0}{Z}\right) + \cdots + \left(1 - \frac{k k_0}{Z}\right)^{n-1} \right]. \tag{8.47}$$

After summing up we have

$$U_n^0 = I_t^0 Z \left[1 - \left(1 - \frac{k k_0}{Z}\right)^n \right]. \tag{8.48}$$

In the limit at $n \to \infty$ under condition $0 < \left(1 - \frac{k k_0}{Z}\right) < 1$, finally $U_t^0 = I_t^0 Z$, i.e., the system provides the target current of remagnetization.

Let us investigate the dynamic properties of the system of remagnetization current stabilization. Let Δ_t be the absolute admissible error of setting a given amplitude of the remagnetization current. Then the relative error will be equal to

$$\delta = \Delta_t / I_t^0. \tag{8.49}$$

Assume that after n steps the relative error of setting the voltage by the system is

$$(U_t^0 - U_n^0)/U_t^0 \leq \delta, \tag{8.50}$$

or

$$\left(1 - \frac{k k_0}{Z}\right)^n \leq \delta. \tag{8.51}$$

Finding the logarithm of the expression (8.51) for the number of cycles for which the specified relative error of setting the amplitude of the remagnetization current is achieved, we obtain

$$n \leq \lg(\delta) / \lg\left(1 - \frac{k k_0}{Z}\right). \tag{8.52}$$

Figure 8.14 presents the dependences of the number of cycles on the relative error of setting a given remagnetization current for some values of the parameter $\xi = k k_0 / Z$, which is determined by the design features of the stabilization system.

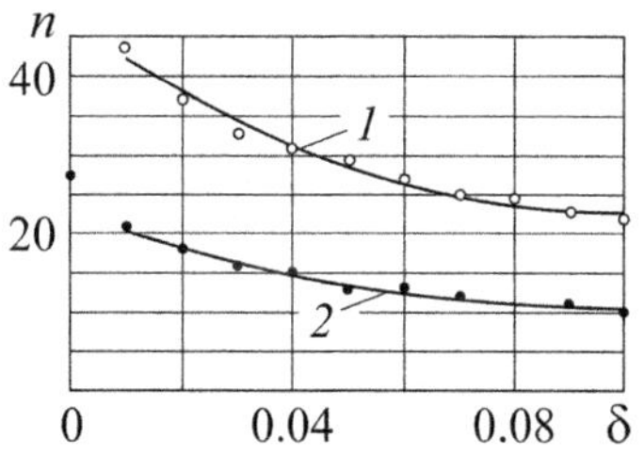

Fig. 8.14 Dependence of the number of cycles on the relative error of setting a given magnetization current: *1*—$\xi = 0.1$; *2*—$\xi = 0.2$

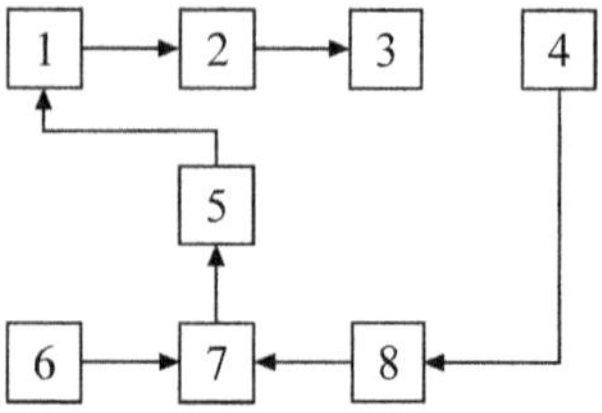

Fig. 8.15 Block diagram of the system for automatic stabilization of the amplitude of the magnetic field induction

Figure 8.15 shows the developed scheme for automatic stabilization of the amplitude of the remagnetization field induction. It consists of a remagnetization signal generator *1*, a remagnetization signal amplifier *2* with a voltage gain k_0, an overhead electromagnet *3*, a measuring coil *4*, a shaping unit of the control signal *5*, a unit for setting the amplitude of the induction of the remagnetization field *6*, a difference measuring unit *7*, a meter of the amplitude of the remagnetizing field induction *8*.

The required value of the induction amplitude of the remagnetizing field is set by program in block *6*. During the n-th remagnetization period in block *8* using the measuring coil *4* the induction amplitude of the remagnetization field $B_n^0 = \Phi_n^0/A$ is measured. Here Φ_n^0 is the amplitude of the magnetic flux in the investigated specimen or in the magnetic circuit of the overhead electromagnet *2* (Fig. 8.11, that is measured using a measuring coil *6*), A is their cross-section. In blocks *7* and *5*, respectively, a difference $(B_t^0 A - \Phi_n^0)$ is found and a control signal $(B_t^0 A - \Phi_n^0)k$ is formed for the remagnetization signal generator *1*. Here, constant k determines the rate at which the system reaches the required value of induction. In the n-th step of the iterative process, the voltage at the output of the amplifier *2* will be equal to

$$U_n^0 = U_{n-1}^0 + (B_t^0 A - \Phi_{n-1}^0)kk_0, \tag{8.53}$$

or taking into account (8.38)

$$U_n^0 = U_{n-1}^0 + \left(B_t^0 A - \frac{N I_{n-1}^0}{\sum_i R_i^m}\right)kk_0. \tag{8.54}$$

Because $I_{n-1}^0 = U_{n-1}^0/Z$, we will get

$$U_n^0 = U_{n-1}^0 + \left(B_t^0 A - \frac{N U_{n-1}^0}{Z\sum_i R_i^m}\right)kk_0, \tag{8.55}$$

and after transformations

$$U_n^0 = B_t^0 Akk_0 + (1-g)U_{n-1}^0, \tag{8.56}$$

where $g = \frac{N}{Z\sum_i R_i^m}$.

The last expression can be written as the sum of n terms of the geometric progression with the denominator $(1-g)$:

$$U_n^0 = B_t^0 Akk_0\left[1 + (1-g) + \cdots + (1-g)^{n-1}\right], \tag{8.57}$$

or in the convolved form

$$U_n^0 = \frac{B_t^0 AZ\sum_i R_i^m}{N}\left[1 - (1-g)^n\right]. \tag{8.58}$$

When $n \to \infty$ under condition $0 < (1-g) < 1$ we finally obtain $U_n^0 = B_t^0 \frac{AZ\sum_i R_i^m}{N}$, i.e., the system provides the desired value of the amplitude of the remagnetization field induction.

For the number of cycles for which the specified relative error of setting the amplitude of the remagnetizing field induction is achieved, similarly to (8.52) we obtain

$$n \le \lg(\delta)/\lg(1-g) \tag{8.59}$$

and the presented in Fig. 8.14 dependences will also describe the dynamics of establishing the required value of the induction amplitude for some values of the parameter $g = N/(Z\sum_i R_i^m)$, which is determined by the design peculiarities of the stabilization system, shown in Fig. 8.15.

Thus, the created model takes into account the typical features of signals: random occurrence in time of individual events (random flux of pulses); limited time, finiteness energy; the randomness of the amplitude. This model allows us except of measuring the traditional parameters of the signals, to use additionally their statistical characteristics the moments and parameters of the functions of the amplitude and time distributions. The amplitude of the MAE signal pulses is distributed according to the exponential law, and, therefore, its parameter λ can be used as informative when diagnosing ferromagnetic objects. The amplitude of induction of the remagnetizing field, the thickness of the non-magnetic layer and the thickness of the specimen affect the amplitude, the shape of the envelope and the duration of the MAE signal. The signal duration decreases with increasing amplitude of field induction, but increases with the thickness of the layer and the material under study at a constant value of the induction amplitude. The sum of signal amplitudes increases with increasing amplitude of the field or thickness of the studied material at a fixed field value and decreases with growth of the non-magnetic layer thickness.

References

Augustyniak M, Augustyniak B, Piotrowski L, Chmielewski M, Sadowski W (2006) Evaluation by means of magneto-acoustic emission and Barkhausen effect of time and space distribution of magnetic flux density in ferromagnetic plate magnetized by a C-core. J Magn Magn Mater 304:552–554

Augustyniak B, Piotrowski L, Chmielewski M (2008) Impact of frequency and sample geometry on magnetacoustic emission voltage properties for two steel grades. J Electr Eng 59:33–36

Buttle DJ, Scruby CB, Yakubovics JP, Briggs JAD (1987) Magnetoacoustic and Barkhauzen emission: their dependence on dislocation in iron. Philos Mag 55(6):717–734
Durin G, Zapperi S (2002) On the power spectrum of magnetization noise. J Magn Magn Mater 242–245:1085–1088
Gaunkar N, Kypris O, Nlebedim IC, Jiles DC (2014) Optimization of sensor design for Barkhausen noise measurement using finite element analysis. J Appl Phys 115:17E512
Nazarchuk Z, Skalsky V, Pochapskyy Ye, Hirnyj S (2012) Application of magnetoacoustic emission for detection of hydrogen electrolytically absorbed by steel. In: Proceedings of the 19th European. Conference on Fracture "Fracture Mechanics for Durability, Reliability and Safety", Kazan, Russia, August 26–31, ID 405, 8
Nazarchuk ZT, Andreykiv OYe, Skalskyi VR (2013) Otsinyuvannia vodnevoi degradatsii feromagnetykiv u magnetnomu poli (Evaluation of hydrogen degradation of ferromagnets in a magnetic field). Naukova Dumka Publishing House
Ng DHL, Lo CCH, Jakubovics JP (1994a) The effects of demagnetizing and stray fields on magnetoacoustic emission. J Appl Phys 75(10):7009–7011
Ng DHL, Lo CCH, Cheng C (1994b) The dependence of magnetoacoustic emission on magnetic induction and specimen thickness. IEEE Trans Magn 30(6):4857–4859
Perevertov O, Stupakov A (2015) Magnetoacoustic measurement on steel samples at low magnetizing frequencies. J Electrical Eng 66:58–61
Pochaps'kyi EP, Mel'nyk NP, Koblan IM (2018) Influence of the conditions of excitation and generation of magnetoelastic acoustic emission signals in ferromagnetic materials. Mater Sci 54(3):444–449
Shibata M, Ono K (1981) Magnetomechanical acoustic emission—a new method of nondestructive stress measurement. NDT Int 227–234
Skal'skii VR, Klim BP, Pochapskii EP (2012) Distribution of the induction of a quasi-stationary magnetic field created in a ferromagnet by an attachable electromagnet. Russ J Nondestr Test 48(1):23–34
Skalsky V, Hirnyj S, Pochapskyy Ye, Klym B, Plakhtiy R, Tolopko Ya, Dolishniy P (2011) The effect of deformation on the parameters of magnetoacoustic signals. Visnyk Ternopilskogo natsionalnogo universytetu (Bulletin of Ternopil National University) Part 1:155–161
Skalskyi VR, Klym BP, Pochapskyi YeP, Simakovych OH, Dolishnij PM (2014) Magnetoakustychnyi metod kontroliu vmistu vodniu v feromagnetykskh (Magnetoacoustic method of controlling hydrogen content in ferromagnets). Fizykokhimichna Mekhanika Materialiv. Spetsialnyi vypusk (Physicochemical Mechanics of Materials. Special issue) 10(2):505–509
Tikadzumi S (1987) Fizika ferromagnetizma. Magnitnyie kharakteristiki i prakticheskiie primeneniia (Physics of ferromagnetism. Magnetic characteristics and practical applications). Mir Publishing House
White S, Krause T, Clapham L (2007) Control of flux in magnetic circuits for Barkhausen noise measurements. Meas Sci Technol 18:3501–3510

Chapter 9
Evaluation of Ferritic-Pearlitic Steels Degradation Under the Influence of Low Concentration of Hydrogen

The results of research on the characteristics of the interaction of low-concentration hydrogen with low-carbon (ferrite-pearlite class) steels using the MAE method are presented in this chapter. It was established that low concentrations of hydrogen in the structure of low-carbon steels do not significantly affect their elastic characteristics. Optimum annealing regimes of steels have been determined to ensure maximum remagnetization after hydrogen charging. The influence of hydrogen on the sum of amplitudes of MAE signals for undeformed and plastically deformed steel was determined, and the effect of carbon content in steel on the manifestation of hydrogen in it was estimated. The growth of MAE activity up to a certain concentration of hydrogen, and then its decrease is illustrated. It is important to take this into account when developing techniques for non-destructive control of the local concentration of hydrogen in such materials.

9.1 Problem State-of-the-Art

Recently, with the development of theoretical and experimental investigations, the physical aspects of the magnetization of the ferromagnet have been described, and the character of changes in domain structure under the influence of the applied (external) magnetic field has been studied (Rudiak 1986; Tikadzumi 1987a, b). However, the physical aspects of the effect of hydrogen on the magnetic properties of such materials have not yet been sufficiently disclosed. In this regard, it makes sense to establish the influence of the hydrogen factor and show the regularities that arise in this case as well as their manifestation at the macrolevel using modern ideas about the domain structure and the character of interactions within ferromagnets.

V. Skalskyi and Z. Nazarchuk, *Magnetoelastic Acoustic Emission*, Springer-AAS Acoustics Series, https://doi.org/10.1007/978-981-99-4032-5_9

Hydrogen, diffusing into the metal and absorbing in it, creates local places of triaxial mechanical tensile stresses, thus promoting the redistribution of dislocations. At high concentrations of hydrogen, pores, micro- and macrocracks are formed. These processes cause deterioration of the operational properties of the ferromagnet at the macrolevel—its hydrogen degradation takes place.

Simultaneously with these processes, there are changes in magnetic anisotropy and the domain structure of the ferromagnet caused by them, as a result of which the contours and dimensions of domains and domain walls change (Tikadzumi 1987a). The latter can obtain a reduction in the fixing points at the dislocations or a decrease in the energy required to separate them from the fixing points. On the other hand, hydrogen can unlock the movement of both individual dislocations and dislocation clusters.

In addition, due to the smallest atomic size among all the chemical elements, hydrogen ions penetrate into the ferromagnet crystal lattice and, at high concentrations there, interact with the electrons of the lattice atoms, reducing their spin magnetic moments. This leads to a decrease in the total magnetic moment of the domains as well as in the spin moment of the atoms in the domain wall. Therefore, it can be argued that there is some decrease in the exchange energy of the ferromagnet magnetic interaction under the hydrogen influence, depending on its concentration. In the external magnetic field at certain levels of intensity, a jump-like movement of the domain walls in the ferromagnet occurs. It causes magnetostriction and is accompanied by MAE emission. Integrally, this manifests itself in different ways, and therefore, by quantifying the magnetoelastic acoustic emission, we can establish the pattern of domain walls behavior under the influence of hydrogen.

Summarizing the above presented, the following model of the influence of the hydrogen factor on the remagnetization of ferritic-pearlitic steels is proposed. The presence of hydrogen of initially low concentration in the ferromagnetic material at first helps to unblock dislocations (unlike plastic deformations, changes in material structure, internal stresses, etc.) because only the primary effects of hydrogen-metal interaction are manifested in this case (Kush 1975). First, hydrogen interacts with the nuclei of the dislocation clusters and accumulates at the grain boundaries. That is, hydrogen in steels seems to displace carbon and nitrogen from dislocations because the energy of interaction of the latter with hydrogen is less than with nitrogen and carbon (Siede and Rostoker 1958; Calland et al. 1969). This facilitates the unblocking of dislocations, and hence their mobility, which in turn increases the number and amplitude of domain wall jumps during the ferromagnet remagnetization in the external magnetic field. This should occur to a certain (typical for this material) hydrogen concentration.

With a further increase of the concentration hydrogen forms clusters and blocks dislocations (especially their accumulation), which leads to a decrease in the intensity and amplitudes of the domain walls jumps. This occurs because the tearing off of the domain walls from the fixation points will be complicated by the increase in the number of dislocations, pore- and microcracking, and will lead to a decrease in the already mentioned indicators of MAE signal generation.

Thus, there should be a phenomenon of dual (first increase, and when reaching the typical value of hydrogen concentration for this material—decrease) changes in the level of amplitudes and the number of jump-like movements of domain walls during quasi-static remagnetization of ferritic-pearlitic steels in the presence of low concentrations of hydrogen.

The term "low concentration" for ferritic-pearlitic steels means its value is up to 3 ppm. In this case, it should be borne in mind that in the exploited steels of this grade, this indicator can reach 7 ppm (Dobrotvorsky and Archakov 1989), and the local concentration of hydrogen in such steels can theoretically exceed the average volume by several orders of magnitude, reaching a ratio of 1:1 (Tkachov 1999).

9.2 Calculation of the Depth of Magnetization in Specimens for Testing

To know what specimen thickness is required to carry out experimental research, it is necessary to establish a depth of penetration of a magnetic field into the material. Such calculations are performed for 15 steel (see Appendix) specimens according to the formula (Nazarchuk et al. 2001):

$$h = \sqrt{2/(\omega\sigma\mu_r\mu_0)} = \frac{1}{\sqrt{\pi f\sigma\mu_r\mu_0}}, \tag{9.1}$$

where μ_r is relative magnetic permeability; μ_0 is magnetic permeability in a vacuum; $\omega = 2\pi f$ is cyclic frequency; σ is the specific electrical conductivity of the material.

Using the magnetization curve (Fig. 9.1a) the dependence of magnetic permeability on the field intensity is constructed (Fig. 9.1b), from which the relative magnetic permeability is found.

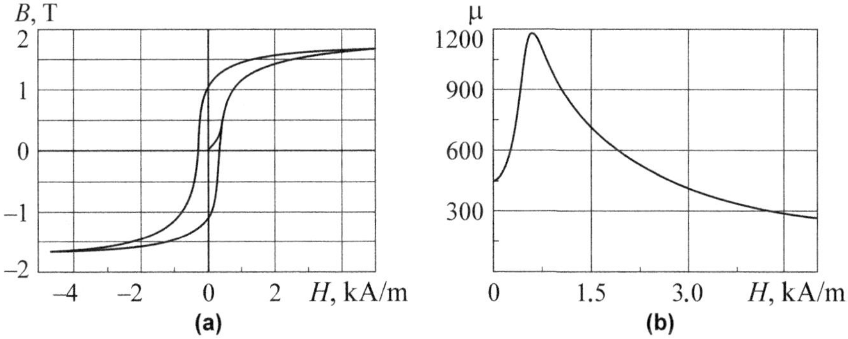

Fig. 9.1 Hysteresis loop (**a**) and the character of the change in magnetic permeability (**b**) for ferritic-pearlitic steels

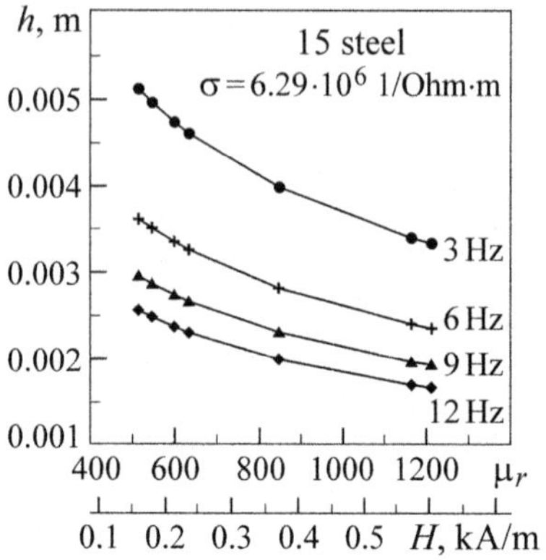

Fig. 9.2 Depth of one-side penetration of the magnetic field for 15 steel at different remagnetization frequencies

Having the necessary data, the depth of the magnetic field penetration for different frequencies of the specimen remagnetization is calculated (Fig. 9.2).

As can be seen from Fig. 9.2, remagnetization of ferritic-pearlitic steels with a frequency of 9 Hz allows us to ensure guaranteed one-side penetration of the magnetic field to a depth of 2–3 mm. By reducing the remagnetization frequency to 3 Hz, such steel is one-side magnetized up to 5–6 mm.

9.3 Technique of Excitation and Recording of MAE Signals

Plane specimens, remagnetized in the solenoid, were used in experimental studies. Preliminary, according to the known dependences (Tikadzumi 1987a), the distribution of the solenoid magnetic field intensity was calculated. The solenoid had 1500 turns of copper wire in five layers, the outer diameter of the coil was 35 mm, and the length was 94 mm.

The magnetoelastic acoustic emission caused by the Barkhausen jumps was converted into an electric signal by a highly sensitive (coefficient of conversion of elastic waves into electric signals not less than 1.6×10^9 V/m) primary piezoelectric transducer of elastic waves with uneven distribution of the conversion coefficient $\pm$ 3 dB in the operating frequency band 0.2–0.6 MHz. The choice of the type of ferritic-pearlitic steels was guided both by the depth of investigation of their physical properties and the practical application of the results to create appropriate applied methods for diagnosing the state of industrial structures made of them. For this purpose, the most common ferritic-pearlitic steels (Ukrainian steels grade: steel 3sp; 08kp; 10; 15; 20; 30; 09G2S; see Appendix) were chosen.

The block diagram of excitation and recording of MAE signals during remagnetization of specimens is presented in Fig. 9.3.

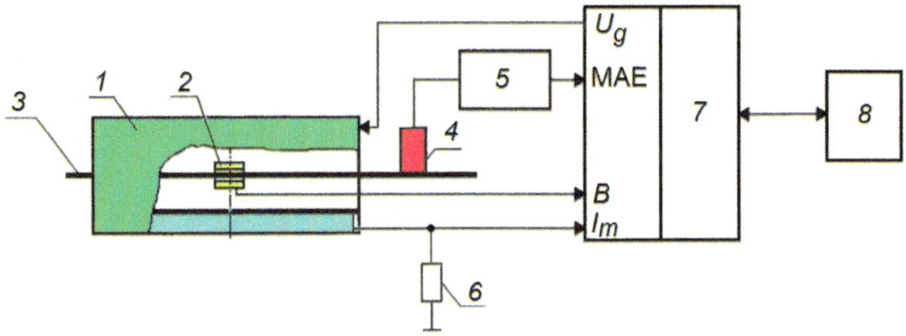

Fig. 9.3 Block diagram of excitation and recording of MAE signals: *1* a solenoid; *2* the measuring coil of the specimen magnetization; *3* the specimen; *4* the MAE piezoelectric transducer; *5* a preamplifier (PP); *6* the reference resistor for estimating the amplitude of the remagnetization current; *7* the MAE unit of signal extraction (USE) and remagnetization signals generation; *8* personal computer

The developed in the Karpenko Physico-Mechanical Institute portable magnetoacoustic system was used to select, record MAE signals, generate signals and estimate the magnetization current, measure the magnetic field induction amplitude. The composition of the system is shown in Fig. 9.3.

To remagnetize the specimens the magnetic field of solenoid *1* was used, by passing through it a quasi-static current from the output of block *7*. The shape of the current change, as well as its magnitude, was set by the program using a PC. To estimate the amplitude of the remagnetization current the reference resistor *6*, which was connected in series with the solenoid winding, was applied. The amplitude of the magnetic field induction in the specimen *3* was determined by the measuring coil *2*.

It has been experimentally established that the sinusoidal remagnetization current is optimal for our case because its shape is not distorted by the solenoid.

From the output of the acoustic emission transducer, the electric signal amplified by the preamplifier was fed to the input of the acoustic channel of the signal measurement unit (SMU), where it was selected by frequency and amplitude, and converted into digital code. From the output of the SMU, the digitized signal passed to PC, where it was statistically processed, visualized and the parameters (amplitude, sum of amplitudes, number of pulses) of MAE signals were calculated.

Prior to the experiments, the optimal frequency of magnetization was determined. To do this, experiments were performed on a steel plate of size 240 × 30 × 2 mm at different remagnetization frequencies for the sinusoidal shape of the solenoid current (Fig. 9.4). In this case it is taken into account that the jumps of the domain walls are proportionally sensitive to the value of $\mathrm{d}H/\mathrm{d}t$, which depends on the frequency. However, the depth of the magnetic field penetration into the ferromagnet also depends on it. Therefore, the frequency of the remagnetization plays an important role, and its value must be optimized.

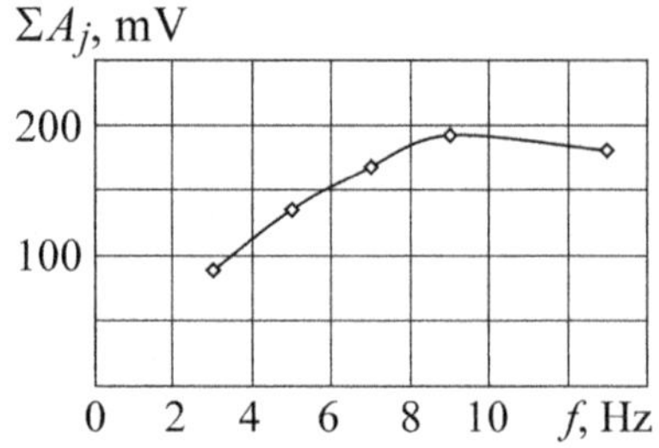

Fig. 9.4 Dependence of the sum of MAE signal amplitudes on the frequency of remagnetization of ferritic-pearlitic steel specimens at frequencies: 3–13 Hz for 3sp steel plate (in the state of delivery) of thickness 2 mm for induction amplitude 0.6 T

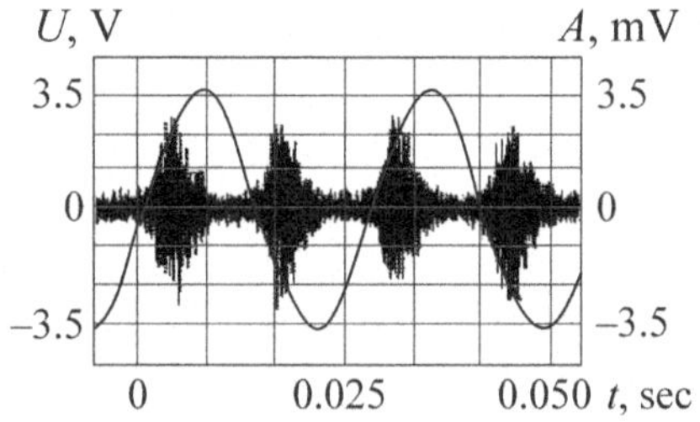

Fig. 9.5 A view of the sinusoid of remagnetization and MAE signals during tests of the steel plate in the magnetic field of the solenoid with intensity $H = 720$ A/m

Based on the experimental results, it has been found that the optimal remagnetization frequency for our MAE measuring conditions is 9 Hz. It is also important that this frequency is not a multiple of the harmonics of the industrial AC power supply network (Fig 9.5).

To construct the dependence of the sum of the MAE signal amplitudes on the magnetic field induction in the specimen, the amplitudes of the pulse envelope were added, which exceeded the level of discrimination, set to be the same in each measurement variant (Fig. 9.6) (Nazarchuk et al. 2017).

9.4 Influence of Heat Treatment on Magnetoelastic Properties of Steels

Structural changes that occur during heat treatment have an influence on the changes in the mechanical and other physical properties of materials. Therefore, it is important to know the relationship between the magnetic parameters of ferritic-pearlitic steels with the peculiarities of the structure. Numerous works testify to the existence of the relationship between magnetic parameters, such as coercive force H_c, residual magnetization I_r, magnetostriction λ, with main structural characteristics, which

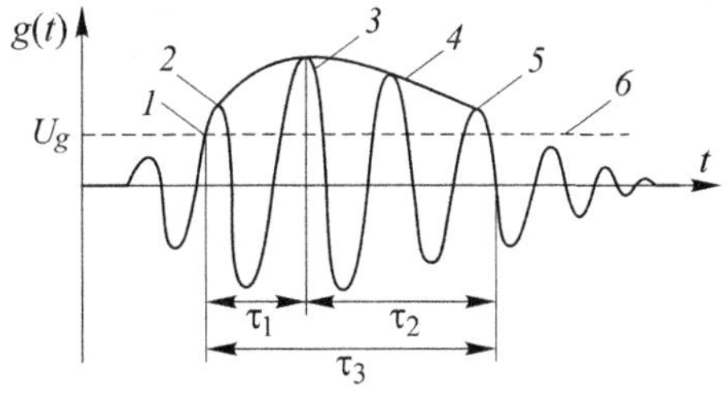

Fig. 9.6 Representation of the response function of the piezoelectric transducer to the event of magnetoelastic acoustic emission: *1* the moment of signal arrival; *2* the first pulse; *3* the envelope amplitude; *4* the envelope; *5* the last pulse; *6* the level (threshold) of discrimination; τ_1 the time of increase of the leading front of the envelope; τ_2 the time of decline; τ_3 the duration of the event

form the basis for creating the magnetic methods of non-destructive testing. It is noted that the increase in the grain size leads to the decrease in coercive force and magnetic permeability, because in addition to inclusions and dislocations in the grain body, the intergranular boundary is the main factor hindering the movement of the domain wall. Under other equal conditions, the minimum values of H_c correspond to the structures consisting entirely or partially of granular pearlite. The presence of intermediate decay products in the structure is accompanied by an increase in the above parameter. For the constant carbon content in steel and for a certain form of carbides, the dependence of the coercive force on the particle size is observed (Tumen State Oil and Gas University 1997).

It is established that the coercive force maximum is observed for the carbide particles diameter of the order of the domain wall thickness. Phase transformations that occur during heat treatment like tempering, cause changes in the magnetic, electrical properties and hardness of the tempered steel specimens due to changes in the density of dislocations, internal stresses, initiation and development of carbide particles.

However, the analysis of literature sources shows that for most structural and alloyed steels, the dependences of the coercive force on the tempering temperature are not unambiguous. Studies on construction of the dependences of residual magnetization, secondary induction, and saturation magnetostriction on the structural state of steels have shown that most of the above parameters change ambiguously with increasing tempering temperature. Published investigations (Tumen State Oil and Gas University 1997) and others on the study of resistance of the residual magnetized state to the action of elastic stresses show a high structural sensitivity of this parameter, as well as its correlation with the values of coercive force and magnetostriction.

Figure 9.7 shows the dependences of the coercive force and magnetoelastic sensitivity of 45 steel specimens on the tempering temperature (Tumen State Oil and Gas University 1997). The dotted line illustrates the calculation curve. The ambiguous character of the change in coercive force with increasing tempering temperature is related with the simultaneous influence of several factors. Firstly, the decrease in

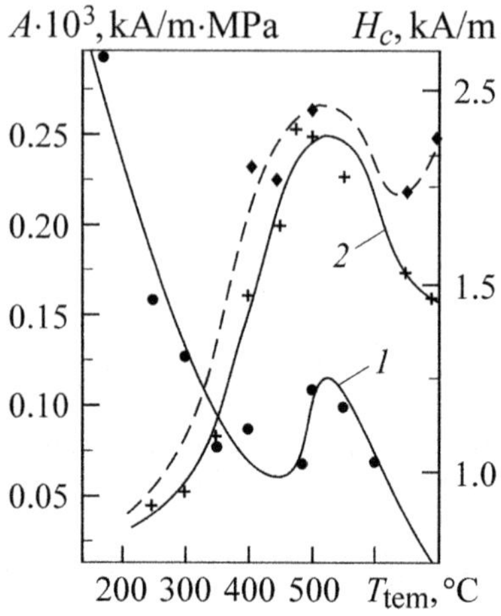

Fig. 9.7 Dependence of coercive force (curve *1*) and magnetoelastic sensitivity (curve *2*) of 45 steel specimens on the tempering temperature

internal stresses with increasing tempering temperature is accompanied by a decrease in coercive force, and on the other hard—an increase in the size of carbide particles that prevent the movement of the domain wall, should lead to increased energy losses for remagnetization, i.e. increase of the coercive force.

It is typical that the curves of the dependence of the magnetoelastic sensitivity on the tempering temperature of ferritic-pearlitic steels have a maximum, which is in the range of tempering temperature, where the dependence of the coercive force is ambiguous. The presence of the maximum on the magnetoelastic sensitivity curve can be associated with the achievement of the "critical size" of carbide particles (when the size of the inclusions is commensurable with the thickness of the domain boundaries), around which an extensive system of small closing domains is formed, which is consistent with the mechanisms of the coercive force change during tempering. The processes of coagulation of carbides that occur during high-temperature tempering, associated with a decrease in the fields of internal stresses and scattering fields, cause a decrease in the coercive force and magnetoelastic sensitivity of the studied steels.

Inclusions with high values of coercive force (cementite, carbides) create a branched system of closing domains around themselves, which exceeds the size of the inclusion hundreds of times.

Under low-temperature tempering, the sizes of carbide particles are small. As the tempering temperature increases, their sizes increase and carbide coagulation processes start in the steel. This leads to the internal stresses decrease, as a result of which the coercive force decreases.

During heating, the ferromagnetic properties of the metal are gradually lost. Curie showed that the complete loss of ferromagnetic properties took place at a certain temperature, which was called the Curie point. The intensity of magnetization gradually decreases with increasing temperature (Fig. 9.8) and the Curie point corresponds to the complete loss of ferromagnetism (Guliayev 1978).

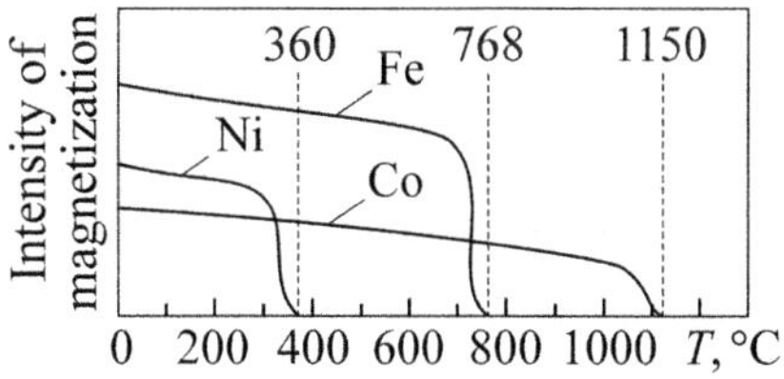

Fig. 9.8 Dependence of magnetic properties of iron, nickel, and cobalt on temperature

Magnetic transformation has a number of features that distinguish it from the allotropic one. First, the magnetic properties gradually decrease when approaching the transformation point and this point does not correspond to a jump-like change in properties. Second, the magnetic transformation has no temperature hysteresis, and increase of the cooling rate does not reduce the transformation temperature. Third, the mechanical and some other physical properties do not change during the transformation. And the most important thing is that the magnetic transformation is not accompanied by recrystallization—the formation of new grains and the change of the lattice. These features significantly distinguish magnetic transformation from the allotropic one, which is characterized by changes in the crystal lattice, recrystallization, and thermal hysteresis of the transformation (Guliayev 1978).

9.5 Estimation of the Influence of Some Physical Factors on MAE Generation

Influence of elastic deformations. Initially, the tensile diagram of the plane specimen (Fig. 9.9) from ferritic-pearlitic 15 steel (material in the state of delivery) was experimentally constructed (Skalsky et al. 2022). The speed of movement of the loading device indenter was 0.05 mm/s.

The obtained result has the shape of a classical strass-strain curve for ductile materials with a "sharp yield point". Based on this, data on its various sections were obtained and the tests to assess the effect of elastic stresses in the ferromagnet on the parameters of the MAE signals were performed. Uniaxial tensile forces were applied to the specimen placed in the solenoid and the dependences of the sum of

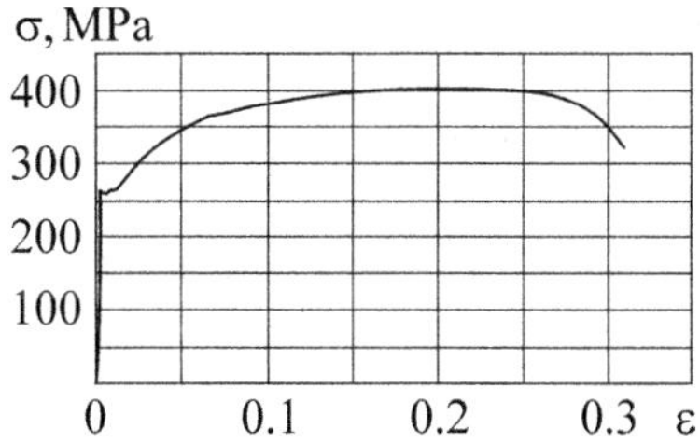

Fig. 9.9 Stress–strain curve for the plane 15 steel specimen

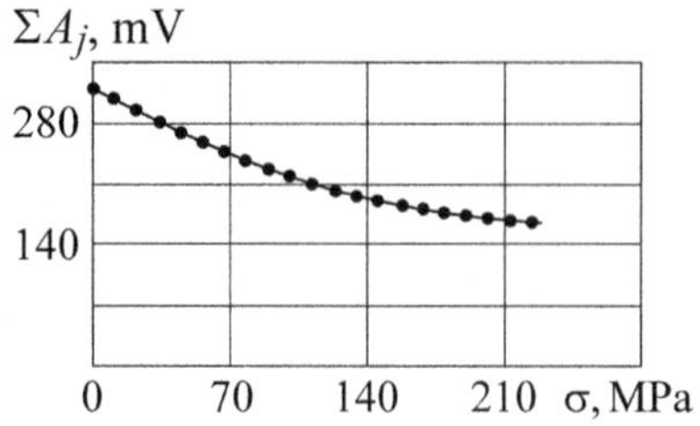

Fig. 9.10 Dependence of the sum of amplitudes of MAE signals on the elastic deformation of the ferromagnet (for the solenoid magnetic field intensity $H = 0.8$ kA/m)

the MAE signals amplitudes caused by the elastic stresses at constant induction of the magnetic field in the specimen under load were written (Fig. 9.10).

From the obtained experimental data it is seen that during the elastic deformations the sum of the amplitudes of magnetoelastic acoustic emission decreases with increasing mechanical stresses in the ferritic-pearlitic steel (confirmation of the Villari effect).

Influence of remagnetization volume. In experiments the data on the influence of the remagnetization volume on the change in the sum of the MAE signals amplitudes are of a considerable interest. The results of such investigations are shown in Fig. 9.11.

Thus, the dependence of the sum of the MAE signals amplitudes increases almost linearly with increasing the volume of the steel magnetization.

Influence of plastic deformations. When stresses in the material reach the values higher than the elastic limit, the deformation becomes irreversible. Removal of the external load removes only the elastic component, and the residual deformation is called plastic. In crystals it occurs by sliding on the corresponding planes and twinning—when the tangential stresses reach some critical values.

Considering the fact that the grains are not oriented in the same way, the plastic deformation, different in volume, occurs in the material structure. Simultaneously with the change in the shape of the grain, the blocks are crushed and the angle of disorientation between them increases. The X-ray diffraction analysis shows that after appropriate deformation some grains and blocks are elastically stressed (internal

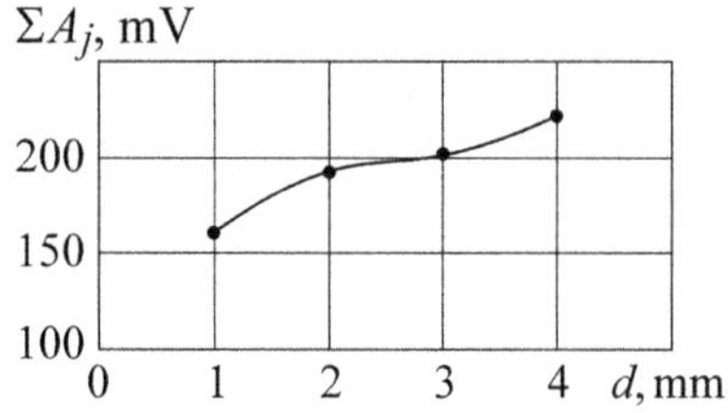

Fig. 9.11 Dependence of the sum of the MAE signals amplitudes on the thickness d of the 3sp steel specimens at their constant length and width (the remagnetization frequency is 9 Hz)

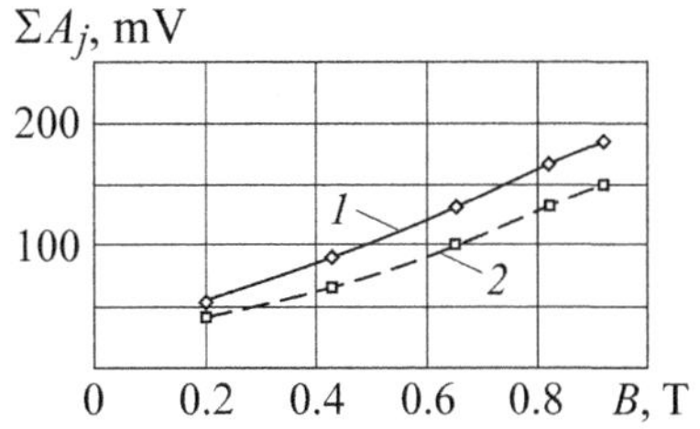

Fig. 9.12 Change in the sum of amplitudes of magnetoelastic acoustic emission signals depending on the magnetic field induction for plastically deformed 15 steel at $\sigma = 350$ MPa (curve 2) and in the state of delivery (curve 1)

stresses of the first kind), and the crystal lattice at the boundaries of grains, blocks and near the sliding planes is distorted (internal stresses of the second kind). This regular orientation of the crystallites relative to the external deforming forces is known as texture. The formation of texture (due to rolling, drawing, etc.) depends on the nature of the metal and the type of plastic deformation and promotes the appearance of anisotropy of its mechanical and other physical properties.

The specimens were tensile plastically deformed on the UME10-TM machine. Figure 9.12 illustrates the dependence of the sum of the MAE signal amplitudes on the magnetic field induction for the original and plastically strained material.

The same experiments were performed for 30 steel with a thickness of $d = 3$ mm. The test results are shown in Fig. 9.13. Typical signals for both grades of steel are presented in Fig. 9.14.

The obtained experimental data show that already at the beginning of plastic deformations the sum of the MAE signal amplitudes decrease. This tendency grows with an increase in the plastic deformation of ferritic-pearlitic steels.

Influence of volume damage of ferromagnet. The experiments were performed on the specimens of ferritic-pearlitic 09G2S steel (sample sizes are given in Fig. 9.15a). The specimens were cyclically loaded to different degrees of damage accumulation in

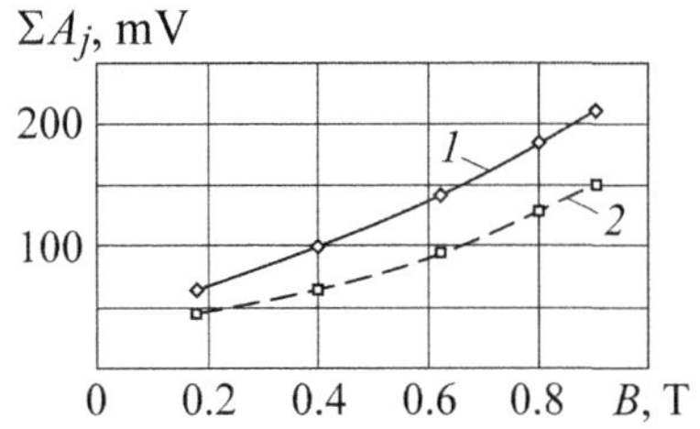

Fig. 9.13 Change of the sum of the amplitudes of the MAE signals depending on the induction of the magnetic field for plastically deformed 30 steel at $\sigma = 360$ MPa (curve 2) and in the state of delivery (curve 1)

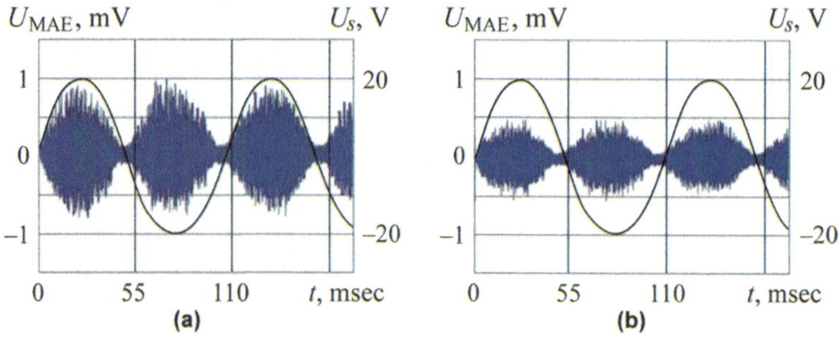

Fig. 9.14 Typical MAE signals during remagnetization of undeformed (**a**) and deformed (**b**) 15 steel specimens ($B = 0.92$ T, $\sigma = 350$ MPa)

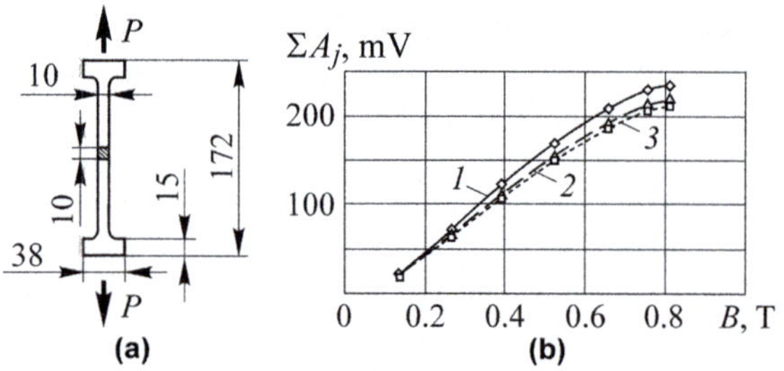

Fig. 9.15 Geometry and dimensions of 09G2S steel specimens (**a**) and change of the sum of amplitudes of MAE signals depending on the magnetic field induction and level of material fatigue damage (**b**) under cyclic loading with frequency $f = 10$ Hz, stress ratio $R = 0.1$: *1* the original material; *2* after $N = 4 \times 10^6$ stress loading cycles $\sigma = 200$ MPa; *3* after $N = 4 \times 10^6$ stress loading cycles $\sigma = 300$ MPa

multi-cycle fatigue. The dependences of the sum of amplitudes of the magnetoelastic acoustic emission on the level of damage of 09G2S steel (Fig. 9.15b) were obtained similarly as in the previous studies.

According to the results of experiments on fatigue volume damage of 09G2S ferritic-pearlitic steel, it can be concluded that the sum of the MAE signal amplitudes is less than for the original material. Thus, it is experimentally established that the level of material damage also reduces this indicator, as well as the elastic stresses and plastic deformations.

9.6 Estimation of the Influence of the Chemical Composition and Structure of Steel on MAE

Two grades of ferritic-pearlitic steels were studied. Plane specimens of 30 steel (see Appendix) after their initial remagnetization with a frequency of 9 Hz with recording the MAE signals, were annealed in an autoclave at a temperature of 830 °C. In the initial state, steel 30 had a fine-dispersed homogeneous ferritic-pearlite structure with a pronounced texture of the near-surface layers (Fig. 9.16a).

After annealing, a medium-grained structure with a clear grain cut and remnants of non-crystallized areas is obtained (Fig. 9.16b). The grain sizes are of 25–70 μm. Figure 9.17 shows the result of investigations of changes in the MAE signal amplitudes depending on the steel structure.

As follows from Fig. 9.17 the amplitudes of the signals increase with increasing external magnetic field intensity H in both annealed and unannealed specimens. However, in the second case, this growth is significantly lower, which is due to the restraint of the domain wall jumps by the available elastic stresses in the material. It

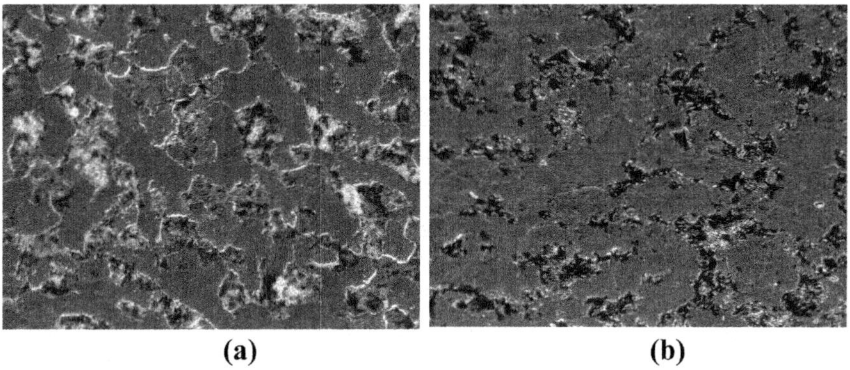

Fig. 9.16 Structure of 30 steel before (**a**) and after (**b**) full annealing (× 1000)

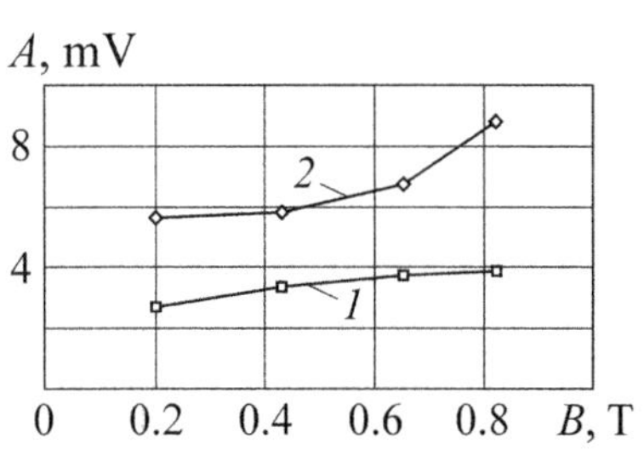

Fig. 9.17 Average amplitudes of the MAE signals for annealed (curve *1*) and unannealed (curve *2*) 30 steel with a change in the magnetic field induction in the specimen

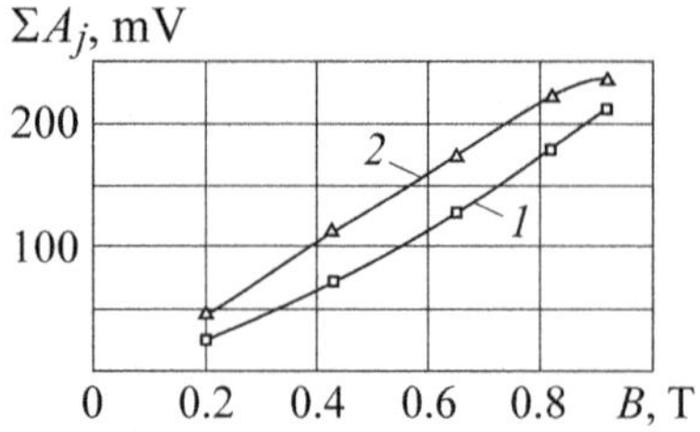

Fig. 9.18 Dependence of the sum of MAE signal amplitudes on the induction of the magnetic field during remagnetization of 15 steel specimens: *1* initial state; *2* annealed at 900–910 °C

is obvious that there are more intergranular boundaries, which are obstacles to such jumps and reduce the activity of the MAE (its total account is significantly reduced).

Similar results were obtained for 15 steel (see Appendix). The specimens were heated to the temperatures of 900–910 °C, followed by cooling together with the furnace. As a result of such heat treatment, the ferrite grains also grew, their boundaries thinned, and parts of the pearlitic phase coagulated, joining into single inclusions. Such heat treatment also helped to reduce the residual stresses caused by plastic deformation during the steel sheets rolling. All these conditions improved the mobility of the domain walls during steel magnetization and thus increased its activity before the MAE generation. Thus, Fig. 9.18 shows that the dependence of the sum of the amplitudes of the MAE signals on the induction of the magnetic field for 15 steel subjected to heat treatment (the curve that corresponds to the full annealing of steel) is characterized by higher MAE activity.

Influence of carbon content in steels. A number of iron-carbon alloys with different carbon content were used for the research (Ukrainian steel grades: 08kp, 15, 3sp, 65G, U8, SCh10 cast-iron; see Appendix). The materials were selected such to study the MAE in ferritic, ferritic-pearlitic, and purely pearlitic structures and with available free carbon (SCh10 cast-iron).

As metallographic studies have shown, the structure of the low-carbon 08kp steel consists of relatively large ferrite grains and a small number of pearlite inclusions, which are located mainly in the form of thin layers along the grain boundaries. Because of the fact that the studied specimens are made of rolled sheets, the ferrite grains are somewhat elongated and their length sometimes reaches 25–30 μm. The microstructure also has a light number of small pores up to 5 μm in diameter, which is obviously due to the method of steel melting.

The microstructure of the low-carbon 15 steel consists of deformed ferrite grains (size 15–20 μm) and pearlite particles, which are placed along their boundaries, or in the form of individual inclusions (size 5–10 μm). A separate group of samples is subjected to full annealing (heating of steel to the temperatures of 900–910 °C and subsequent cooling with a furnace). In the process of heating, their structure is transformed into austenite, and the subsequent slow cooling contributes to the phase recrystallization and equilibrium of the ferritic-pearlitic structure. After such heat

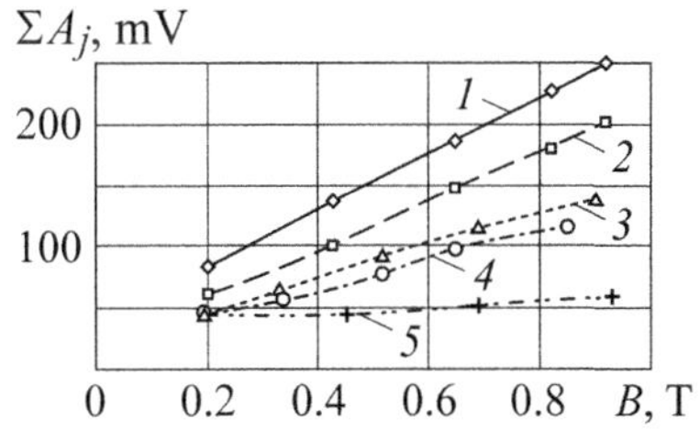

Fig. 9.19 Dependence of the sum of the amplitudes of the MAE induction of the magnetic field on the remagnetization of ferromagnetic alloys with different carbon content: *1* 08kp steel; *2* 15 steel; *3* 65G steel; *4* U8 steel; *5* SCh-10 steel

treatment, the coarsening of equilibrium grains of ferrite with sizes of 20–30 μm and pearlite, which precipitates in the form of inclusions (size 10–15 μm), is obtained.

High-carbon 65G steel consists of a large amount of pearlite and ferrite, which is much less compared to low-carbon steel. Ferritic carbides with manganese are also present in the solid solution. To obtain a pure pearlitic structure a full annealing of eutectoid U8 steel (0.8% C), which consists of heating to the temperatures of 800–810 °C and subsequent cooling with a furnace, is carried out. Under such equilibrium conditions, austenite decomposes during cooling, forming an eutectoid mixture of ferrite with secondary cementite.

Figure 9.19 illustrates the dependence of the sum of the MAE signal amplitudes on the induction of the magnetic field for ferromagnetic alloys with different carbon content. It can be seen that the highest MAE activity is observed for 08kp steel, as its microstructure contains coarse ferrite grains with an insignificant number of pearlite inclusions at the grain boundaries. Thus, during the remagnetization of this material, the jump-like movement of the domain walls is not affected by the grain boundaries or the presence of carbide inclusions, or other phases.

The MAE signals with slightly smaller amplitudes are generated during the remagnetization of 15 steel. This is explained by a slightly finer steel structure and the presence of grain boundaries along which the pearlite inclusions are located. As noted above, the fine-dispersion of the structure and the presence of phases at the grain boundaries complicates the jump-like movement of the domain walls during the material magnetization, and thus reduces the activity of elastic waves generation, caused by these jumps.

A sharp decrease in the activity of MAE generation was observed under remagnetization of high-carbon 65G steel specimens. In this case, the movement of the domain walls was blocked by a large amount of pearlite and small carbide inclusions available in the microstructure, which were present not only along the boundaries but also in the solid solution of the grain body.

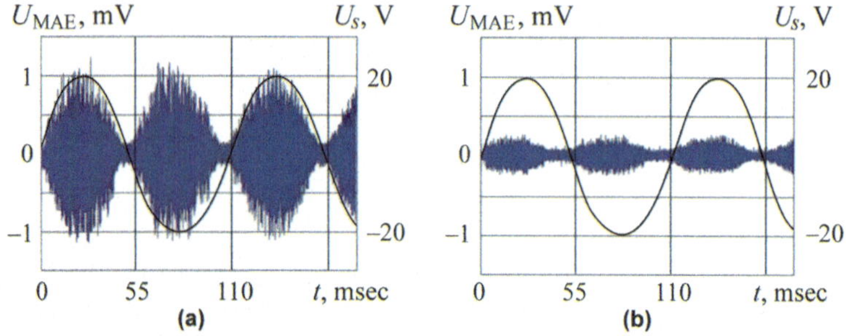

Fig. 9.20 Typical MAE signals during remagnetization of 08kp steel (**a**) and SCh10 steel (**b**) in the magnetic field of the solenoid ($B = 0.92$ T)

In the case of remagnetization of eutectoid U8 steel specimens, the structure of which after full annealing consisted of fine-dispersed pearlite, MAE signals with amplitudes lower than those of the above-mentioned pre-eutectoid steels were observed. Apparently, this effect was caused by the fragmentation of the structure and the presence of a large number of boundaries of pearlite grains, which were an "obstacle" to the rotation of the domain walls.

Gray cast-iron, the microstructure of which consists mainly of pearlite, graphite inclusions and a small amount of ferrite, is characterized by the lowest activity of MAE generation (Fig. 9.20).

Thus, the increase of carbon in iron alloys leads to a decrease in the activity of MAE during their remagnetization in the external quasi-static magnetic field.

Influence of heating temperature (incomplete annealing) during hydrogen charging of ferritic-pearlitic steel. Since in the experiments, the specimens are hydrogenated mainly from the gas phase, the question arises to evaluate the effect of heating of the ferritic-pearlitic steel to elevated temperatures on the generation of MAE. Here it is necessary to choose such a heating temperature which would promote remagnetization of the specimens after its hydrogen charging as much as possible.

After analyzing the curves of the change in coercive force and magnetoelastic sensitivity during heating of steels, shown in Fig. 1.17, the conclusion is drawn that these requirements are best met by heating the sample to a temperature of 550 °C and cooling it together with the furnace, that is, incomplete annealing at this temperature.

After that, experiments were conducted to determine the effect of such heat treatment on the change in the sum of the amplitudes of the MAE signals. The research results are shown in Fig. 9.21.

Note that both annealings were carried out in air. Mode of the first one: heating temperature was 550 °C with holding at this temperature 3.6 ks; the second annealing was performed at 550 °C for 14.4 ks. The time interval between annealing's was 259.2 ks at room temperature, the environment was air. As one can see, the curves

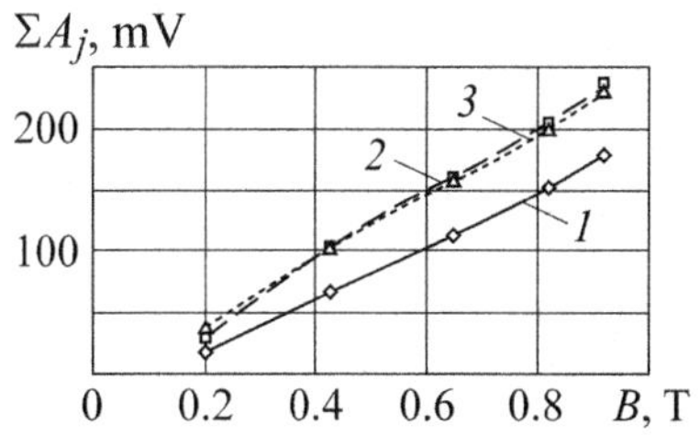

Fig. 9.21 Change of the sum of the amplitudes of the MAE signals depending on the induction of 3sp steel specimens at the annealing temperature of 550 °C: *1* the state of delivery; *2* the first annealing; curve *3* the second annealing (fine hatching)

2 and *3* almost coincide, which indicates no effect of reannealing on the change of structure (magnetic including) and magnetic properties of steel during secondary annealing at the same temperature and different holding times. The same was observed after several experiments on the same specimen of the material.

To assess the effect of the heating environment, a second group of the same specimens (550 °C; 3.6 ks) was similarly annealed in an inert argon gas at a pressure of 1.2 MPa. The obtained results (Fig. 9.22) indicate their good agreement with the results of specimens annealing in the air (Fig. 9.21). The same character of the change of curves and almost the same increase in the sum of the amplitudes of the MAE signals after annealing is observed.

Thus, for the induction range of 0.2–0.6 T, the increment in the sum of the amplitudes of MAE signals for both steels is 30–50%, and for the range of 06–0.9 T the increment is in the range of 20–30%.

Analysis of the metallographic data of the 3sp steel structure before and after annealing demonstrates that on the surface of the material specimens in the initial state a slight texture in the form of slightly elongated in the direction of rolling ferrite grains is observed. Their size ranges from 10 to 40 μm. Small pearlite grains 5–7 μm in size are located mainly at the joints of three ferrite grains or in narrow layers elongated along their boundaries.

The texture on the sheet surface, which was annealed at 550 °C for 3.6 ks before annealing, disappeared, the size of polyhedral ferrite grains decreased to 30 μm. The

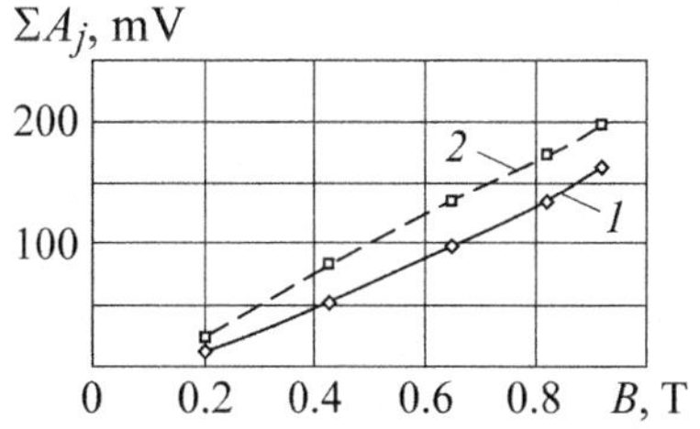

Fig. 9.22 Change of the sum of the MAE signal amplitudes depending on the induction of 3sp steel specimen under annealing 550 °C in argon: *1* the state of delivery, *2* after annealing

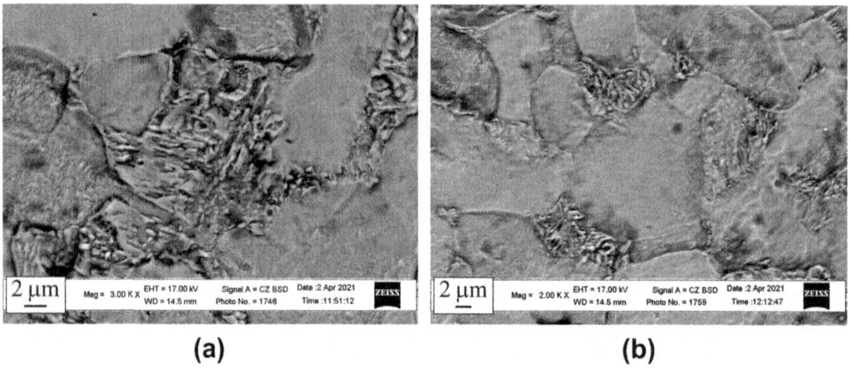

Fig. 9.23 Structure of the surface layers of steel 3sp before (**a**, × 3000) and after (**b**, × 2000) annealing

lamellar structure of pearlite in the steel structure in the initial state changed due to the spheroidization of cementite particles after annealing (Fig. 9.23).

Thus, although the domain structure of the ferromagnet changes with the change of crystal structure (Vonsovsky and Shur 1948; Becker and Döring 1939), it is obvious that a significant improvement in the dislocation situation is the dominant factor, internal residual stresses reduce, which greatly facilitates the movement of domain walls.

Summing up, note that the annealing of the ferritic-pearlitic steel at a temperature of 550 °C changes the morphology of its grain structure, improves magnetic characteristics, heals dislocations, and relieves residual internal stresses. As a result, the activity and jumps of the domain walls increase during steel magnetization, in contrast to the influence of the physical factors described above: elastic and plastic deformations, volume damage and increased carbon content. The presence of air during annealing does not affect the increase in the sum of the amplitudes of the MAE signals.

9.7 Influence of Chemical Composition and Heat Treatment on Magnetoelastic Properties of Steels and Alloys

It is known that most metal structures are made of different steels, the microstructure and properties of which often differ greatly. Therefore, in order to build an effective diagnostic technique, it is necessary to study the influence of the chemical composition and heat treatment on the magnetoelastic properties of structural steels, and therefore on the features of the generation of MAE. It should be noted that the structural changes that occur during heat treatment have an influence not only on the mechanical but also on the other physical properties of materials, so it is important to study the relationship of magnetic parameters of structural steels with their peculiarities.

Numerous works by well-known foreign scientists—Vonsovsky S. V., Shura Ya. S., Kondorsky E. I., Neel L., Mikheev M. N., Kulev V. G., Gorkunov E. S., and others indicate the relationship between magnetic parameters, such as coercive force H_c, residual magnetization I_r, magnetostriction λ with basic structural characteristics, which became the basis for the creation of magnetic methods of non-destructive testing. It is noted that the increase in grain size leads to a decrease in coercive force and magnetic permeability because in addition to inclusions and dislocations in the grain body, the main factor hindering the movement of the domain wall is the intergranular boundary. Under other similar conditions, the minimum values of H_c correspond to structures consisting entirely or partially of granular pearlite. The appearance of lamellar pearlite causes an increase in the coercive force. The presence of intermediate decay products in the structure is accompanied by an increase in the above parameter. At a constant carbon content in steel and for a certain shape of carbides, the dependence of the coercive force on the particle size can be traced (Tumen State Oil and Gas University 1997). It is established that coercive force maximum is observed for the diameter of carbide particles of the order of the domain wall thickness. Phase transformations that occur during heat treatment of the tempering type, cause changes in the magnetic, electrical properties and hardness of tempered steel samples due to changes in the density of dislocations, internal stresses, initiation and development of carbide particles.

However, the analysis of literature sources shows that for most structural and alloy steels, the dependences of the coercive force on the tempering temperature are not unambiguous. Studies on the construction of dependences of residual magnetization, secondary induction, and saturation magnetostriction on the structural state of steels have shown that most of the above parameters change ambiguously with increasing tempering temperature. Published research (Tumen State Oil and Gas University 1997) and others on the study of the resistance of the residual magnetized state to the action of elastic stresses demonstrate the high structural sensitivity of this parameter, as well as its correlation with the values of coercive force and magnetostriction.

In Tumen State Oil and Gas University (1997), the mechanisms of quasi-inverse change of the residual magnetization of 30KhGSA, 60G structural steels, 45 steel and Fe–Co–V alloys during loading in the elastic region of the tensile diagram were studied. The authors constructed residual magnetization hysteresis loops using standard specimens for mechanical tests of the studied steels after ten tensile-compression cycles. The value of the magnetoelastic sensitivity $\Lambda = \Delta I_r / \Delta \sigma$, which is defined as the tangent of the loop inclination of the stress axis, was chosen as a characteristic of the residual magnetization change.

The increase in the residual magnetization under the tension of materials with positive magnetostriction (positive piezoelectric effect) proceeds because the residual magnetized magnet has a magnetic texture—a greater number of magnetic moments oriented along the axis of magnetization. In tension, the percent of transverse magnetic moments that are reoriented along the direction of the initial residual magnetization will be greater than in the opposite direction, which causes an increase in the residual magnetization of the specimen.

Under compression, the magnetic moments rotate in the direction perpendicular to the load axis. In this case, due to the available texture, the decrease in direct magnetization will be greater than the inverse one, which will reduce the residual magnetization in general.

The negative piezoelectric effect ($\lambda_{100} > 0$, decrease in the residual magnetization of steel during tension and, consequently, increase in compression) may be caused by the presence of areas with inverse magnetization relative to the $I_{r\sigma}$ vector due to heterogeneity of the material properties. Inclusions with high values of coercive force (cementite, carbides) form a branched system of closing domains around themselves, which exceeds the size of the inclusion hundreds of times. They can play the role of sources of the magnetizing field. The magnetic force lines of the scattering fields of rigid areas must be oriented opposite to the direction of the residual magnetization $I_{r\sigma}$. The authors of Tumen State Oil and Gas University (1997) suggest that in these areas there is a magnetization in the opposite direction of the magnetosoft sections. This fact reduces the total magnetization of the specimen and the associated magnetostatic energy. During tension of the specimen, the magnetization of the magnetorigid regions increases, increasing the scattering fields formed by them. At the same time, the volume fraction of the magnetized areas is growing. If, in general, the increase in the inverse magnetization is greater than that of the direct one, a decrease in the residual magnetization under tension, i.e., a negative piezoelectric effect, is observed. The opposite effects are observed under compression.

Figure 9.24 illustrates the microstructures of 45 and 30KhGSA steels at different tempering temperatures. One can see that during the medium-temperature tempering, areas of carbide accumulation are clearly visible, which can act as sources of scattering fields, causing a negative piezoelectric effect. Under high-temperature tempering, the heterogeneity of the structure is significantly reduced and the piezoelectric effect becomes positive.

To study the character of the negative piezoelectric effect, the authors of Tumen State Oil and Gas University (1997) investigated the effect of the body geometry on the change in the residual magnetization of cylindrical specimens of K52F7 alloy in the state of delivery ($H_c \sim 600$ A/m). The specimens were of the same diameter but of different lengths.

An increase in residual magnetization (negative piezoelectric effect) was observed when compressing a specimen with a diameter of 4 mm and a length of 15 mm. For specimens with a length of 30 and 45 mm and a constant diameter, the residual magnetization first increases during compression and then decreases. For 60 mm long specimens, the residual magnetization during compression only decreases (positive piezoelectric effect). The authors explain this phenomenon as follows. Due to the high demagnetizing factor, areas of inverse magnetization appear at the specimen faces.

During compression, the value of the inverse magnetization decreases, which in general leads to the increase in the residual magnetization of short specimens. Obviously, the number of remagnetized areas near the ends changes slightly when

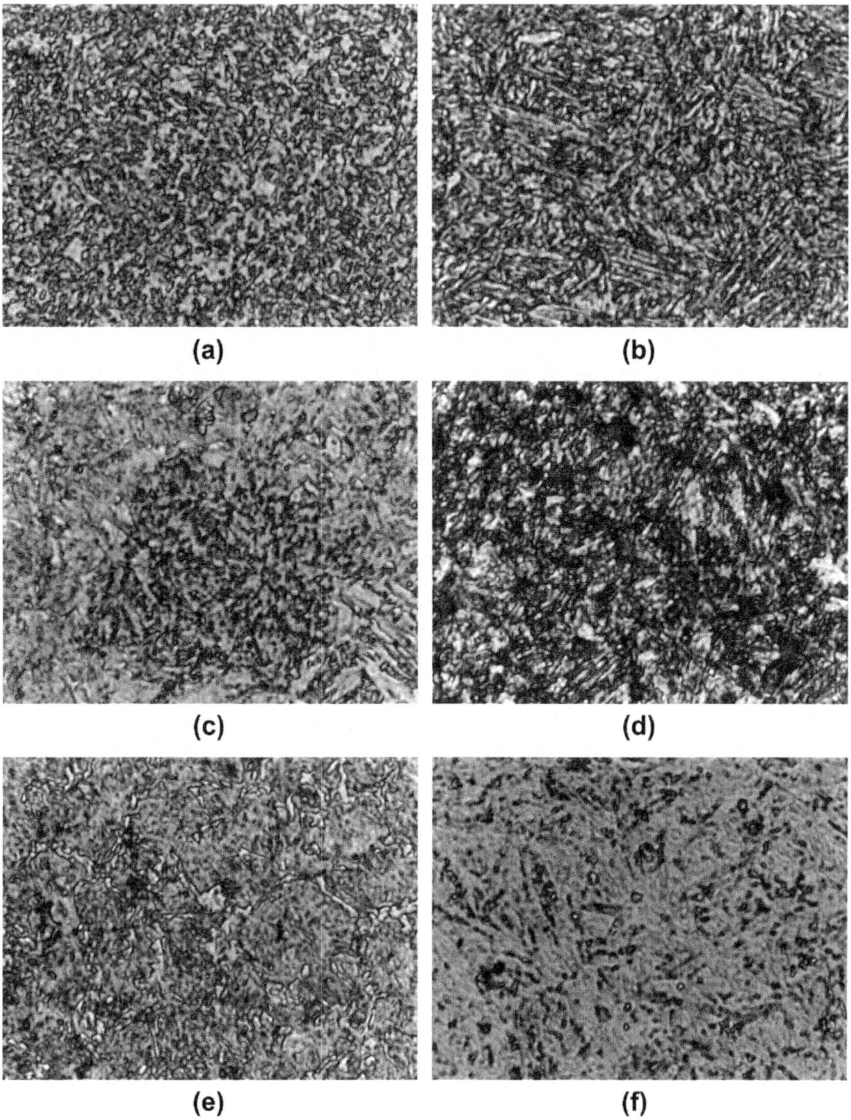

Fig. 9.24 Microstructures of 45 and 30KhGSA steels after hardening in water and subsequent tempering at different temperatures: **a** 45 steel, $T_{tem} = 700$ °C; **b** 45 steel, $T_{tem} = 480$ °C; **c** 45 steel, $T_{tem} = 250$ °C; **d** 30KhGSA steel, $T_{tem} = 700$ °C; **e** 30KhGSA steel, $T_{tem} = 530$ °C; **f** 30KhGSA steel, $T_{tem} = 400$ °C ($\times$ 1000)

the linear dimensions of the specimen change, while as the length of the specimen increases, the relative volume fraction of the areas of inverse magnetization should decrease. As a result, the piezoeffect changes this sign to positive. Similar results were obtained by the authors on the 3kp steel specimens in the state of delivery ($H_c \sim 400$ A/m).

To reduce the edge effects of closing magnetic field lines, the authors of Tumen State Oil and Gas University (1997) investigated specimens with a rigid surface layer, namely cylindrical specimens with a diameter of 6 mm and a height of 15 mm from 3kp steel and 45 steel, which were cemented in solid carburizer (charcoal) for temperatures 910–930 °C. As a result, a carbon-saturated cemented layer was formed on the surface, which has a higher magnetic rigidity compared to the core. All the samples were also thermally quenched in water at temperatures of 810 °C and subsequent tempering for 1 h. It is found that the non-cemented and subjected to low tempering specimens of 3kp steel and 45 steel are characterized by a negative piezoelectric effect, while the cemented—by a positive one. Thus, the increase in length is due to the reduction of edge effects (Fig. 9.25) (Shved 1985). The opposite effects are observed under high tempering. The positive piezoelectric effect is typical of non-cemented and tempered at 600 °C specimens of 3kp steel. Apparently, this is due to the fact that the material in which the high-temperature tempering takes place becomes magnetosoft and, as a result, the coercive force decreases and the edge effects weaken. The presence of a magnetorigid cemented layer leads to the appearance of remagnetized areas inside the specimen as a result of the closure of the magnetic field lines of the rigid shell. Therefore, a negative piezoelectric effect is observed in the cemented specimens.

During the transition from low-temperature to medium and high tempering, cemented specimens are characterized by the wide loops of magnetoelastic change in residual magnetization. This is due to the different mechanisms of rearrangement of the domain structure under the direct and reverse action of the magnetic field. It is established (Tumen State Oil and Gas University 1997) that during compression of specimens as a result of irreversible restructuring of the domain structure, a change in the residual magnetization takes place, and under the specimen unloading the character of restructuring of the domain structure changes. The processes of domain structure restructuring are also influenced by both carbide inclusions and high internal stresses caused by the difference between the coefficients of linear expansion of the non-cemented matrix and the carbon-saturated outer layer.

Thus, the change in the sign of the piezomagnetic effect of the residual magnetized state during the transition from non-cemented to cemented specimens confirms the influence of the scattering fields of rigid carbide inclusions on the change of the residual magnetization during loading. During low-temperature tempering, when the sizes of carbide particles are small and, consequently, small areas of inverse magnetization, the piezomagnetic effect created by them is positive. As the tempering temperature increases, the size of carbide inclusions grows, scattering fields increase, which cause a negative piezoelectric effect. Also at high temperatures, the processes of coagulation of carbides begin in steel, which leads to a decrease in the internal

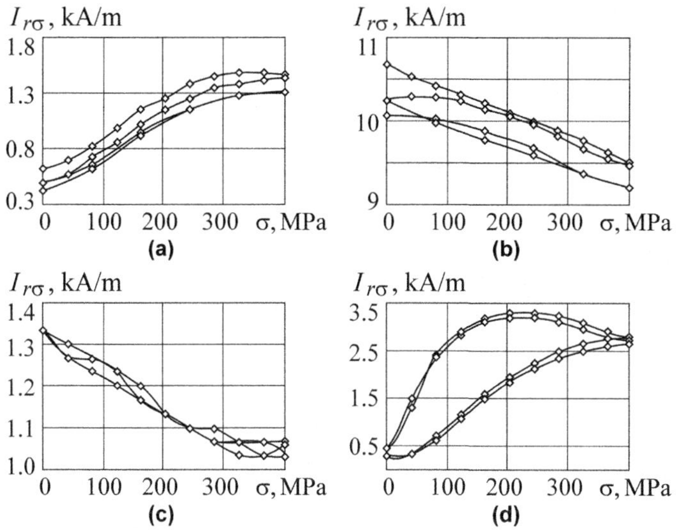

Fig. 9.25 Magnetoelastic change of residual magnetization during compression of 45 steel, $T_{tem} = 200$ °C (uncemented) (**a**); 45 steel, $T_{tem} = 200$ °C (cementation time 1 h) (**b**); 3kp steel, $T_{tem} = 200$ °C (uncemented) (**c**); 3kp steel, $T_{tem} = 600$ °C (cementation time 1 h) (**d**)

stresses and scattering fields. As a result, the coercive force decreases and a positive piezoelectric effect is observed.

In Tumen State Oil and Gas University (1997) the peculiarities of the piezoelectric effect due to the rotation of the spontaneous magnetization vector of domains in the Fe–Co alloy (52%)–V (5, 7, 9%) specimens, which have only a positive and much larger magnetostriction than structural steels, are investigated. It is established that the magnetic structure of these alloys under a certain heat treatment is usually single domain. Therefore, only one mechanism of change of the residual magnetization under load can be identified for these alloys, namely the rotation of the spontaneous magnetization vector of domains.

Figure 9.26 presents the microstructures of Fe–Co–V (K52F7 and K52F9) alloys subjected to different heat treatment.

It is found that increase in the vanadium content leads to a decrease in the grain size. It is obvious that the grain boundaries of these alloys are a "difficult obstacle" for the domain wall. The authors have established that the smaller the grain size, the more pronounced is the effect of rotation processes on the change in residual magnetization.

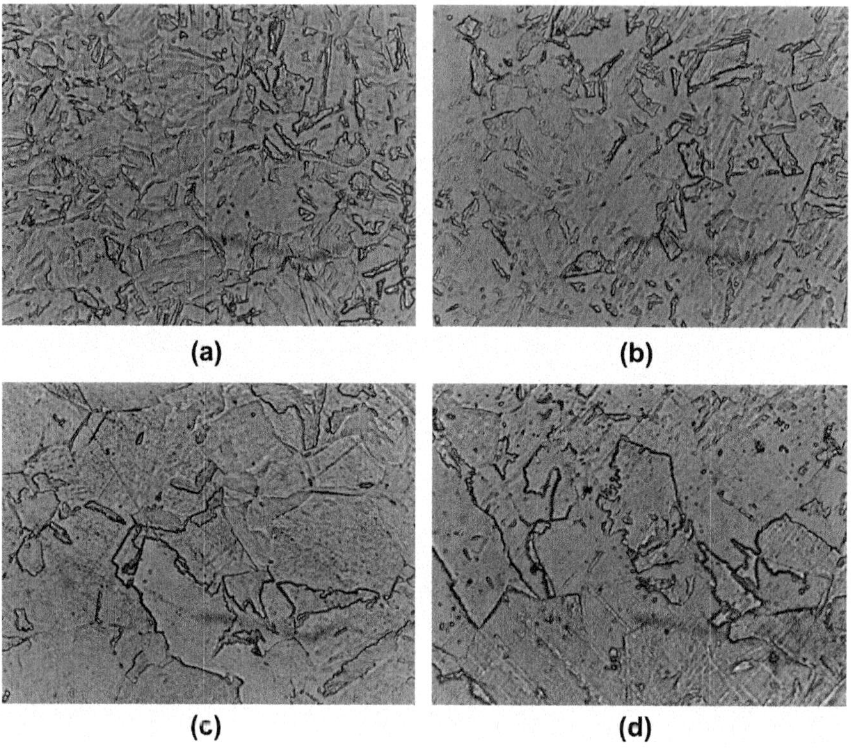

Fig. 9.26 Image of the microstructure of Fe–Co–V alloys after quenching and the following tempering: **a** K52F9, $T_{\text{tem}} = 400$ °C; **b** K52F9, $T_{\text{tem}} = 600$ °C; **c** K52F9, $T_{\text{tem}} = 200$ °C; **d** K52F7, $T_{\text{tem}} = 600$ °C (× 400)

9.8 Influence of Hydrogen on the Jump-Like Movement of Domain Walls in Steels

Taking into account the main ideas and results of previous studies on the influence of different physical factors, it is concluded that for an unambiguous interpretation of the effect of hydrogen on the jump of domain walls in ferritic-pearlitic steel, it is necessary to conduct experiments only on identical specimens of the same steel grade, modes of hydrogen charging and record of MAE signals.

So, the specimens with dimensions of $240 \times 30 \times 2$ mm^3 were prepared from one sheet of 3sp steel in the state of delivery (Skalskyi et al. 2023). The mechanical characteristics of the alloy are given in Table 9.1, and its chemical composition—is in Appendix. Chemical analysis of steel shows the presence of such elements (wt.%); Co—0.001; Sn—0.005; V < 0.001; W < 0.01.

These specimens are used as a basis for further research. Other grades of ferritic-pearlitic steels are also used to compare or confirm the test results.

Table 9.1 Mechanical characteristics of 3sp steel

Material	σ_{02} (MPa)	σ_b (MPa)	E (MPa)	δ (%)	ψ (%)
3sp steel	230	390	210×10^3	32	56

Known experimental investigations of iron and its alloys in gaseous hydrogen (Tkachev et al. 1999; Bilby and Hewitt 1962; Romaniv et al. 1972) at atmospheric pressure have shown that the most significant changes in the mechanical properties occur at room temperature, when hydrogen in the metal does not dissolve. It is also found that gaseous hydrogen has the greatest effect on the mechanical properties of metals and alloys at temperatures close to room temperature (Romaniv et al. 1972).

Therefore, all the experiments on remagnetization of the specimens were performed in air at room temperature and atmospheric pressure. In order to improve the reliability of experimental results, all the specimens before hydrogen charging were subjected to annealing at a temperature of 550 °C with a holding time at this temperature of 14.4 ks. The time between annealing and hydrogen charging from the gas phase was approximately 61.2 ks.

Hydrogen charging of specimen from the gas phase. The specimens were hydrogenated in a specially designed chamber in a gaseous hydrogen (Fig. 9.27). Taking into account theoretical calculations, hydrogen charging modes are chosen in each experiment so as to bring the concentration of hydrogen in the metal as close as possible to the real one, which is in long-term operated ferritic-pearlitic steels in the range of low concentrations (Dobrotvorsky and Archakov 1989; Dmytrakh et al. 2020; Smiyan 2018). After completion of the hydrogen charging, the specimens were retested in the solenoid.

Based on the obtained test results of 3sp steel specimens, Fig. 9.28 presents the dependences of the sum of the amplitudes of the MAE signals on the change in the

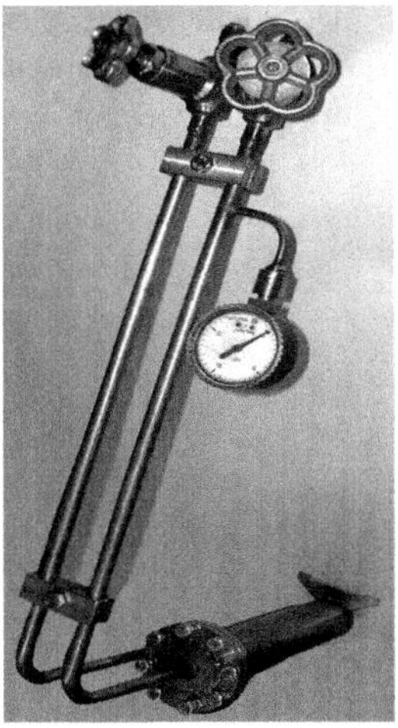

Fig. 9.27 General view of the chamber for specimens

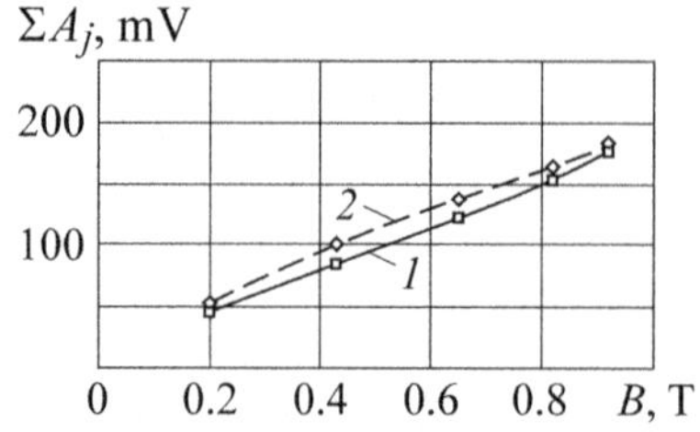

Fig. 9.28 Change in the sum of the amplitudes of the MAE signals depending on the induction of the magnetic field of 3sp steel specimens: *1* the state of delivery; *2* after hydrogen charging (hydrogen concentration in the metal $C = 2.7$ ppm)

magnetic field induction for the initial state (the state of delivery) and after hydrogen charging.

The specimens were hydrogenated at a temperature of gaseous hydrogen of 550 °C, the pressure in the chamber was 0.5 MPa, and the exposure time was 14.4 ks. The concentration of hydrogen in the steel was determined using an ELTRA H-500 gas analyzer.

The results show that the sums of the amplitudes of the MAE signals increase with increasing magnetic field induction in both non-hydrogen-charged and hydrogen-charged specimens. However, in the second case they are significantly larger (Figs. 9.28 and 9.29) for both steel grades. This can be explained only by the increase in the jump-like displacements of domain walls under the influence of available hydrogen both in amplitudes (increase in the length of mechanical displacements during a single jump) and in quantity. A similar result was observed in the case of annealing of ferritic-pearlitic steels (see Figs. 9.21 and 9.22). Therefore, it is necessary to determine the contribution of each of these two competing factors to the total amplitude of the MAE signals during the remagnetization of the hydrogen-charged steel specimen. To do this, proceed as follows.

During the experiments, MAE signals were first recorded for the specimens in the state of delivery (original material), and then—for annealed. After that, they were hydrogenated in a gaseous hydrogen, and then remagnetized and MAE signals were

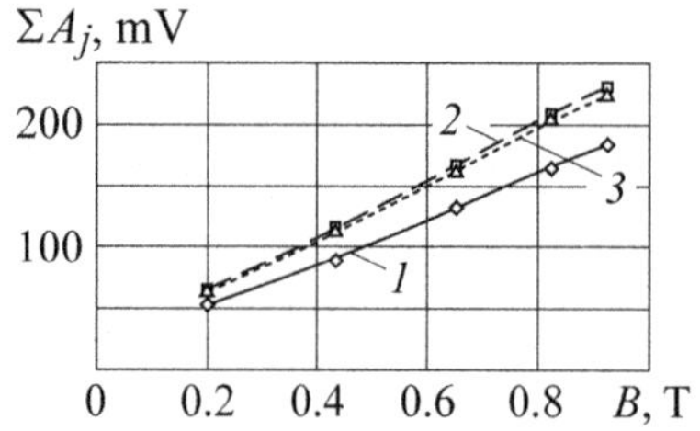

Fig. 9.29 Dependence of the sum of amplitudes of MAE signals on magnetic field induction for non-hydrogen-charged (*1*) and hydrogen-charged (*2, 3*) 15 steel plates ($C = 2.2$ ppm); *3* repeated measurements after 43.2 ks of aging in air

recorded. Incomplete annealing was performed at a heating temperature of 550 °C for 3.6 ks; then they were remagnetized in the solenoid with recording of MAE signals. They were then hydrogenated in gaseous hydrogen at the same temperature with a chamber holding time of 14.4 ks and a gas pressure of 0.5 MPa. The cooled samples were also remagnetized, recording the MAE signals. The results of the experiments are shown in Fig. 9.30. It follows that the presence of hydrogen as well as annealing increases the activity of MAE signals generation during the remagnetization of ferritic-pearlitic steels. Comparing the contribution of each of these factors, one can see that the contribution of the influence of the hydrogen in the sum of the amplitudes of the MAE signals is dominant.

Hydrogen charging of the specimens in gaseous hydrogen at a pressure of 0.5 MPa at temperatures of 550 °C for 4 h caused additional grinding of ferrite grains on the 3sp steel surface to 7–20 μm. Even at the low resolution of the microscope, the formation of damage along the grain boundaries between ferrite and pearlite, the appearance of which was facilitated by high-temperature hydrogen, is obvious.

As a result of metal hydrogen charging, the size of polyhedral ferrite grains decreased not only on the steel surface, but also in the center of the cross-section of the specimens in the range from 7 to 12 μm. Although the effect of hydrogen in the

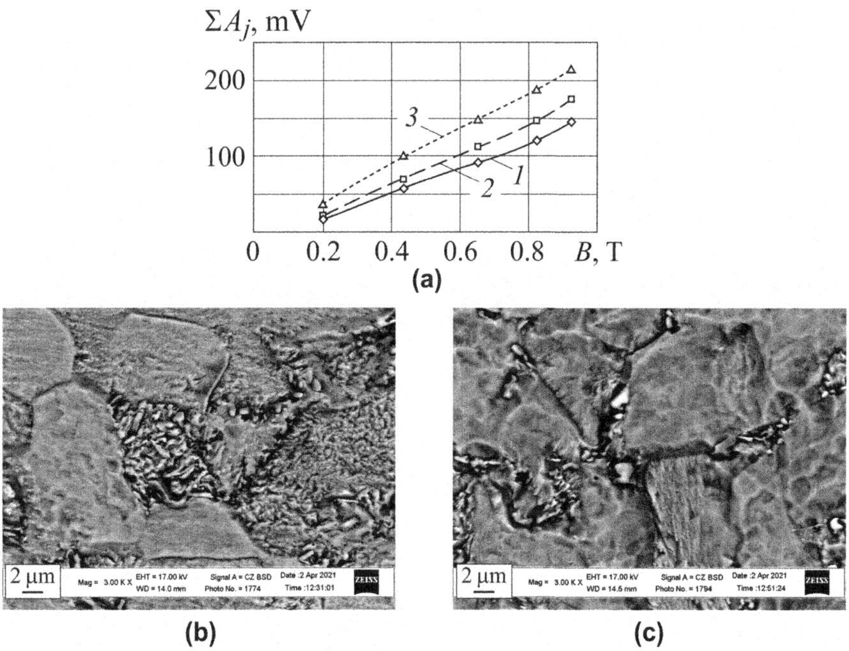

Fig. 9.30 Change in the sum of amplitudes of MAE signals depending on magnetic field induction of 3sp steel specimens: *1* the state of delivery; *2* after annealing; *3* after hydrogen charging (hydrogen concentration in the metal $C = 0.181$ ppm) (**a**) and metallography of their structure in the center of the cross-section (**b**, × 3000) and on the surface (**c**, × 3000)

center of the cross-section is less (compared to the structure in the initial state and after annealing) than on its surface, the presence of this effect is obvious. Therefore, it can be argued that the mechanism of action of hydrogen from the gas phase on the morphological features of the grain structure of the surface layers of 3sp steel and the diffusion-mobile hydrogen that penetrated into the metal through the surface, is the same, although they differ in efficiency.

Analysis of the structural features of three variants of 3sp steel (see Fig. 9.10) at higher resolutions made it possible to reveal the character of the damage observed along the interface between ferrite and pearlite grains at low resolution. First of all, a change in the morphology of cementite particles within pearlitic grains is found. In particular, the lamellar structure of pearlite in the structure of steel in the initial state changes due to the spheroidization of cementite particles—both after annealing and after hydrogen effect. In addition, there is a massive precipitation of elongated or small rounded carbide particles along the grain boundaries. This was most pronounced in the structure of steel after hydrogen charging (Fig. 9.30b, c).

The most important fact is that cementite and ferrite interfaces in pearlite grains and grain boundaries (even between ferrite grains) were clearly weakened by defects along these boundaries, which was especially evident after hydrogen action. In this case this feature was shown most clearly in the center of the cross-section. This suggests that the interfaces serve as energy-saving traps for hydrogen, which, falling into them, contribute to the decohesion between the structural components, which intensify the formation of scattered damage, including the specimen cross-section.

Thus, although the domain structure of the ferromagnet changes with the change of crystal structure, it is obvious that the significant improvement in the dislocation situation is a dominating factor, and the internal residual stresses are reduced, which greatly facilitates the movement of domain walls.

Figure 9.31 presents the MAE signals, which are typical for 3sp steel before and after hydrogen charging. The gain factor of the equipment measuring path is 78 dB.

Thus, annealing of ferritic-pearlitic 3sp steel at a temperature of 550 °C together with hydrogen charging at the same temperature changes the morphology of its grain structure, improves magnetic characteristics, heals dislocations, and relieves

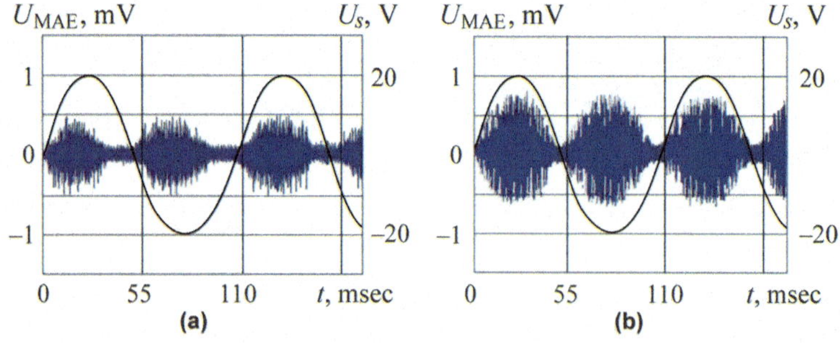

Fig. 9.31 Typical signals of MAE during remagnetization in the magnetic field of the solenoid (B = 0.92 T) for the 3sp steel specimen in the initial state (**a**) and hydrogen-charged (**b**)

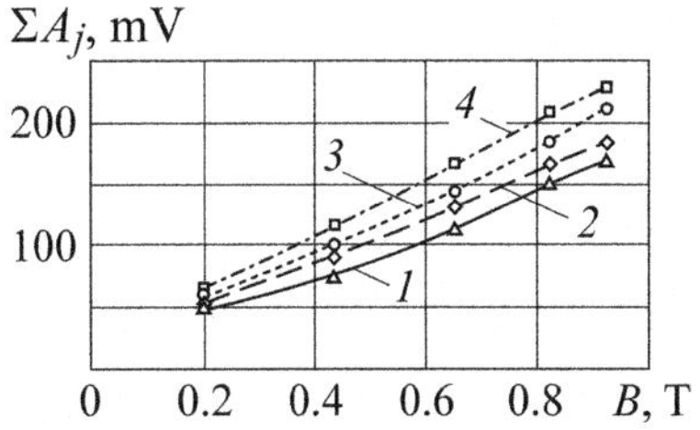

Fig. 9.32 Change of the sum of the 3sp steel specimens: *1* after deformation; *2* the state of delivery, *3* after hydrogen charging of the deformed material (hydrogen concentration $C = 0.482$ ppm); *4* after steel hydrogen charging in the state of delivery ($C = 0.373$ ppm)

residual internal stresses. As a result of hydrogen action, the activity and value of the domain walls jumps during the magnetization of steel increases even more, which significantly affects the increase in the sum of the amplitudes of the MAE signals.

Influence of hydrogen on MAE signal generation by plastically deformed steel. The sensitivity of the MAE method to the presence of plastic deformations is shown above. Based on this, the question arises about the possible manifestation of the hydrogen factor in this state for the ferritic-pearlitic steels. For this purpose, the specimens were remagnetized, tensioned in the stress range $\sigma_{02} < \sigma < \sigma_b$, MAE signals of the plastically deformed specimen were recorded, hydrogenated and MAE were recorded again. Test data are shown in Fig. 9.32.

From experimental results one can see that even in plastically deformed steel in the presence of hydrogen, MAE is generated more actively than in the case of its hydrogen charging in the initial state. This forms the basis for the creation of new methods for real-time analysis of equipment in the production conditions in order to identify the sites of local hydrogen damage. Similar results are obtained for 15 steel.

Thus, the influence of hydrogen in plastically deformed ferritic-pearlitic steels during their remagnetization by an external quasi-static magnetic field occurs by increasing the activity of generating MAE signals, which is manifested by an increase in the sum of their amplitudes.

Influence of carbon content on the manifestation of hydrogen presence in ferritic-pearlitic steels. The above presented in paragraph 9.6 stipulated testing of how the effect of hydrogen on the remagnetization of ferritic steels with different carbon contents is manifested. Specimens for experiments were hydrogenated according to the scheme described above, analyzing the sum of the amplitudes of the MAE signals (Fig. 9.33).

We can see that, as in all previous cases, hydrogen significantly increases the activity of MAE during the remagnetization of ferritic-pearlitic steel compared to its state of delivery. In this case, the less carbon, the greater the impact of hydrogen.

As was mentioned above, during aging of the hydrogenated material it is degassed. This takes place even at room temperature and atmospheric pressure. However, even

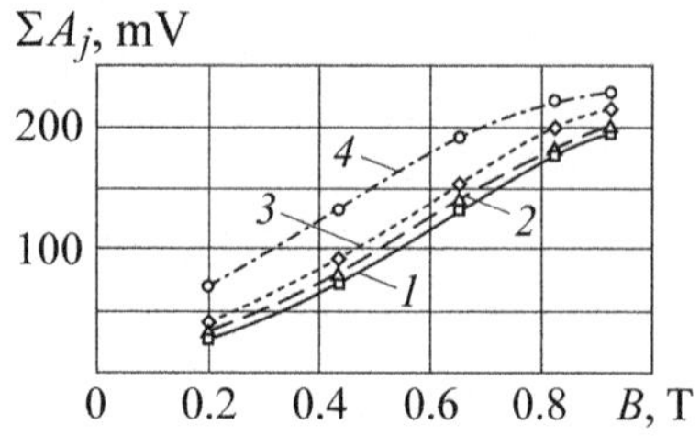

Fig. 9.33 Change in the sum of amplitudes of MAE signals depending on the magnetic field induction in the ferritic-pearlite steel specimens: *1* the state of delivery of steel 15; *2* the state of delivery of 3sp steel; *3* after 15 steel hydrogen charging (hydrogen concentration $C = 0.332$ ppm); *4* after 3sp steel hydrogen charging ($C = 0.361$ ppm)

in this case, the MAE method shows a high sensitivity to the detection of hydrogen factor. This is confirmed by the research results, as shown in Figs. 9.34 and 9.35.

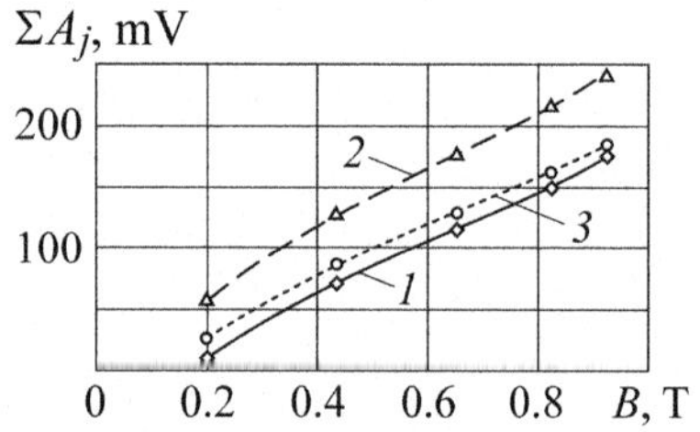

Fig. 9.34 Change of the sum of the amplitudes of the MAE signals depending on the induction of the magnetic field of specimens of ferritic-pearlitic 3sp steel: *1* the state of delivery; *2* after hydrogen charging from the gas phase; *3* after aging the hydrogenated material for 4.84 Ms (56 days) in air at room temperature and atmospheric pressure

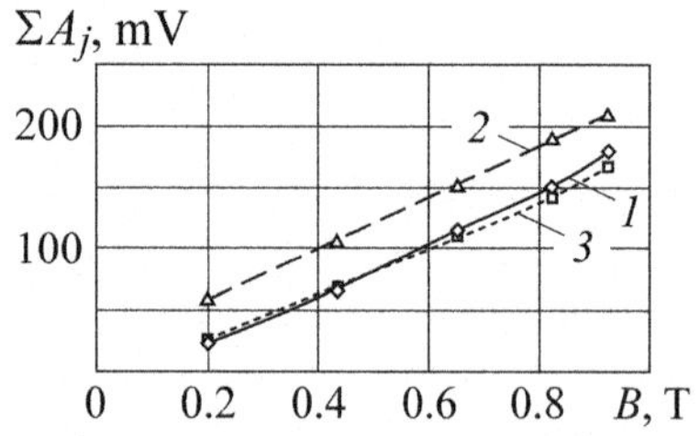

Fig. 9.35 Change of the sum of the amplitudes of MAE signals depending on the induction of the magnetic field of specimens of ferritic-pearlitic 3sp steel: *1* the state of delivery; *2* after hydrogen charging from the gas phase; *3* after aging the hydrogenated material for more than 9 years in air at room temperature and atmospheric pressure

9.9 The Phenomenon of Dual Growth of MAE

In modern literary sources there are a large number of publications describing concepts, theories, mathematical models, experimental results, etc. on hydrogen issues. Based on the most common of them, on the basis of research we make the following statement about the domain walls jumps during quasi-static magnetic field remagnetization of the ferritic-pearlite steels with available low-concentrated hydrogen in them.

At the beginning of hydrogen charging, hydrogen protons with a high diffusion rate (Barrera et al. 2018) penetrate into the metal, intensively concentrating at the boundaries of grains, inclusions, filling pores and other discontinuities. Decrease in the number of dislocations in steel unblocks the attachment points of the domain walls and stimulates their jump-like movement, which is manifested in an increase in the number and magnitude of jumps. This can last up to a certain value of hydrogen concentration. Then the diffusion of hydrogen into the crystal lattice begins. In the crystal lattice of metal, it is energetically advantageous not to distribute dissolved hydrogen uniformly, but to localize it in clusters (Tetersky et al. 1964; Besnarg 1966). These clusters have the form of an almost periodic lattice of the hydrogen subsystem, the period of which coincides with the parameter of the metal lattice. The latter means that the atomic concentration of hydrogen in such hydrogen-enriched zones is close to unity (Tkachov 1999). The effect depends on the electron density of the metal and the temperature. The clusters become unstable and dissolve at temperatures of about 500 K, i.e., about the time when the effects of hydrogen embrittlement of metals disappear (Tetelman and Robertson 1966).

Further, hydrogen causes the fact that hydrogen clusters in metal not only reduce the interaction energies of metal atoms, packing defects, and lattice deformations (Dobrotvorsky and Archakov 1989; Vavrukh and Solovian 1985; Ostash and Vytvytsky 2011; Robertson 2001; Kikuta 1976; Yukhnovskyi and Tkachev 1987), which also facilitates the movement of dislocations, but also causes an increase in dislocations and relaxation stresses (Robertson 2001; Shved 1985; Nibur et al. 2006; Lu et al. 1997). Hydrogen-metal clusters can cause the formation of new dislocations, point defects or specific phases in metals at high hydrogen concentrations (Tkachev et al. 1999).

The appearance of new dislocations and other damage of steel begins to impede the movement of domain walls. This, in particular, does not contradict the existing models of hydrogen inhibition of the dislocations movement of so-called atmospheres of Cotrell (Thompson and Bernstein 1985). The free movement of dislocations is exhausted, their number increases, as a result of which they accumulate near structural barriers (grain boundaries, secondary phases, etc.). And it is hydrogen that contributes to this (Andreikiv and Hembara 2008; Gerberich et al. 2009). Thus, the activity of the MAE decreases.

So, we see that there should be a duality in the character of MAE generation: first an intense growth with increasing hydrogen concentration in steel, and then at a certain, typical for a particular steel grade hydrogen concentration—an insignificant

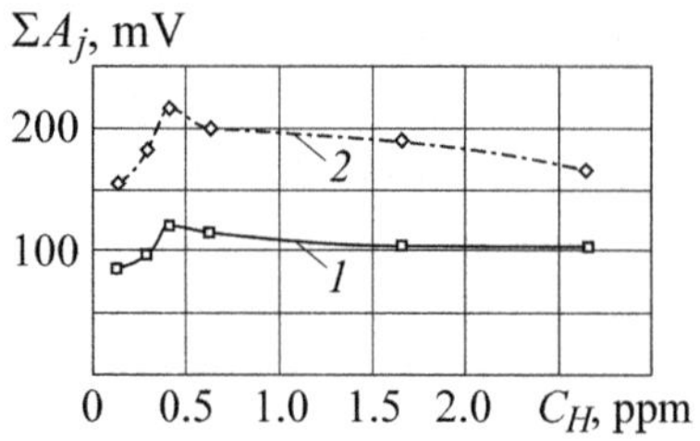

Fig. 9.36 Dependence of the sum of the amplitudes of the MAE signals on the hydrogen concentration in the specimens of ferritic-pearlitic 3sp steel under induction of the magnetic field in the specimens *1* 0.4 T; *2* 0.8 T

decrease. This will represent the processes and mechanisms described above. In Fig. 9.36 the results of the corresponding experimental investigations are given.

As one can see the research results fully confirm the above statements.

References

Andreikiv OY, Hembara OV (2008) Mekhanika ruinuvannia ta dovgovichnist metalevykh materialiv u vodenvmistkykh seredovyshchakh (Fracture mechanics and life time of metallic materials in hydrogen-containing environments). Naukova Dumka Publishing House

Barrera O, Bombac D, Chen Y (2018) Understanding and mitigating hydrogen embrittlement of steels: a review of experimental, modelling and design progress from atomistic to continuum (review). J Mater Sci 53:6251–6290

Becker R, Döring W (1939) Ferromagnetismus. Springer-Verlag

Besnarg S (1966) Influence de la haute pukete'du fer sur son aptitude au chargement en protons. Ann Chim 6(3):245–283

Bilby BA, Hewitt J (1962) Hydrogen in steel—the stability of microcracks. Acta Metall 10(6):587–600

Calland I, Azou P, Bastien P (1969) Comportement de l'acier doux sous contrainte en présent d'hydrogènem. Comp Rend L'acad Sci 268:27–32

Dmytrakh IM, Syrotiuk AM, Leshchak RL (2020) Ruinuvannia ta mitsnist trubnykh stalei u vodenvmistkykh seredovyshchakh (Fracture and strength of pipe steels in hydrogen-containing environments). Prostir Publishing House

Dobrotvorsky AM, Archakov YI (1989) Teoreticheskoie issledovaniie vliyaniia vodoroda na mekhanicheskiie svoistva zheleza (Theoretical study of the influence of hydrogen on the mechanical properties of iron). Fiz-Khim Mekh Mater (Physicochem Mech Mater) 3:37

Gerberich WW, Stauffer DD, Sofronis P (2009) A coexistent view of hydrogen effects on mechanical behavior of crystals: HELP and HEDE, effects of hydrogen on materials. In: Somerday BP, Sofronis P, Jon R (eds) Proceedings of the international hydrogen conference. ASM International, Ohio

Guliayev AP (1978) Metallovedeniie (Metal science). Metallurgiya Publishing House

Kikuta EI (1976) Izucheniie vodorodnoi khrupkosti i rol vodoroda v mikrostrukture (Study of hydrogen brittleness and role of hydrogen in microstructure). Yese Gakkaisi 45(2):1016–1021

Kush H-G (1975) Wassrstoffeinfluß auf Streckgrenze und Ludersdehnung von Eisen bei Raumtemperatur. In Korroziya pod napriazheniyem i vodorodnoie okhrupchivaniie (Stress corrosion and hydrogen embrittlement. In: Scientific symposium), Dresden, pp 71–84

Lu H, Li M, Zhang T, Chu W (1997) Hydrogen-enhanced dislocation emission, motion and nucleation of hydrogen induced cracking for steel. Sci China (Ser E) 40:530–538

Nazarchuk ZT, Koshovyi VV, Skalskyi VR (2001) Mekhanika ruinuvannia i mitsnist materialiv. T. 5: Neruinivnyi control tekhnichna diagnostyka (Fracture mechanics and strength of materials. Vol. 5: non-destructive testing and technical diagnostics). PhMI Publishing House

Nazarchuk Z, Skalskyi V, Serhiyenko O (2017) Some methodological foundations for selecting and processing AE signals. In: Acoustic emission. Methodology and application (foundations in engineering mechanics). Springer, Cham, pp 107–159

Nibur KA, Bahr DF, Somerday BP (2006) Hydrogen effects on dislocation activity in austenitic stainless steel. Acta Mater 54:2677–2684

Ostash OP, Vytvytsky VI (2011) Dvoistist dii vodniu na mekhanichnu povedinku stalei i struknurna optymizatsiia ih vodnetryvkosti (Dual action of hydrogen on mechanical behavior of steels and structural optimization of their hydrogen resistance). Fiz-Khim Mekh Mater (Physicochem Mech Mater) 4:5–19

Robertson IM (2001) The effect of hydrogen on dislocation dynamics. Eng Fract Mech 68:671–692

Romaniv AN, Tkachev VI, Krypiakevych RI (1972) Malotsyklovaia ustalost stali 2X13 v srede gazoobraznogo vodoroda (Low-cycle fatigue of 2X13 steel in gaseous hydrogen medium). Fiz-Khim Mekh Mater (Physicochem Mech Mater) 1:102–104

Rudiak VL (1986) Protsessy perekliucheniia v nelineinykh kristalakh (Switching processes in nonlinear crystals). Nauka Publishing House

Shved MM (1985) Izmeneniie ekspluatatsionnykh svoistv zheleza i stali pod vliianiiem vodoroda (Changes in the operational properties of iron and steel under the influence of water). Naukova Dumka Publishing House

Siede A, Rostoker W (1958) On the problem of hydrogen embrittlement of iron. Trans Am Inst Min Metall Petrol Eng 212:27–32

Skalsky V, Nazarchuk Z, Stankevych O, Klym B, Selivonchyk T (2022) Effect of operational factors on magnetoacoustic emission of low-carbon steels. Int J Press Vessels Pip 199:104744

Skalskyi V, Nazarchuk Z, Stankevych O, Klym B (2023) Influence of occluded hydrogen on magnetoacoustic emission of low-carbon steels. Int J Hydrogen Energy 48(15):6146–6156

Smiyan OD (2018) Voden i ruinuvannia metalu obiektiv tryvaloi ekspluatatsii (Hydrogen and fracture of metal of long-term operating objects). Naukova Dumka Publishing House

Tetelman AS, Robertson WD (1966) Hydrogen embrittlement of iron and steels. Trans Am Inst Min Metall Pet Eng 236(4):78–84

Tetersky VA, Soshko AI, Karpenko V (1964) K voprosu o vliyanii radiatsionnogo izlucheniia v gazovykh sredakh na mekhamicheskiie svoistva stali. V knige: Vliayniie rabochikh sred na svoistva materialov (To the question of the influence of radiation in gaseous media on the mechanical properties of steel. In: Influence of working environments on the properties of materials). Naukova Dumka Publishing House

Thompson AU, Bernstein IM (1985) Rol metllurgicheskikh faktorov v protsessakh razrusheniia s uchastiyem vodoroda. Dostizheniia nayki o korozii i tekhnologii zashchity ot neie. Korrozionnoie rastreskivaniie metallov (The role of metallurgical factors in destruction processes with the participation of hydrogen. Achievements of corrosion science and corrosion protection technology. Corrosion cracking of metals). Metallurgy Publishing House

Tikadzumi S (1987a) Fizika ferromanrtizma. Magnitnyie svoistva veshchestva (Physics of ferromagnetism. Magnetic properties of matter). Nauka Publishing House

Tikadzumi S (1987b) Fizika ferromagnitizma. Magnitnyie kharakteristikii praktycheskiie primeneniia (Physics of ferromagnetism. Magnetic characterristics and practical applications). Nauka Publishing House

Tkachev VI, Kholodnyi VI, Levina VN (1999) Rabotosposobnost stalei i splavov v srede vodoroda (Workability of steels and alloys in a hydrogen environment). Vertikal Publishing House

Tkachov VI (1999) Mechanism of the reversible effect of hydrogen on the mechanical properties of steel. Mater Sci 35:477–484

Tumen State Oil and Gas University (1997) Materialy regionalnoi nauchno-tekhnicheskoi konferentsii (Proceedings of the regional scientific-technical conference). Tumen State Oil and Gas University, Tumen, p 184

Vavrukh MV, Solovian VB (1985) Lokalizatsiia primesei vodoroda v metalle (Hydrogen localization in metal). Fiz-Khim Mekh Mater (Physicochem Mech Mater) 4:26–29

Vonsovsky SV, Shur YA (1948) Ferromagnetizm (Ferrmagnetism). State Publishing House of Theoretical and Technical Literature

Yukhnovskyi PI, Tkachev VI (1987) Sostoyaniie vodoroda v metalle (State of hydrogen in metal). Fiz-Khim Mekh Mater (Physicochem Mech Mater) 4:107–108

Chapter 10
Evaluation of Absorption of Electrochemical Hydrogen by MAE Parameters

A solenoid electrochemical cell was prepared to study the effect of hydrogen absorbed from the electrolyte on the magnetoelastic acoustic emission (MAE) parameters. Cylindrical specimens made of cold-rolled 15 steel (see Appendix) were investigated. The electrochemical cell is filled with 0.1 N NaOH solution, which provides high electrical conductivity, and does not promote the release of electrolysis by-products other than hydrogen and oxygen. As with gas-phase hydrogen charging, during electrolytic hydrogen charging, a significant increase in the power of MAE signals is noticeable for hydrogen-charged samples compared to non-hydrogen-charged ones. Comparing the obtained results for electrolytic hydrogen charging with the results for hydrogen charging from the gas phase, it is possible to note a significantly higher sensitivity of MAE parameters during cathodic hydrogen charging. This can be explained by the role of diffusion-mobile hydrogen in the electrolysis conditions and its absence in experiments with gaseous hydrogen charging. The results of the research are described in detail in this chapter of the monograph.

10.1 Hydrogen as a Factor of Structural Material Degradation

Aging of structural materials, which is revealed at the macrolevel by decreasing plasticity and toughness, is a concern in various industries: oil and gas, petrochemical, pipeline transport, construction, energy, etc. (Nazarchuk et al. 2017; Batchelor et al. 2011). Degradation of the physical properties of materials begins with submicrostructural changes, caused, of course, by the dynamics of dislocations and diffusion processes, resulting in irreversible segregation (either to grain boundaries or to microstructural defects within grains) of certain chemical elements, especially if they have a thermodynamic tendency to form a separate phase with physical properties significantly differing from the matrix. For example, the formation of ion-covalent

V. Skalskyi and Z. Nazarchuk, *Magnetoelastic Acoustic Emission*, Springer-AAS Acoustics Series, https://doi.org/10.1007/978-981-99-4032-5_10

compounds such as oxides, phosphides, sulfides, nitrides, hydrides, carbides, and combined compounds, even in the early stages of initiation, when there is a high degree of coherence with the matrix, as in the case of Gignier–Preston zones, can cause a significant change in the material sensitivity to the initiation and propagation of microcracks in its volume. This occurs when the newly formed physical inhomogeneity, which may cause modification of the internal stress field, reaches such dimensions and is oriented in the stress field in such a way that conditions for the development of fracture arise.

Degradation of carbon and low-alloy steels—the most common grades of structural materials and ferromagnets—is a well-documented phenomenon, especially in the steam-generating equipment of power plants, gas and oil pipelines. Among the main factors affecting the aging process are variable loads and temperatures as well as hydrogen absorbed by the metal (Okura et al. 1998; Kurzydlowski and Lunarska 2008). However, if the effects of mechanical and thermal fatigue are studied in sufficient detail, including at the level of microstructural changes in the metal, the understanding of the role of hydrogen still remains unclear, despite more than half a century of research by scientists and engineers in this field.

Hydrogen gets into structural materials already at the manufacturing stage. Hydrogen enters into the molten metal, for example, from heat-resistant ceramics and ferroalloys, during drawing—from lubricants C_xH_y, due to annealing in hydrogen-containing atmospheres; before galvanization—from acids (HCl, H_2SO_4); during galvanization; due to corrosion; at the stages of storage and transportation; due to welding, etc.

Today, for many categories of metal items the requirements concerning maximum hydrogen concentration are put forward, which is carefully controlled either by extracting hydrogen from a molten metal fragment (total hydrogen content) or by hot extraction and thermodesorption of solid metal specimens (hydrogen content of diffusion-mobile and desorbed from the reverse traps). The analysis of the amount of hydrogen is carried out on the principle of changing the thermal conductivity of the gas mixture and using infrared and mass spectrometry.

It is known that the absorption of hydrogen by metals is possible both from hydrogen-containing gases (H_2, H_2S, NH_3, etc.) and from liquids (H_2O, H_2SO_4, HCl, etc.), and occurs in several stages. The creation of the conditions of increased activity of neutral atomic hydrogen on the surface of the metal/medium interface, part of which can get inside the metal is a necessary condition for hydrogen charging.

The process of absorption of atomic hydrogen by metal from the electrolyte by electrochemical reduction of hydrogen ions begins with electrosorption of hydrogen-containing compound, according to a simplified scheme, using one of the equations (Protopopoff and Marcus 2007):

$$H^+ + Me + e^- = H_{ads}(Me), \tag{10.1}$$

$$H_2O + Me + e^- = H_{ads}(Me) + OH^-, \tag{10.2}$$

depending on the activity of the proton in the electrolyte, i.e., the pH of the latter. For the increased proton activity (acidic electrolytes), the reduction according to the scheme (10.1) dominates. In another case, the dominant scheme is (10.2). It should be noted here that Eq. (10.1) does not illustrate the hydrating state of the proton in the electrolyte and the dehydrating stage in the H^+ proton reduction scheme.

As a result of the reduction process a chemisorbed hydrogen atom $H_{ads}(Me)$ is formed, and electrons for electrosorption are supplied either from an external source of electric current (for underground or underwater metal-environment systems, this may be, for example, cathodic protection or stray currents from electric transport systems) or from electrochemical oxidation of the metal according to equation:

$$Fe = Fe^{2+} + 2e^{-}. \quad (10.3)$$

After the formation on the metal surface of neutral absorbed atomic hydrogen $H_{ads}(Me)$, the further behavior of the latter is determined by one of the following two processes: recombination-desorption with the formation of molecular hydrogen H_2 and absorption of atomic hydrogen by metal. Depending on the activation energy of the process, desorption can be chemical, i.e., occur by direct formation of molecular hydrogen from two adsorbed atoms:

$$2\,H_{ads}(Me) = H_2 + 2Me, \quad (10.4)$$

or electrochemical, during which the discharge of the proton of the electrolyte is involved with the transfer of the electron from the metal. Depending on the activity of the proton in the regions of the electrolyte adjacent to the metal surface, this occurs according to one of the following schemes:

$$H_{ads}(Me) + H^{+} + e^{-} = H_2 + Me, \quad (10.5)$$

$$H_{ads}(Me) + H_2O + e^{-} = H_2 + OH + Me. \quad (10.6)$$

The formed hydrogen molecules have significant dimensions: the bond length in the hydrogen molecule H_2 is 0.74 Å, the van der Waals radius of each atom is 1.17 Å, as a result of which the effective van der Waals length of the hydrogen molecule exceeds 3 Å. Considering the energy of the interatomic bond in the hydrogen molecule, which is 458 kJ/mol, it is clear why molecular hydrogen at room temperature is chemically virtually inert and is not dangerous for iron, nickel, or other metals from which structural materials are prepared and in which interatomic distances are shorter: for iron and nickel, these are 2.48 Å and 2.49 Å, respectively.

The hydrogen molecules formed on the surface of the metal combine to form a gas phase in the form of bubbles, overcoming the surface energy at the electrolyte-gas interface, and, reaching the appropriate size, are pushed out of the electrolyte by Archimedean force. This happens to the vast majority of hydrogen atoms reduced on the metal surface.

However, a small part of the chemisorbed atomic hydrogen $H_{ads}(Me)$ on the metal surface undergoes absorption with the formation of a solid solution of $H_{ads}(Me)$ according to

$$2H_{ads}(Me) = H_2 + 2Me. \tag{10.7}$$

The kinetics of the process (10.7) depends on numerous factors, including primarily the chemical state of the films on the metal surface, their coherence, general and local at the site of absorption, chemical nature of the metal, crystallographic orientation of the grain surface absorbing hydrogen atom, power of micro- or nanogalvanic vapors existing in the vicinity of the hydrogen absorption site, the composition of the electrolyte, the density of the electric current applied to the metal (Fermi level of electrons), temperature, pressure, etc.

The increase in the chemical activity of surface atomic hydrogen, and thus subsurface hydrogen absorbed by the metal, causes a gradient of chemical potential, which becomes the driving force of hydrogen diffusion in the metal volume. In the literature, one can find a significant number of publications devoted to the study of hydrogen diffusion parameters in metals, including many reviews on this topic, such as the review of Kiuchi and McLellan (1983). The diffusion coefficients of hydrogen D(H) published in these sources demonstrate huge differences—more than four orders of magnitude for room temperature. This inconsistency of data is caused by significantly different levels of defects in the studied materials. Even the smallest change in their concentration will affect hydrogen capture and therefore affect diffusion parameters. This applies to both 0D, 1D, 2D, or 3D defects in ultrapure iron, as well as metallic and non-metallic impurities in the metal, which are usually present even in relatively pure iron in concentrations commensurable or even much higher than dangerous levels of absorbed hydrogen. For such investigations, even heat treatment by annealing in a quartz furnace can cause silicon absorption into the metal, which in turn also affects the diffusion of hydrogen. Such elements as Cr and Mo can have a significant influence on the values of the diffusion coefficients D(H) (Kim et al. 1981).

In an isotropic medium, the diffusion of hydrogen in metal is described by Fick's laws, while in real materials the picture can be more complicated because there is the action of factors such as the field of internal deformations, determined not only by anisotropic properties of the crystal structure of the material and applied external stress vectors, but also by the modification of this field by defects and chemical inhomogeneities, as well as tetragonal distortion of the crystal structure of the metal by the absorbed hydrogen, not to mention the presence of other phases, interfacial and intergranular boundaries, generation and movement of dislocations, and their interaction with absorbed hydrogen.

Investigation of the state and dynamics of hydrogen in the metal, despite their significant amount, has not yet clarified many issues, including the physical nature of hydrogen in the metal. This is caused by the limitations of analytical tools in detecting the lightest of all existing atoms, as well as in the case of metals such as iron, by extremely low solubility in the volume-centered crystal lattice of metal, which is about three H atoms per 10^8 Fe atoms (Bernstein 1970). In real structural materials, which are both chemically and by the defect level, significantly different from ultrapure iron, the level of "metallurgical" hydrogen (i.e., hydrogen contained in the metal sold by metal producers) can easily be 1 ppm, and dangerous hydrogen levels—from tenths to several ppm units.

Such investigations are complicated due to the ferromagnetic properties of iron, which is the basis of most structural materials. Such materials are characterized by magnetocrystalline anisotropy and the relationship between elastic parameters and magnetostrictive properties of the metal. Since the hydrogen absorbed by the metal shows a significant positive partial volume and causes tetragonal distortion of the crystal lattice, it cannot but cause mutual hydrogen-magnetic interaction, especially regarding the interaction of both hydrogen and domain walls with the defects of the microstructure (Mu et al. 1989; Ruščák and Perng 1993; Sanchez et al. 2005).

10.2 Theories of Hydrogen Embrittlement of Metals and Alloys

With the development of investigations on the effect of absorbed hydrogen on the physical properties of the metal, ideas about its possible mechanisms have also been developing. Various physical models of hydrogen embrittlement have been proposed, although none of them can claim universality, and in some cases, several mechanisms of hydrogen degradation have been observed simultaneously. In such cases, the dominance of one of the embrittlement mechanisms is possible, which under certain conditions may change to the dominance of another mechanism. Let us briefly consider several main mechanisms of hydrogen embrittlement.

Theory of pressure. The theory of planar pressure—perhaps the oldest of all theories—was developed in the works of Zapffe and Sims (Zapffe and Sims 1941; Zapffe 1947; Sims et al. 1948; Sims 1950). According to it, hydrogen in the metal accumulates in the sites of increased free volume, which the authors call cavities, or "substructural disjunctions" forming a gas phase, the pressure in the middle of which increases, contributing to or causing fracture of the metal. Deformation promotes the increase of the size of these cavities and the pressure reduction. However, further desorption of hydrogen into these cavities again leads to an increase in pressure. If according to the authors of the theory, the rate of hydrogen desorption into the cavity does not provide maintain or increase the pressure due to its decrease caused by the corresponding deformation rate of the metal, one should expect reduction of the hydrogen embrittlement effect. Temperature, influencing directly on gas pressure, increases the embrittlement effect, so there must be a critical temperature and a critical cooling rate, below which the pressure in the cavities will not increase due to hydrogen desorption, and therefore the embrittlement effect will also decrease.

Petch and Stables extended the Orowan's idea about the delayed fracture of glass to the phenomenon of hydrogen embrittlement (Orowan 1944; Petch and Stables 1952), noting that the release of hydrogen into microcrack cavities reduces the critical stress for crack propagation in the Griffiths formula, namely crack propagation should occur step by step. Since the embrittlement process is controlled by hydrogen diffusion, the metal must exhibit plastic behavior at high deformation rates. As the temperature decreases, the diffusion of hydrogen decreases, since it is described by the Arrhenius relation:

$$D = D_0 \, e^{-\frac{Q}{RT}}, \tag{10.8}$$

and therefore the ductility of the metal should increase.

This theory proves to be important to explain the hydrogen blistering of steels, but it cannot be extended to other effects of hydrogen degradation of metals.

Theory of decohesion or hydrogen strength reduction. This mechanism, which is often cited to explain hydrogen embrittlement, was proposed by Troiano and developed by Oriani (Troiano 1960; Oriani 1972, 1978). According to this theory, hydrogen dissolved in the crystalline structure of metal reduces the strength of interatomic bonds, causing a decrease in the level of stresses at which the metal fracture occurs. The amount of hydrogen dissolved in the metal lattice is generally very low and should therefore be the cause for the concentration of hydrogen in the vicinity of the fracture site. In the case of intergranular cracking, this may be due to the accumulation of hydrogen in traps at the grain boundaries or at the boundaries of the secondary phase. In the case of cross-grain cracking, the increase in the hydrogen concentration in crystalline structure is possible due to higher levels of its deformation as a result of triaxial stresses that occur in the area in front of the crack tip.

The theory of decohesion is considered as the dominant mechanism in the case of internal hydrogen cracking of high-strength steels and alloys in which hydride phases are not formed, although the theory itself, even in the case of high-strength steels and alloys, ignores plasticization of the zone in front of the crack tip (Thomson 1978).

Theory of surface energy or adsorption reduction of strength. The theory proposed by Petch and Stables (Petch and Stables 1952; Petch 1956) considers the effect of hydrogen adsorbed at the crack tip and on the metal surface inhomogeneities, on the reduction of surface energy, which, in turn, decreases the threshold values of the stress intensity factors which are necessary for the development of brittle fracture. This theory applies to cases where the adsorption of hydrogen occurs at the same time as the event of fracture.

The theory of adsorption strength reduction was developed in the studies of Coleman and co-authors on stress corrosion cracking (Coleman et al. 1961), where it was proposed that in the corrosive environment the adsorption of various atoms or compounds, both neutral and charged (ions), which reduced the threshold values of stresses, or their concentration factors, took place.

Formation of the hydride phase. Early studies of the effects of hydrogen (the 1960s–70s) revealed hydrogen embrittlement due to the formation of hydrides or other relatively brittle phases (Sherman et al. 1968; Owen and Scott 1972). Based on these and his own studies on hydride formation in niobium, Birnbaum was the first to formulate the theory of the initiation and growth of the hydride phase (with significantly different mechanical properties) as the cause of hydrogen embrittlement (Birnbaum and Baker 1972; Gahr and Birnbaum 1976, 1978; Gahr et al. 1977; Birnbaum 1976; Grossbeck and Birnbaum 1977; Makenas and Birnbaum 1980).

This theory was confirmed on other materials with a high content of strong hydride-forming agents, such as Ti, V, or Zr (Cannelli and Cannelli 1981; Koike and Suzuki 1981; Puls 1981; Hardie and McIntyre 1973), and was later extended to all cases of phase transformations induced by absorbed hydrogen, because the formation of brittle phase (hydride, martensite, etc.) and its subsequent fracture was the main reason here (Benson et al. 1968; Narita and Birnbaum 1980; Rozenak et al. 1969). The main requirements are the stabilization of the newly formed phase by dissolved hydrogen in the metal and the stress field in the front of the crack tip (Westlake 1969; Narita et al. 1982; Flannagan et al. 1981) and the brittleness of this phase (Birnbaum 1984).

This mechanism can also include systems in which pseudohydrides are formed that do not have an ordered hydrogen substructure. Such systems include Ni and its alloys, Pd and its alloys, metals of group V–B (Nb, Ta, V) at high temperatures and austenitic stainless steels/alloys in which pseudohydrides are formed at high levels of hydrogen charging.

This mechanism of hydrogen embrittlement is considered to be well studied with a significant base of experimental evidence. Hydrides have mechanical properties that differ significantly from the properties of the matrix, and therefore crack initiation and propagation occur along these brittle phases. In the case of iron, in which the solubility of hydrogen is extremely low, a stable hydride phase cannot be formed at normal temperatures, so this mechanism can practically not be extended to structural steels and alloys.

Theory of hydrogen-enhanced localized plasticity. This theory was first proposed by Beachem (1972) on the basis of a detailed fractographic examination of steel specimens. It was ignored for many years until Birnbaum took it up and developed it on the basis of extensive experimental research. The obvious contradiction of the terms "embrittlement–plasticity" can be explained as follows. The stress field in the metal strongly influences local distribution of the hydrogen dissolved in the crystal structure. In the sites of its increased concentration, the yield strength of the metal decreases, causing strongly localized plastic deformation and ductile fracture of the metal, while at the macrolevel the deformation remains insignificant (Matsumoto et al. 1981; Birnbaum and Sofronis 1994). This theory is confirmed by direct observations of the hydrogen-intensified dislocations movement using a penetrating electron microscope with a controlled atmosphere. It is important that the phenomenon of hydrogen-enhanced localized plasticity is confirmed for a wide range of materials, such as Fe (Tabata and Birnbaum 1983, 1984), Ni (Eastman et al. 1982; Robertson and Birnbaum 1984, 1986), Al (Bond et al. 1988), etc., proving that this mechanism is universal and extends to different types of crystal structures and different chemical composition of the metal.

The causes of the intensification of the movement of dislocations under the influence of hydrogen absorbed by the metal are still unclear, although the hypothesis of "elastic shielding" of dislocations by dissolved hydrogen in the metal has been proposed (Birnbaum 1994). According to it, the hydrogen accumulation can move together with dislocations in the temperature range in which hydrogen embrittlement is observed, influencing the interaction of elastic defects at close distances, without manifesting itself at large distances.

So, hydrogen absorbed by the metal can cause an embrittlement effect in different scenarios, often competing, depending on the thermodynamic and kinetic aspects of fracture. Having an influence on the field of internal stresses, hydrogen must inevitably affect the processes of magnetostriction, and when interacting with structural defects—the dynamics of domain walls, and therefore should affect the parameters of remagnetization of ferromagnets and their elastic properties. So, it is expected that the hydrogen absorbed by the metal will affect the magnetoelastic acoustic emission, and thus there must be correlations between the activity of hydrogen in the ferromagnet and different physical parameters related to the magnetic and magnetoacoustic properties of the latter.

10.3 Relationship Between Absorbed Hydrogen and Magnetic Properties of Metals

The study of the relationship between hydrogen, absorbed by the metal, magnetic field parameters and the effects associated with these factors is in its initial stage due to the lack of experience in both experimental and theoretical fields. Although some correlations between hydrogen and magnetic factors have been experimentally revealed, it is still too early to speak about the existence of at least some established patterns, as these studies are very fragmentary and involve a small group of materials, among which there are almost no typical structural materials-ferromagnets which undergo the greatest degradation under the influence of absorbed hydrogen. Nevertheless, the experimental data accumulated for the last decades still give hope to the development of magnetic and magnetoacoustic approaches to hydrogen materials science and their actual implementation for non-destructive testing of structural materials.

Some aspects of the hydrogen-magnetic relations for metals can be analyzed from the literature, which have a long history of experimental research and a more developed theoretical basis. For example, the effect of a magnetic field on the mechanical properties of ferromagnetic materials has been studied since the 1950s. The studies of the influence of hydrogen on the mechanical properties of these materials are no less old (Matsui et al. 1979; Moriya et al. 1979). For example, it has been established that defects in the crystal structure affect the parameters of the hysteresis curve, and therefore plastic deformation or fatigue has an influence on these parameters (Jiles 1988; Jiles and Utrata 1988; Devine et al. 1992; Thompson and Tanner 1994). The Barkhausen effect, which depends on the interaction between the domain walls and the inhomogeneities of the crystal lattice, is also sensitive to plastic deformation and other mechanical properties of the metal (Ng et al. 2003). Hydrogen also interacts with structural defects, changing their energy parameters, which inevitably affects the characteristics of plasticity or Barkhausen noise.

Examples of such studies include the establishment of a synergistic effect of absorbed hydrogen and magnetic field on the mechanical properties of high-strength low-alloy 2.25Cr–1Mo steel, sensitive to hydrogen embrittlement (Ruščák and Perng 1993). For high-strength pipe steels X52, X70, and X80 it is found that the residual magnetization causes an increase in the concentration of hydrogen absorbed by the metal by more than 60% (Sanchez et al. 2005), and we speak here about a residual magnetization that can be caused by non-destructive magnetic methods such as magnetic flux scattering, which is used for internal diagnostics of existing main gas pipelines. However, for high-strength steel AISI 4340, which has a significant tendency to hydrogen embrittlement, no effect of magnetic field on this process is observed, although the development of fracture is monitored by the sensitive acoustic emission method (Thompson and Tanner 1994). These and other similar studies (Mu and Chu 1989; Ramanathan et al. 2013) give only episodic ideas about the influence of the magnetic field on hydrogen embrittlement, and therefore the study of these phenomena is still waiting for its researchers.

As for the publications devoted to the influence of absorbed hydrogen on the magnetic properties of materials, their number is greater, although they are, of course, devoted to the latest materials of special purpose. For example, the influence of absorbed hydrogen on the magnetic properties of rare-earth metal compounds (Buschow and Sherwood 1978; Fish et al. 1979), magnetic ordering in ferromagnetic intermetallics Th_6Mn_{23} and Y_6Mn_{23} (Malik et al. 1977), $GdNi_2$ (Malik and Wallace 1977), $GdFe_2$ (Aoki et al. 1987), YFe_2 (Paul-Boncour and Percheron-Guegan 1999; Singh and Gupta 2004), and other intermetallides is studied (Wallace 1979; Shenoy et al. 1983). However, there are not many studies on the effect of absorbed hydrogen on the magnetic properties of iron-based structural materials. For example, studying the effect of hydrogen and deformation on the stability and phase transformations of the austenitic phase in stainless steel, the authors found no changes in the magnetic properties (Narita et al. 1982). In Frank and Ferman (1965), which studied the change in the magneto-mechanical properties of Fe–Si steels, the absorbed hydrogen is considered exclusively as a factor that determines the internal stresses in the metal. From this point of view, a number of later studies have been conducted (Smith and Birchak 1968, 1969, 1972; Rothenstein and Anghel 1970). In other papers, the effect of absorbed hydrogen on the magnetic properties of iron in the form of nanoparticles has been established (Novakova et al. 2001a, b). The influence of hydrogen absorbed in relatively large quantities in the microcavities of nickel and iron on the magnetic properties (coercive force, residual nonmagnetization) of materials is given in Govindaraju et al. (1996). The same category includes the study of the effect of the distribution of hydrogen absorbed in amorphous alloys with a high content of Fe on their magnetic properties (Alcalá et al. 2000). It has been suggested that for samples with a high coefficient of magnetostriction, internal stresses and magnetic anisotropy are caused by the distribution of absorbed hydrogen, rather than its total content in the metal. Similar studies have been carried out on amorphous samples of other chemical compositions (Kovaľaková et al. 2005; Palit et al. 1977). All these works are quite diverse and concern only certain aspects of the issue, and therefore cannot be generalized.

Regarding studies on the effect of absorbed hydrogen on the micromagnetic properties of materials, there are very few studies and they concern mainly materials for special purpose or new materials that do not yet have any purpose application. These include a series of publications on the instability of amorphous metallic materials under the influence of absorbed hydrogen. Although this phenomenon has been widely studied for decades on various alloys (Savyak et al. 2004), the use of micromagnetic methods in such studies is very rare (Weissman 1996). For example, Mohanta, Mitra, and Chattoraj have shown that for an amorphous iron-based alloy, the processes of metal hydrogen charging-dehydrogen charging cause corresponding reversible changes in the Barkhausen signal intensity and coercive force (Mohanta et al. 2003). According to the authors, the internal stresses induced by the absorbed hydrogen are responsible for these changes. In Ceniga et al. (1999), Skryabina et al. (2000), Ceniga (2001), Ceniga and Kováč (2001, 2002), Novák et al. (2002), Skryabina and Spivak (2003), several more examples of successful use of the Barkhausen effect in the investigation of the effect of absorbed hydrogen on the stability of various amorphous alloys based on Fe, Co, Ni are shown. Another example is the application of the Barkhausen effect to study the influence of absorbed hydrogen on the stability of the sigma phase of Fe–Cr (Cieslak et al. 2009). In the mentioned studies, the Barkhausen noise method is used exclusively as a detector, which indicates sudden changes in the magnetization of the material, without presenting the nature of the hydrogen influence on the remagnetization processes. It can be concluded that hydrogen plays a secondary role, stimulating phase transformations in metastable metals.

With regard to structural materials with ferromagnetic properties, there are very few sources in the literature on the impact of their hydrogen charging on the processes of remagnetization, in particular the Barkhausen effect and magnetoacoustic emission. Considering the above said and the rapid development of acoustic emission diagnostics both in the field of basic research and its instrumental base, there is a motivation for these studies—to establish the influence of hydrogen on the parameters of magnetoacoustic emission.

10.4 Method of Hydrogen Charging from Electrolyte

To study the effect of hydrogen absorbed from the electrolyte on the MAE parameters, a solenoid electrochemical cell was prepared, the external and schematic view of which is shown in Fig. 10.1. The solenoid winding is wound on a polyvinyl chloride pipe with a diameter of 50 mm, which served as the body of the electrochemical cell. The base of the solenoid was closed on both sides with covers and appropriate holes for filling and draining the electrolyte, sealing for the centering specimen pin, specimen output, electronic thermometer, and anode inputs.

Cylindrical specimens with a diameter of 12 mm and a length of 260 mm made of cold-rolled 15 steel (see Appendix) were investigated. This ferritic-pearlite steel has a high content of α-Fe, which provides its high (compared to pure iron) ferromagnetic properties. The lower part of the specimen was completely located in the electrochemical cell volume, while the upper insignificant part of it protruded above the upper cover of the specimen for electrical connection and for the installation of the primary acoustic emission transducer.

The electrochemical cell is filled with 0.1 N NaOH solution, which is a convenient electrolyte that provides high electrical conductivity, does not promote the release of electrolysis by-products other than hydrogen and oxygen, and protects the steel specimen surface from excessive corrosion during periods of cathode current interruption.

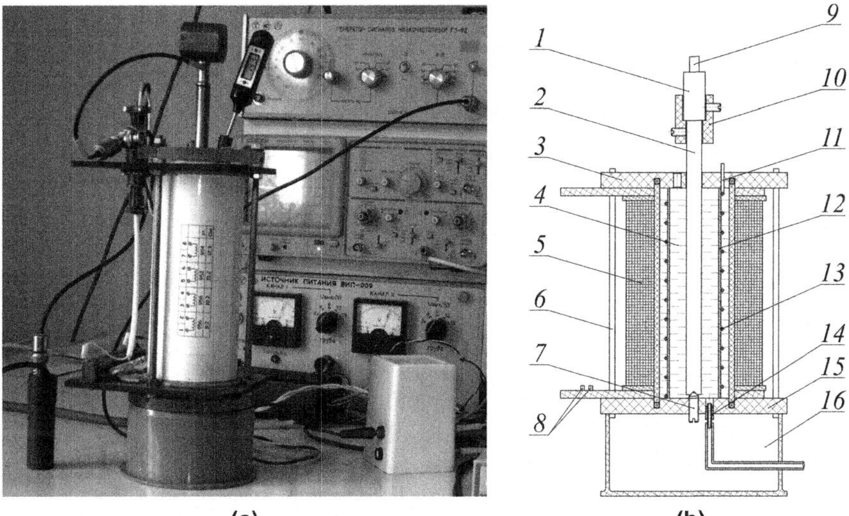

(a) (b)

Fig. 10.1 External (left) and schematic (right) view of the solenoid-electrochemical cell to study the effect of metal-absorbed hydrogen on the parameters of MAE signals: *1* is the primary AE signal transducer; *2* is the specimen; *3* is the upper cover; *4* is the electrolyte; *5* is the solenoid winding; *6* is the bracing wire; *7* is the support; *8* are terminals; *9* is the HF cable; *10* is the nozzle; *11* is the anode output; *12* is the frame; *13* is the platinum electrode; *14* is the insulating base; *15* is the lower cover; *16* is the pallet

A platinum anode is installed coaxially with the solenoid and the specimen (cathode) through a digital measuring device UT101, operating in the mode of measuring the electrochemical cell current, which connected to a stabilized current source Etalon EP.10010.1.3. With the help of these devices, the average current density on the surface of the working electrode 0–20 mA/cm^2 was ensured.

The advantage of the electrochemical hydrogen charging of metal compared to hydrogen charging from the gas phase is that the metal is not subjected to heat effect, which can cause not only relaxation of internal residual stresses in it, to which, as known, magnetic properties of ferromagnets are sensitive, but also changes in material microstructure—phase transformations, segregation of the secondary phase, grain growth, recrystallization, etc. In addition, under conditions of electrolytic hydrogen charging, the activity of reduced hydrogen on the metal surface is easily controlled by the current. Under such conditions, it is not difficult to achieve a significant concentration of atomic hydrogen on the metal surface, as well as simply to control the low activity of reduced atomic hydrogen on the metal surface. This is an important point, because the solubility of hydrogen in α-Fe is very low (Ovchinnikov and Rassada 1990), the concentrations of hydrogen are also small and critical, that cause irreversible fracture of the microstructure, which is tenths or even hundredths of ppm for pure iron (Kolachev 1985) and not more than 1 ppm for pipe steels (Tkachev et al. 1999).

To excite the magnetic field inside the specimen, the solenoid winding is connected to a PCG10/8016 digital functional AC alternator, controlled by computer, via a low frequency amplifier and a current measuring system. The frequency of the AC alternator can be changed in the range of 0.01 Hz–1.0 MHz. The signal is monitored using a computer oscilloscope module PCS500. During the research, the influence of different forms of AC supply voltage of the solenoid: sinusoidal, meander, sawtooth, triangle-like, pulse, is tested.

To the upper end of the specimen with the help of a special holder is connected a broadband high-sensitive (coefficient of conversion of elastic waves into electrical signals not less than 1.6×10^9 V/m) primary piezoceramic transducer. It converts elastic oscillations of the metal surface into an electrical signal due to the propagation of magnetoacoustic emission waves. The non-uniformity of the conversion factor of the primary transducer is ± 7 dB in the operating frequency band 200–1000 kHz. The metal surface is lubricated with vaseline at the point of connection of the transducer. The transducer is connected to the preamplifier using a standard BNC-BNC cable.

The MAE signal from the primary transducer through the pre-amplifier SAA-06 with a gain of 40 dB arrives at the input of the information-measuring system MAE-1L, which amplifies, processes and records the MAE signal. The measuring channel of the unit is designed for amplification, filtering of the signal in the range of 200–1000 kHz, and its digitization. Developed in the DELPHI environment, the MAE-1L software allows us to adjust the signal gain, the magnitude of its discrimination threshold, the speed of digitization and the sampling size of the signal. Through the parallel port, this unit is connected to a computer that stores digitized data and analyzes MAE signals using MAESTAT, MS EXCEL, MATLAB, etc. software. The MAE-1L information and computing unit is synchronized with the signal of the functional generator. This means that the beginning of the MAE signal recording occurs at the same phase delay of the first quarter of the period, i.e., at the rise of the sine wave.

The elastic MAE waves, which cause the displacement of the metal surface, arise due to the abrupt movement of the domain walls during the remagnetization of steel in an alternating magnetic field. The dynamics of domain walls movement can represent on the one hand the defectiveness of the metal structure, and on the other—the energy distribution of defects in the microstructure. Both of these metal characteristics depend on the amount of hydrogen absorbed by the metal, which, in turn, depends on the current and time of electrolysis and other physicochemical parameters. With this in mind, the dependences of the MAE signals parameters on: *a*—the magnitude of the cathode current passing through the specimen surface; *b*—magnetic field intensity and *c*—hydrogen charging time are studied.

10.5 Results of the Research

In this part of the research, two series of experiments were conducted. In the first, the MAE signals recorded by the measuring system were subjected to typical post-processing, using an algorithm of threshold values that extracted low-amplitude pulses from the calculations. In the second series of experiments, the recorded MAE signals were subjected to another method of processing during which only the MAE pulse values shifts caused by the principle of the measuring system were removed and the low-amplitude pulses were not filtered to preserve the completeness of the useful MAE signal.

For the first series of experiments, Fig. 10.2 shows the dependence of the parameter $\sum A_i$ (in thousands of a. u.) of the MAE signal on the time of electrolytic hydrogen charging of the specimen with a cathode current $I_c = 50$ mA. As can be seen from the figure, the initial ($t_H = 0$) value of $\sum A_i$ is about 100 and this level of MAE signal intensity lasts more than 5 min.

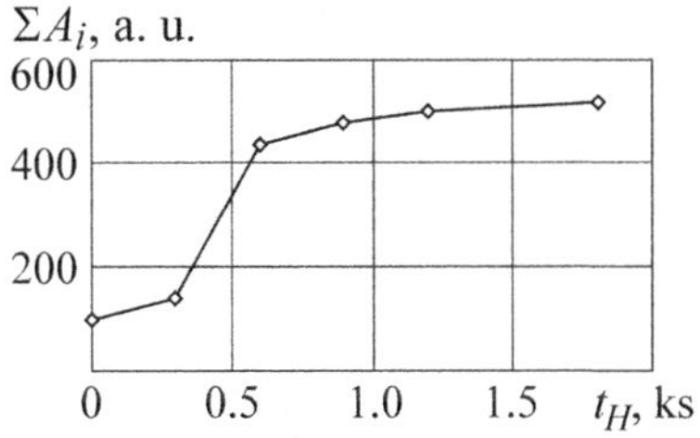

Fig. 10.2 Dependence of the parameter $\sum A_i$ on time of electrolytic hydrogen charging t_H of the specimen. Cathodic current $I_c = 50$ mA, the amplitude of magnetic field intensity $H_a = 8.7$ kA/m

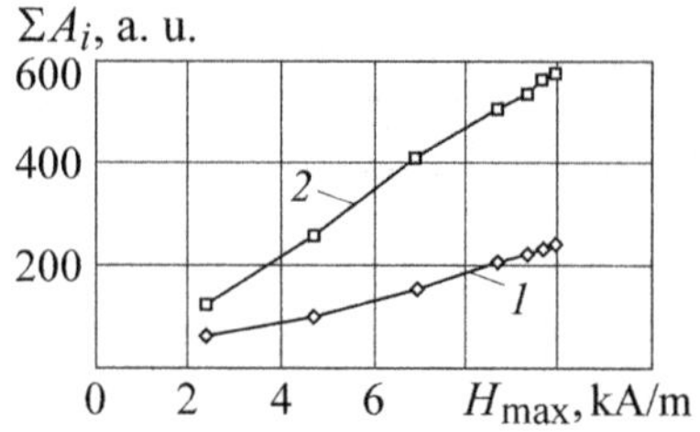

Fig. 10.3 Dependence of the parameter $\sum A_i$ on the magnetic field intensity H_a for the specimen before (*1*) and after (*2*) electrolytic hydrogen charging with cathode current $I_c = 50$ mA for 1.8 ks

It is likely that such time is necessary, on the one hand, for the reduction of surface oxide-hydroxide films, and on the other hand, for the penetration of atomic hydrogen in sufficient quantity and depth to affect the integral parameter of MAE signals $\sum A_i$. After 600 s since the beginning of cathodic hydrogen charging, the sum of the amplitudes $\sum A_i$ exceeds 400 and continues to increase, reaching the value of $\sum A_i = 500$ after 72 min of cathodic current action.

The change in the sensitivity of the MAE signal parameter ΣA_i on the amplitude of the magnetic field intensity H_a due to electrolytic hydrogen charging was estimated. Figure 10.3 shows two such dependences of $\sum A_i$ on H_a: one for the specimen before electrolytic hydrogenation and the other are after it for the specimen with cathode current $I_c = 50$ mA for 1800 s.

The amplitude of the magnetic field intensity does not exceed 10 kA/m, which is lower than the hysteresis loop bend prior to the magnetic saturation of the metal. Again, as in Fig. 10.2 there is a huge difference in the power of the MAE signals for non-hydrogen-charged and hydrogen-charged specimens. The approach of the curves shape to the linear ones does not have a significant physical interpretation, because the parameter $\sum A_i$ is distorted by the stress ratio of the recorded signal relative to the abscissa axis and the lost part of the low-amplitude pulses. The relative increase in the value of the parameter $\sum A_i$ as a result of hydrogen charging of the metal, although less than in Fig. 10.2, is still more than 100%, which indicates an extremely high sensitivity of the method to electrolytic hydrogen charging and the research prospects in this direction.

It is important to evaluate how the intensity of MAE signals changes with increasing cathode current applied to the specimen, and hence the activity/pressure of atomic hydrogen on the metal surface. Figure 10.4 illustrates the dependence of the parameter $\sum A_i$ on the time t_H of electrolytic hydrogen charging for cathode current $I_c = 150$ mA at the same amplitude of the magnetic field intensity $H_a = 8.7$ kA/m as in Fig. 10.2.

Two fundamental differences between Figs. 10.2 and 10.4 can be noted here. Firstly, for the higher cathode current only 300 s are enough for the parameter $\sum A_i$ to reach its almost maximum value, while for $I_c = 50$ mA after 300 s this parameter remains at the noise level. Secondly, in the case of lower cathode current, the parameter $\sum A_i$ has a constant tendency to slightly increase during the hydrogen charging

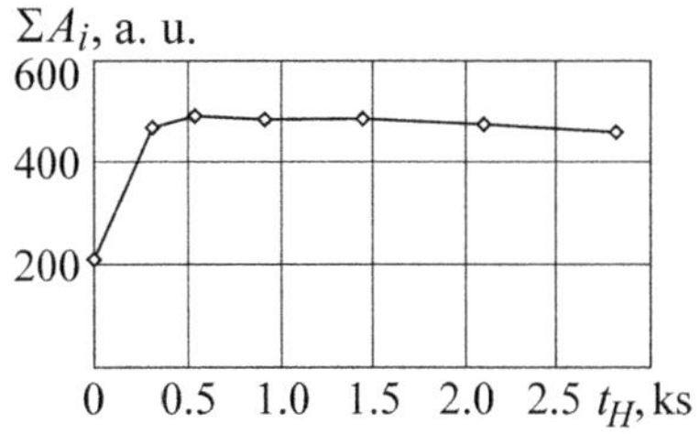

Fig. 10.4 Dependence of the parameter $\sum A_i$ on time t_H of electrolytic hydrogen charging of the specimen. Cathodic current $I_c = 150$ mA, amplitude of magnetic field intensity $H_a = 8.7$ kA/m

period, while for $I_c = 150$ mA this parameter of MAE signals reaches its maximum after 10 min of hydrogen charging, then gradually decreases by ~ 8% to the end of experiment.

The effect of specimen hydrogen charging at the same higher cathode current on the dependence of $\sum A_i$ parameter on the magnetic field intensity H_a is determined. Figure 10.5 shows such dependences for the hydrogen charging period of 1.8 ks. Note that the curves are quite like those shown in Fig. 10.3. Besides, the levels of $\sum A_i$ in the case of a higher cathode current are not larger, but slightly lower, although for a hydrogen-charged specimen the values of $\sum A_i$ are twice as high as for a non-hydrogen-charged one. The fact of insignificant decrease of the parameter $\sum A_i$ requires further study.

A further increase in the cathode current to the level $I_c = 200$ mA does not fundamentally change the observed tendencies (Fig. 10.6): for the hydrogen-charged specimen the parameter $\sum A_i$ is about twice as large as for the non-hydrogen-charged one, although compared to previous hydrogen charging modes it slightly decreases.

After obtaining the above dependences, the second series of tests is performed with similar physical remagnetization conditions, however with slightly lower levels of cathode currents and other modes of processing the measurement results. Figure 10.7 presents the dependences of the $\sum A_i$ parameter on the amplitude of the magnetizing field intensity H and the cathode current of hydrogen charging. If we approximate the obtained dependences of $\sum A_i$ on H parabolically (approximation curves are not shown), then for the least squares method, we obtain the determination coefficients

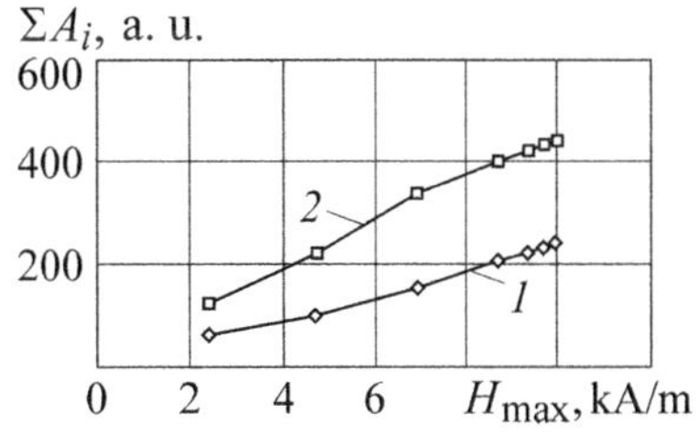

Fig. 10.5 Dependence of the parameter $\sum A_i$ on the magnetic field intensity H_a for the specimen before (*1*) and after (*2*) electrolytic hydrogen charging with cathode current $I_c = 50$ mA for 1800 s

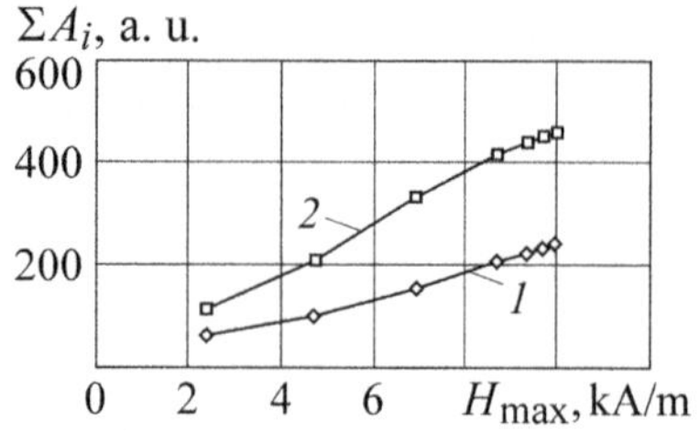

Fig. 10.6 Dependence of the parameter $\sum A_i$ on the magnetic field intensity H_a before (*1*) and after (*2*) electrolytic hydrogen charging of the specimen with cathode current $I_c = 200$ mA for 1800 s

within the range $R^2 = 0.9982 - 0.9999$, which indicates the functional dependence of the variables. The parabolic character of these dependences means that during the experiments the induction level does not reach the saturation of the ferromagnet. Therefore, the main source of the MAE elastic waves is jump-like movements of 90° domain walls due to changes in the directions of magnetization vectors from one axis of the volume-centered Fe crystals to another. Note that the vertices of the parabolic approximations are located at the levels of recorded noise $\sum A_i = 58 \pm 1$ a. u. within the scatter of experimental data.

As can be seen from Fig. 10.7, under the electrolytic hydrogen charging of the ferromagnet specimen, the MAE signal intensity increases significantly, especially for cathode hydrogenating currents of 50 mA and more. However, as in the previous series of tests, in these experiments the increase in the MAE signals intensity with

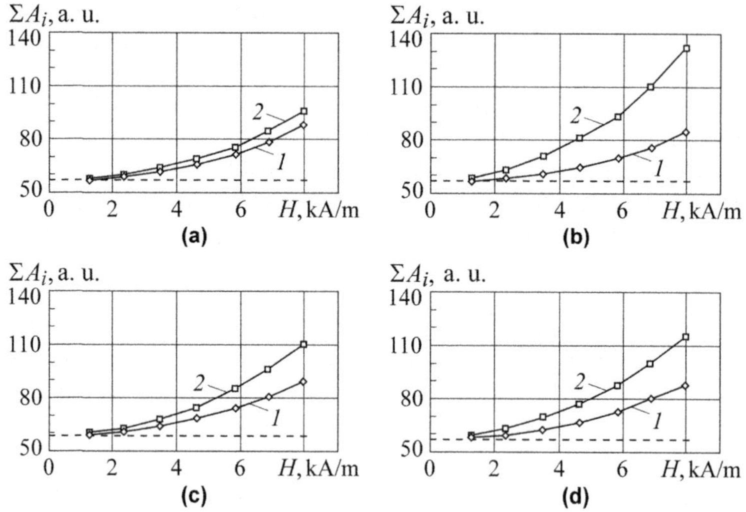

Fig. 10.7 Influence of the remagnetizing field intensity H on the parameter $\sum A_i$ for specimens before (curve *1*) and after electrolytic hydrogen charging for 1.8 ks (curve *2*) with a current of 25 mA (**a**), 50 mA (**b**), 100 mA (**c**), and 150 mA (**d**). The noise level is indicated by a dotted line

cathode current is recorded in the sequence 25 → 100 → 150 → 50 mA, i.e., asymmetric, reflecting, apparently, the simultaneous influence of several competing factors. The use of the cathodic hydrogenating current of 25 mA causes only a slight increase in the MAE signal intensity above the level generated before the cathode current switch on, i.e., before the hydrogen charging process. Therefore, considering the relatively high reproducibility of MAE signals for the original specimens (before hydrogen charging), this current can be considered to be the sensitivity limit of the method for cathodic hydrogen charging of the metal.

In another series of experiments presented in Fig. 10.8, the time variation of the $\sum A_i$ parameter after switching on the specimen hydrogen charging current for different values of the cathode currents is studied. In this case the amplitude of the magnetizing field for all cases is the same $H_a = 6.8$ kA/m. At the moment of switching on ($t = 0$), the value of the parameter $\sum A_i$ is in the range of 76.6–80.6 a. u., and after 0.4–0.8 ks it reaches its maximum value.

On the obtained dependences (Fig. 10.8) the duration of the transient process increases with the current magnitude. Note that for a cathode current of 25 mA during the first 0.47 ks, the MAE signal level is unchanged, similar to the signal obtained when there is no cathodic hydrogen charging. Then, $\sum A_i$ gradually increases, reaching a stationary value in ~ 2.0 ks after the current is witched on. As the magnitude of the current increases, the increase in the parameter $\sum A_i$ to a steady-state level accelerates. It should be noted that the reproduction of the MAE signal level for "stationary" hydrogen charging deteriorates, which coincides with the results given

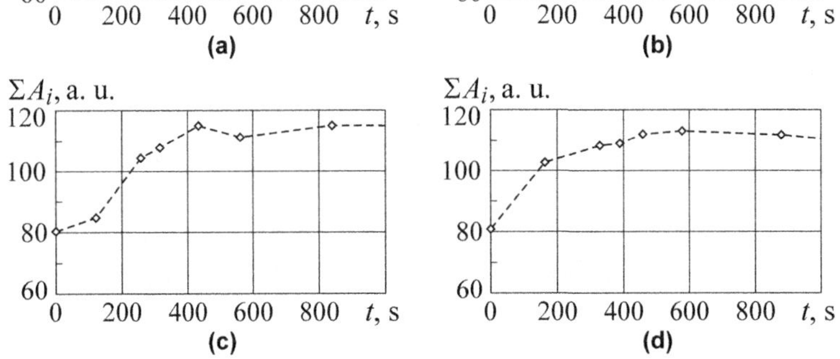

Fig. 10.8 Dependence the parameter $\sum A_i$ on time t under electrolytic hydrogen charging with a current of 25 mA (**a**), 50 mA (**b**), 100 mA (**c**), and 150 mA (**d**)

above. Moreover, for current $I_c = 150$ mA, the MAE signal level decreases slightly after reaching its maximum value (at time $t = 0.58$ ks), again confirming the results obtained earlier.

The delay in the process of $\sum A_i$ increase after switching on the cathode current, apparently, is due to the restoration of the surface oxyhydroxide films and the creation of conditions for high activity of atomic hydrogen on the metal surface. The time required to reach a steady-state MAE signal level is significantly shorter than that required to hydrogenate a 12 mm diameter specimen volume. This may mean, for example, that only the near-surface layers of the metal are responsible for the increase in the MAE signal, or another phenomenon, that suppresses the signal level during hydrogenation of the metal volume comes into operation. As shown above, the MAE signal is significantly reduced due to the elastic deformation of iron due to magnetocrystalline anisotropy and under plastic deformation caused by the increased defectiveness of its structure.

Comparing the obtained results for electrolytic hydrogen charging with the results for hydrogen charging from the gas phase, it is possible to note a significantly higher sensitivity of MAE parameters during cathodic hydrogen charging. This can be explained by the role of diffusion-mobile hydrogen in the electrolysis conditions and its absence in experiments with gaseous hydrogen charging. Combining these two methods, it is possible to distinguish the influence of mobile and stationary hydrogen on the MAE signal parameters, but such issues require further detailed study.

Since the metal surface state is important for all processes related to hydrogen absorption by the metal, some aspects of the surface impact should be considered in the case of the MAE method under electrolytic hydrogen charging of the metal. Firstly, the degree of magnetization of ferromagnets depends on the depth is the highest in the near-surface layers and fades exponentially with depth. This means that the MAE is more pronounced in the near-surface volume of the metal, the depth of which is inversely proportional to the root of the magnetic field frequency. On the other hand, with decreasing frequency (when the demagnetization depth increases) the number of acoustic waves generated in this volume per unit time decreases, and therefore there is an optimal remagnetization frequency for each setting, at which the MAE signal intensity is the highest, so the method becomes the most sensitive, as shown above.

Secondly, the surface of the metal can be acoustically more "active" than its volume, because the surface Rayleigh and Love waves can transfer higher energy than the longitudinal waves. Although the redistribution of powers of different types of waves depends on the geometric characteristics of the object under study, density of the material, its mechanical characteristics, stress fields, in any case, the contribution of surface waves can be significant.

Thirdly, there is the opposite effect—the influence of surface elastic waves on the kinetics of all stages of chemical and electrochemical processes occurring on the metal surface, including electron transfer processes, dissociation of polyatomic formations (molecules, ions, including hydrated), recombination (formation of molecular hydrogen), as well as diffusion processes, including the diffusion of hydrogen in the near-surface layers of the metal.

Fourthly, the near-surface layers of the metal are the sites of maximum residual stresses caused by the specimen manufacture conditions, or due to absorbed hydrogen. If hydrogen affects the stress field in the near-surface layers, this can have an influence on the intensity of the elastic waves generated during metal magnetization, as previously established.

Surely, the state of the metal volume is also important, including the chemical and phase composition at the macro-, micro-, and nanolevels, anisotropic parameters of the material (e.g., predominant orientation due to casting, rolling or heat treatment), defect distribution, concentration, etc. All these parameters must affect the magnetic and acoustic properties, as well as the interaction of hydrogen with the metal, and hence the parameters of the MAE signals. However, further research is needed to determine these effects.

The obtained results on the influence of hydrogen on the MAE signal parameters do not contradict each other, but on the contrary—mutually confirm and complement each other, presenting a range of data and dependences stipulating for further study of this phenomenon. If the found relationships are confirmed for other hydrogen charging conditions and other materials, and the methodological basis for applying the MAE method for structural materials is developed, it can be widely used to detect hazardous hydrogen concentrations in critical equipment prior to the early stages of hydrogen damage of the metal.

The high sensitivity of the MAE method to the process of electrolytic hydrogen charging of steel revealed in the experiments, due to the extremely low solubility of hydrogen, forms a good basis for further research, during which it will be necessary to find answers, in particular, to the following questions:

(1) Why $\sum A_i$ changes unsystematically with increasing cathode current and why for large cathode currents $\sum A_i$ passes through the maximum and then slowly decreases, while such a tendency is not observed for low-current hydrogen charging modes?
(2) How do the parameters of MAE signals and the concentration of hydrogen in the metal correlate with each other and what threshold values of hydrogen concentration can be detected by the MAE method?
(3) What should be the parameters of the alternating magnetic field to achieve maximum sensitivity of the MAE method to hydrogen absorbed by the metal?

Note that the studied densities of cathode currents for electrolytic hydrogen charging of metal (0.4–2.4 mA/cm^2) are, at first glance, low, as they can be compared with the cathodic protection currents of underground or underwater metal structures. On the other hand, such currents can be dangerous, because the work of cathodic protection under certain conditions causes the formation of aggressive carbonate-bicarbonate environment under the delaminated protective coating of main gas and oil pipes, causing corrosion cracking of the latter. In addition, stray currents can cause similar current densities on pipe surfaces operating in local gas and water distribution networks. Similar values of cathode currents can also be caused by the work of galvanic pairs, macro- and mega-cells, which are formed in large-scale metal structures. The last ones are subjected to significantly different physical and chemical effects (temperature, oxidant concentration, soil porosity, nature contacting liquids, etc.) in different areas. These current densities have almost no influence on the properties of short-term operation materials, but during long-term operation in the presence of mechanical and thermal cycling can cause degradation of the strength characteristics of structural materials. If the MAE method is sensitive to such levels of cathode currents, it opens up prospects for its application to identify the preconditions of the accelerated fracture of ferromagnetic materials.

10.6 Diagnostics of Ferromagnetic Materials by the MAE Signals Parameters

Objects made of ferromagnetic materials undergo essential degradation during long-term operation, which leads to a significant change in their physicochemical particularly magnetic properties. Therefore, modern methods of technical diagnostics should be based on new approaches to creating a methodology for detecting these changes.

Based on the results of experimental studies on the influence of mechanical and physicochemical factors on the change of information parameters of MAE signals, a method for estimating the stress state of ferromagnetic structural elements is proposed. This makes it possible to detect the regions of residual stress during technical diagnostics of large-scale objects by the MAE method.

Investigation of steel corrosion. The MAE method is used for this purpose because of its high sensitivity to changes in the structure and the stress–strain state of the material (Nazarchuk et al. 2013; Fisher and Lally 1968). The intensity of corrosion processes can be tested by the parameters of MAE signals, because the degraded material changes its domain structure, thus affecting the change of the signal parameters.

For experiments, the specimens were prepared of 15 steel (plate dimensions 230 × 30 × 2 mm) and 2ps steel (plate dimensions 230 × 30 × 6 mm) (see Appendix) with holding them for 173 h in NACE solution (5% $NaCl$ + 0.5% CH_3COOH + H_2S (sat.)) (NACE Standard MR-0175-96 1996). Later, the specimens were washed with water, corrosion products were removed, washed in acetone and dried.

Figure 10.9 shows the specimens' surface after experiment. It is damaged by blisters, the formation of which is initiated by hydrogen (Radkevych and Pokhmurs'kyi 2001; Khoma 2010).

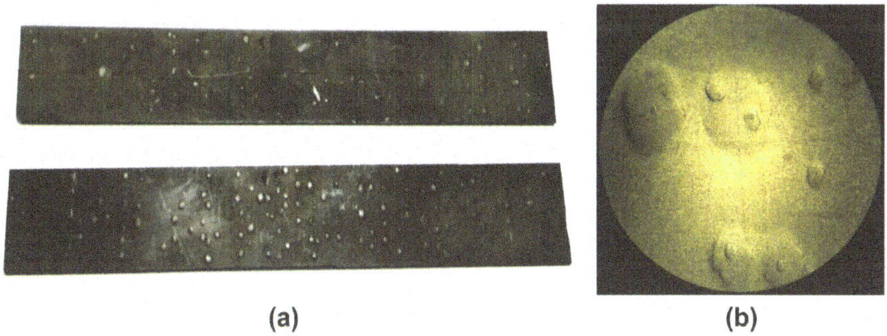

Fig. 10.9 General view of the surface of plate 15 steel specimens (**a**) and surface bubbles in 2ps steel (× 16) (**b**) developed in NACE solution

The appearance of blisters is most likely to occur in the areas of accumulation of coagulated vacancies, in the presence of slag inclusions, micro- and macrocavities. Analysis of the gas composition in the bubbles shows that it can contain up to 99.5% of hydrogen and the gas pressure can reach several hundred atmospheres (Radkevych and Pokhmurs'kyi 2001).

The influence of the hydrogen sulfide medium on the parameters of the MAE signal as a result of remagnetization of the specimens is established (Fig. 10.10).

The obtained dependences show the difference between the sums of the MAE signals amplitudes for the original sample and the one exposed to hydrogen sulfide medium. In particular, for a specimen of 15 steel with a thickness of 2 mm, damaged by aggressive media, the sum of the amplitudes is approximately by 50 a. u. lower than for the original one, at a constant induction of the magnetic field in the specimen $B = 1.3$ T.

With an increase in the thickness of the specimen to 6 mm for 2ns steel the difference increases almost twice to 90–100 a. u. at $B = 0.4$ T, which is explained by its larger remagnetized volume. The decrease in the sum of the MAE signal

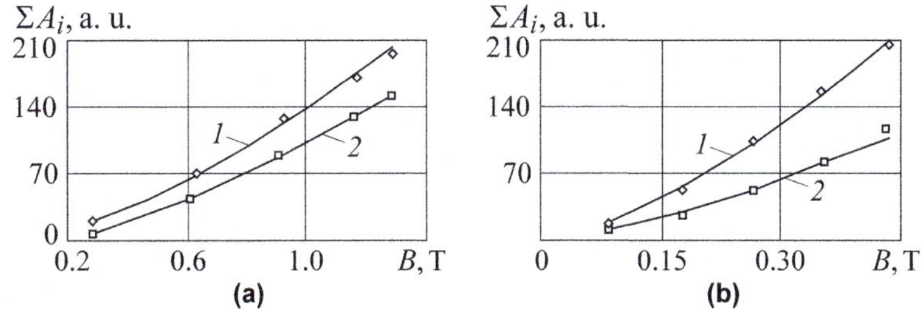

Fig. 10.10 Dependences of the sum $\sum A_i$ of the MAE signal amplitudes on the amplitude of induction B of the remagnetizing field for the original (*1*) and exposed to hydrogen sulfide (2) plate specimens of 15 steel with a thickness of 2 mm (**a**) and 2ps steel 6 mm thick (**b**)

amplitudes in the damaged specimens is explained by degradation of the material structure, which causes a decrease in the mobility of 90° domain walls due to an increase in the number of centers of their attachment.

As a result of experimental studies, the sensitivity of the MAE signal parameters to damage of ferromagnetic materials caused by hydrogen sulfide corrosion is confirmed. In particular, it is found that when the ferromagnetic samples are magnetized, the sum of the amplitudes of the MAE signals decreases significantly for the damaged specimens compared to the original ones.

Diagnostics of long-term operated pipe steels. Under real operation conditions of ferromagnetic constructions at the industrial enterprises it is not possible to avoid the influence of working environments, action of loadings or deformations on the material. The results of experimental studies of the influence of carbon content, structure of ferromagnetic material, heat treatment, hydrogen and corrosion degradation, stress state on the change of the MAE signal parameters are presented in the previous sections.

Each of the above factors occurs during the process of joints formation by welding. The main problems after the crystallization of welded joints (WJ) are the presence of different microstructures in all its zones and the introduction of additional stresses due to high temperatures.

The safe operation of structures with welded joints, in particular, pipelines, is provided first of all by control of their serviceability and diagnostics. Particular attention should be paid to WJ, taking into account their features, such as structural heterogeneity in the WJ cross-section, the accumulation of non-metallic inclusions in the weld metal, the presence of residual post-welding stresses, hot and cold cracks during formation of WJ. In the case of long-term operation, these factors under the influence of the transported product, changes in operating pressures and environmental factors over time can contribute to the formation of microcracks and, consequently, the material fracture. So it is important to perform timely diagnostics of the structure elements state to ensure their safe operation.

Influence of structural inhomogeneity of different zones of pipe steels on the change of the MAE signals amplitude. During long-term operation of main or industrial pipelines, the structure of metal and its mechanical properties undergo changes, corrosion damaged and stress sections of pipelines appear (Chernov et al. 2002; Poliakov et al. 2010). At the same time, structural metal has not yet exhausted its residual life, and the replacement of such equipment in most cases is not technically and economically reasonable. In this case, to extend the life of safe operation of pipelines and their elements, it is important to provide technical diagnostics and monitoring of the most dangerous areas.

Dimensions of the oil pipeline are: outer diameter is 720 mm, and pipe wall thickness is 10 mm. Dimensions of the gas pipeline are: outer diameter is 1000 mm, wall thickness of the pipe is 10 mm. The chemical composition (in percentage) of material for long-term operated oil pipeline (19G steel) is given in Table 10.1, and for long-term operated gas pipeline (17G1S steel)—in Table 10.2.

Table 10.1 Chemical composition of material for oil pipeline

C	Si	S	P	Mn	Cr	Cu	Fe (remaining)
0.18	0.49	0.035	0.03	0.95	0.11	0.19	98.16

Table 10.2 Chemical composition of material for gas pipeline

C	Si	S	P	Mn	Cr	Cu	Fe (remaining)
0.16	0.51	0.04	0.035	1.33	0.32	0.14	97.49

The influence of the pipeline material degradation on the parameters of the MAE signals was studied on the workpieces made of pipe walls (Skalskyi et al. 2018). They were conventionally divided into three zones: the inner (*1*), which was about one third of the thickness and included the inner surface of the pipe, the middle—the next one-third of the thickness (2) and outer (*3*), which included the outer surface of the pipe. As one can be seen from Fig. 10.11, the orientation of the workpieces of material for specimen preparation was along the axis of the pipe. Prismatic specimens with a size of 240 × 30 × 2 mm were made from the workpieces.

During cutting out of workpieces for specimens, and hence during their manufacture, such processing modes are used, in which minimal hardening and mechanical stresses are formed (blanks are cut out after removing the thermal-affected zone caused by melting during their preparation by melt cutting), and also minimal changes in the structure and phase state occur.

The MAE is recorded by a highly sensitive primary piezoelectric transducer with an operating frequency band of 0.2–0.6 MHz. The MAE signals at different external magnetic field induction B_{max} are investigated: 0.21 T; 0.45 T; 0.7 T; 0.9 T; 1.03 T.

Note that the microstructure of the degraded (Fig. 10.12a–c) metal of the oil pipe consists of ferritic-pearlite columnar crystals. From metallographic studies it is found that the exploited metal is damaged by a large number of pores with a size of 1–3 μm in the cross-section of the pipe wall. A 10–20% increase in the number of pores in the metal thickness of the pipe from the outer (Fig. 10.12a, d) to the inner (Fig. 10.12c, f) layer is observed. The pores formed during operation are of the correct spherical

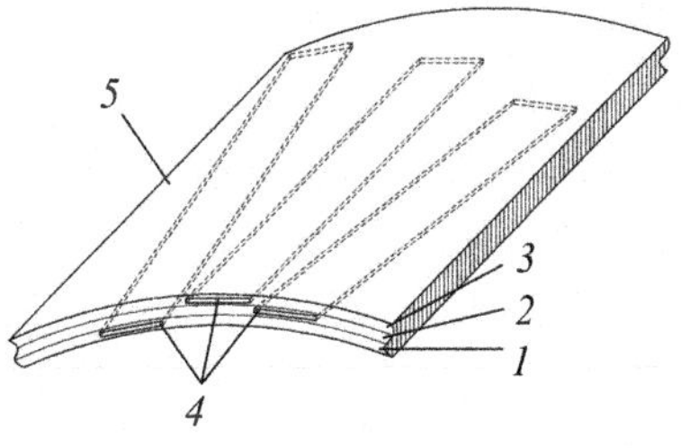

Fig. 10.11 Scheme of cutting workpieces for preparing specimens from long-term operation pipes (*1–3* are inner, middle and outer layers of the pipe wall, respectively; *4* is places of cutting specimens from the wall; *5* is the pipe wall)

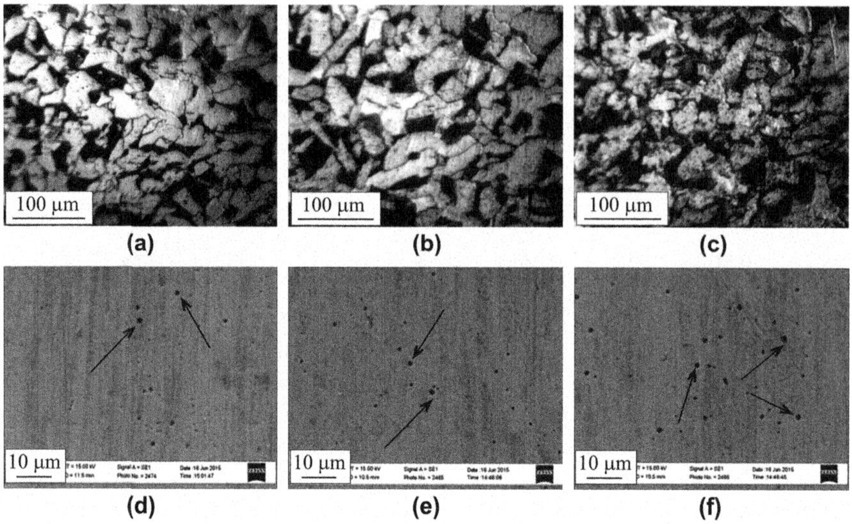

Fig. 10.12 Microstructure of the oil pipeline material from different layers of the pipe wall: **a** is outer; **b** middle; **c** inner layers; **d**, **e**, **f** are pores caused by the presence of gases

shape (Fig. 10.12d, f), which is typical for defects caused by the presence of gases in the metal structure.

Figure 10.13 shows the microstructure of the pipeline material for each of the layers of the pipeline wall. The metal structure of the pipe consists of ferrite and pearlite grains (approximately 65% ferrite and 35% perlite), is fine-grained with tape structure, typical of tubular steels, which are mainly made by rolling. Average grain sizes are: ferrite is 30–35 μm; pearlite is 20–25 μm.

Technological defects, the appearance of which is caused by the process of steel production are also revealed. Figure 10.13d illustrates the non-metallic inclusion of phosphide formed during final deoxidation of the liquid metal. Shrinkage shells (Fig. 10.13e) caused by metal shrinkage and a small number of gas pores (Fig. 10.13e, f), which appear as a result of the release of gases during cooling of the metal to the solid phase, are also recorded.

The average values of the microhardness of the metal of different layers according to the pipe wall thickness specimen are shown in Fig. 10.14a. Figure 10.14b illustrates the results (averaged for layers) of measuring the microhardness of the gas pipeline material from different layers of the pipe wall. The highest microhardness is in the inner wall layer—185 HV, and of the outer and middle layers—in the range of 170–175 HV.

According to the literature (Makarenko et al. 2001) a decrease in toughness and hardness takes place in oil and gas pipes as a result of their long-term operation, which is a phenomenon of its operational degradation and is due to the development of scattered damage.

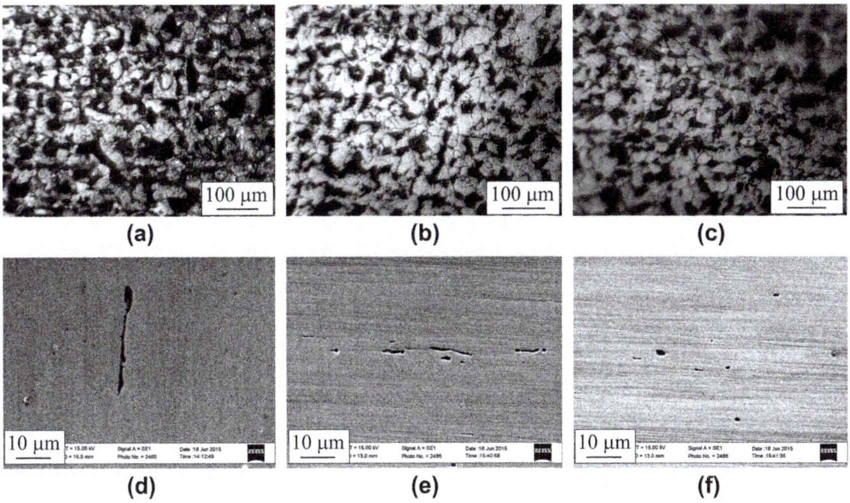

Fig. 10.13 Microstructure of the gas pipeline material from different layers of the pipe wall: **a** is inner; **b** middle; **c** outer layers; **d** is phosphide inclusions; **e**, **f** are rolling defects

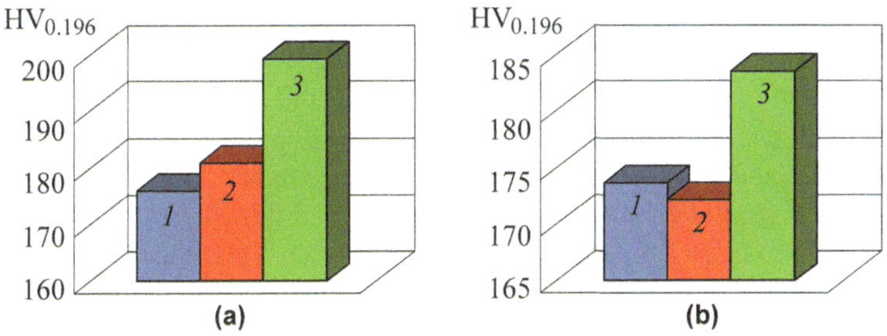

Fig. 10.14 Microhardness of metal of different layers by the pipe wall thickness: **a** is oil pipeline and **b** is gas pipeline (*1* outer, *2* middle, *3* inner layers)

The hardness of the inner surface layers of the unexploited pipe is significantly higher than of the outer one. For long-term operated pipes, this difference remains, although it is significantly reduced (Fig. 10.15).

The results obtained in our investigations correlate with those known in the literature and indicate, on the one hand, the operational damage of the metal in the pipe wall volume, while the hardness gradient by its thickness means the negative role of hydrogen in steel degradation.

According to measurements of macrohardness of metal of different layers of the pipeline wall, a slightly higher Rockwell and Brinell macrohardness have the specimens from the inner layer of the pipe wall—92–93 HRB and 192–197 HB, respectively, while the specimens from the outer and middle layers of the pipe have the same Rockwell and Brinell macrohardness—91 HRB and 187 HB, respectively.

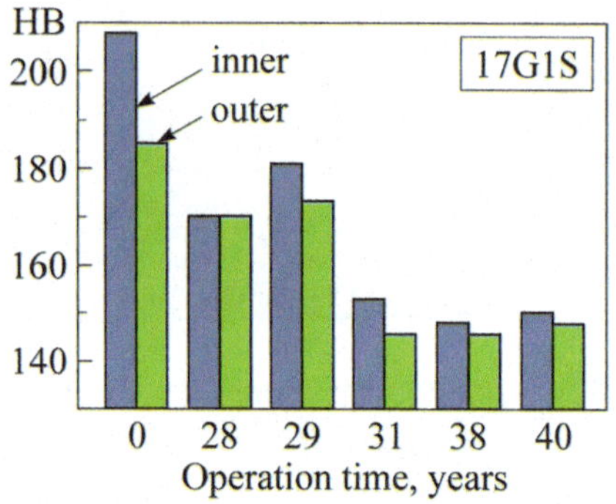

Fig. 10.15 Hardness HB for 17G1S steel after different operation time (inner and outer surface of the pipe wall)

According to the results of studies of the microhardness of oil pipeline material, it is found that the highest value of hardness (~ 200 HV) is recorded in the inner layer, the lowest (170 HV)—in the outer, which should be represented in the parameters of MAE signals during their remagnetization. By measuring the macrohardness of the metal of different layers of the oil pipeline wall, it is found that in all layers Rockwell as well as Brinell macrohardness is the same—90 HRB and 187 HB.

Figure 10.16 shows the results of experimental studies of the dependence of the sum of the MAE signal amplitudes on the magnetic field induction in the specimens for different layers of the walls of long-term operated pipes.

The obtained dependences indicate the difference in the values of the sum of the amplitudes of the MAE signals for different layers of the oil and gas pipeline wall. The largest sum of the amplitudes of the MAE signals is recorded in the specimens from the inner layer of the pipe wall, and the smallest one in the specimens from the outer layer. With increasing induction of the magnetic field, this feature is more pronounced.

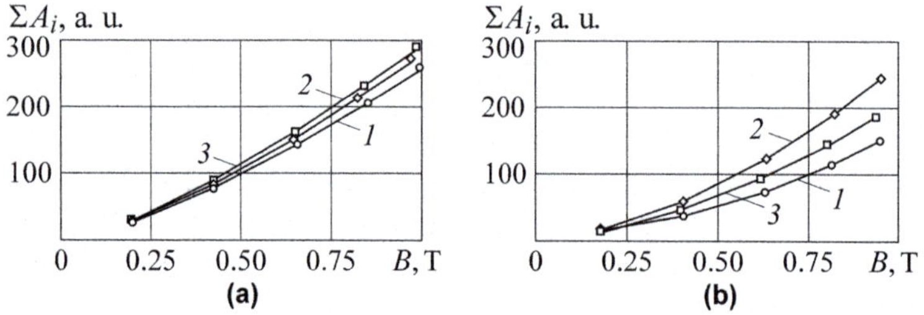

Fig. 10.16 Dependence of the sum of amplitudes of MAE signals on the magnetic field induction in specimens from different layers of the pipe wall: **a** oil pipeline; **b** gas pipeline (*1* external, *2* middle, *3* internal)

As a result of the long-term operation of steels of main and industrial pipelines under the influence of aggressive environments their structure and mechanical characteristics change. The possibility of hydrogenating steels has also a significant effect on this process. For example, when measuring the amount of occluded hydrogen by vacuum extraction, it is found that the amount of hydrogen in the material from the lower part of the pipe is twice as high as in the original material.

It is also found that the hydrogen desorption rate is much lower in the exploited material compared to the original one due to its more intensive trapping. The difference in the properties of the exploited and unexploited materials is also confirmed by measurements of the rate of hydrogen penetration through the membrane. It is found that the effective diffusion coefficient of hydrogen is much lower in the exploited material than in the original. This means that the efficiency of trapping hydrogen in the exploited material from the lower part of the pipe is much higher than in the original one, due to the increasing defectiveness of the material. That is, the degraded material has much more structural defects that reduce the transport of hydrogen, which is also confirmed by a much lower rate of its desorption. After long-term operation of steel due to the growth of metal defects, the intensity of hydrogen capturing in traps increases significantly.

Thus, based on the obtained results, using the MAE signals of long-term operated steels, it can be concluded that the dominant factor in increasing the activity of the MAE signals generation is the presence of hydrogen in the ferromagnet, which during the long-term operation of steels deeply penetrates into the structural defects—traps, thereby facilitating the domain walls jumps during the remagnetization of the specimens. The difference in the activity of MAE in different layers of the pipe wall confirm the different degree of their defectiveness, and hence the different values of the effective diffusion coefficients in them and different degrees of hydrogen charging.

The specific feature of these pipe steels is their good weldability. According to the welding technology, the main requirement is the uniformity of the welded joint with the base metal and the absence of defects in the weld (Makarenko et al. 2001). In this case, the mechanical properties of the weld metal and the near-joint zone must not be lower than the lower limit of the mechanical properties of the base metal. When welding such steels, the welds must be free of cracks, leaks, pores, cuts, etc. The geometric dimensions and shape of the welds must meet the welding process standards. When welding low-alloy steels, the required number of alloying elements in the weld metal is also obtained by their transition from the base metal. The welding mode also affects the factors responsible for the formation of cold cracks. Therefore, it is especially important to follow the technological requirements during the manufacture and welding of pipeline elements from these steels. The working environment must be considered when selecting the material for the pipes (Makarenko et al. 2001).

To evaluate the influence of mechanical properties of welded joints on the change of MAE parameters, the microstructure of pipe metal in diametrical cross-section using prismatic specimens of dimensions $10 \times 30 \times 60$ mm, cut out from longitudinal welds made of 19G oil and 17G1S gas pipeline, is investigated (Skalskyi et al. 2021).

The specimens are polished and etched with 2% nitric acid solution. The NU-2 optical microscope is used. The influence of the heterogeneity of the structure of different zones of the pipeline on the MAE signal parameters is studied on specimens of 240 × 10 × 3 mm in size, cut out from the three main zones of the longitudinal welded joint. The specimens are remagnetized using overhead electromagnet with 1260 windings of copper wire on each leg of the magnetic circuit. The amplitude of the magnetic field induction in the specimen is measured using a coil (300 windings) with an active resistance of 14 Ω.

Metallographic studies show that the studied materials belong to the class of ferritic-pearlitic steels (Figs. 10.17 and 10.18). The structure of welded joints is characterized by greater dispersion than in the base metal, which has a certain tape structure due to rolling. In the thermal-affected zones the main area is the area of overheating with coarse-grained ferrite-pearlite structure and the available ferrite-pearlite.

The authors of Guyot and Cagan (1993) analyze the influence of different microstructural factors on the domain structure of the material, therefore, considering structural heterogeneity in the cross-section of the WJ, a number of experimental studies have been performed to establish relationships between the informative parameters of the MAE method and the magnetic field induction B for specimens cut out from different zones of pipe welded joint.

According to the results of the experiment, the dependences are constructed (Fig. 10.19). There is a noticeable difference between the values of the sum of the amplitudes of the MAE signals for the specimens cut out from three welded joint zones.

According to the literature (Guyot and Cagan 1993; Ng et al. 1994, 1996), 90° domain walls, which are the main source of MAE, are located at the grain boundaries. So, in the absence of other factors in the structure of metal with a higher degree of dispersion, and, accordingly, a greater total length of the boundaries, the number of MAE signals will increase and, as a result, increases their sum of amplitudes.

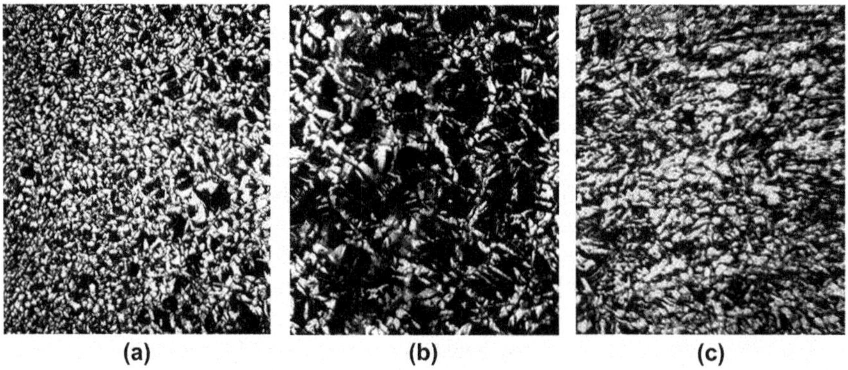

Fig. 10.17 Images of the microstructure of the metal of the welded joint of 19G steel of oil pipeline: **a** base metal; **b** the heat-affected zone; **c** welded joint (× 125)

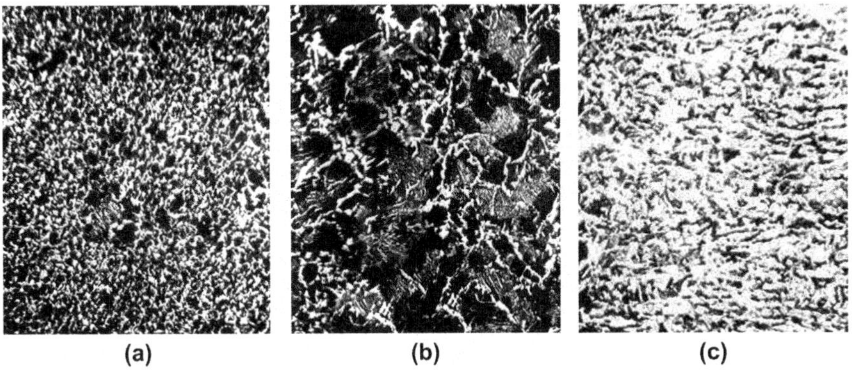

Fig. 10.18 Images of the microstructure of the metal of the welded joint of 17G1S steel of gas pipeline: **a** base metal; **b** heat-affected zone; **c** welded joint (× 125)

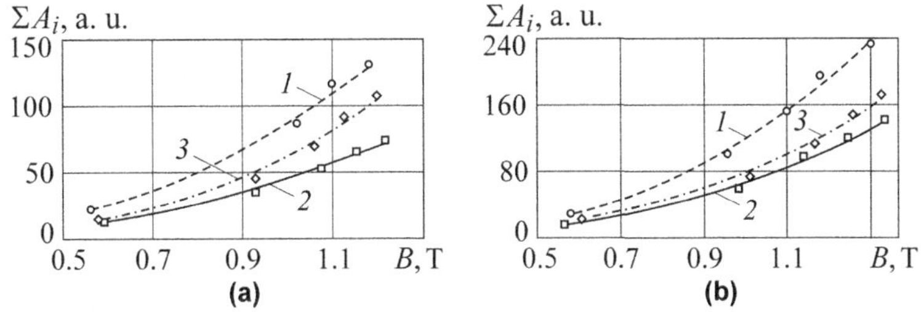

Fig. 10.19 Dependences of the sum of amplitudes of MAE signals on the amplitude of magnetic field induction B for specimens from different zones of the pipe welded joint: **a** gas pipeline; **b** oil pipeline (*1* weld metal, *2* heat-affected zone, *3* base metal)

At a constant amplitude of the induction B of remagnetizing field, the largest values of the sum of the amplitudes $\sum A_i$ are recorded for specimens from the welded joint (curve *1*). The lowest values of $\sum A_i$ (curve 2) are recorded in the specimens from the heat-affected zone. This effect is explained by structural transformations in the ferromagnet under the influence of high temperatures during welding, which cause the restructuring of the material domain structure (Nazarchuk et al. 2013).

Thus, the MAE method is sensitive to the changes in the metal structure due to long-term operation of ferromagnetic materials. This fact allows us to build, on the basis of the MAE phenomenon, the method of non-destructive testing of long-term operation steel damage, and thus for diagnosing volumetric damage of structural elements made of such ferromagnetic materials.

Estimation of the influence of residual stresses in the vicinity of the welded joint and stress regions of long-term operated pipe steels on the parameters of MAE signals. To assess the influence of residual post-welding stresses in the vicinity of the welded joint and stress areas of ferromagnetic structural elements of long-term

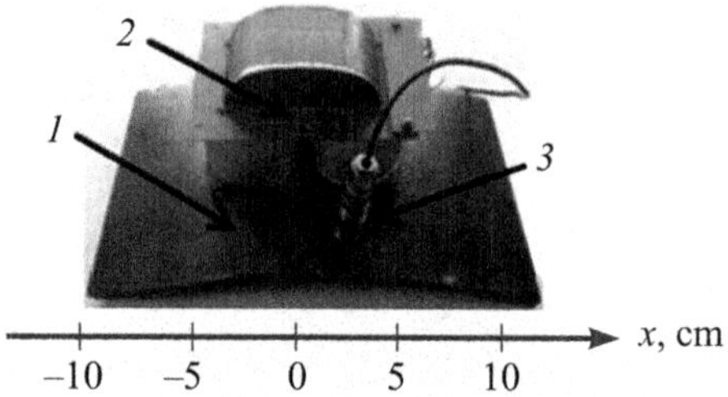

Fig. 10.20 Location of overhead electromagnet on a fragment of a pipe with a longitudinal welded joint: *1* a pipe fragment, *2* electromagnet, *3* acoustic emission transducer

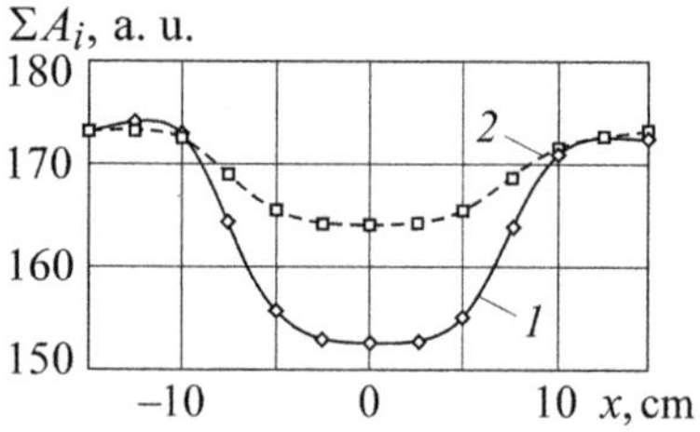

Fig. 10.21 Dependence of the sum of amplitudes of MAE signals on the location of overhead electromagnet on the pipe fragment: *1* initial state, *2* after annealing

operation on the parameters of MAE signals, a fragment of 19G steel pipe with a diameter of 1020 mm, wall thickness 10 mm, after 48 years of operation in oil with longitudinal WJ was investigated (Fig. 10.20).

To relieve stress, annealing was performed for 2 h at $T = 550$ °C, after which the plate was cooled together with the furnace. Based on the experimental results the dependences of the sum of the amplitudes of the MAE signals on the location of the overhead electromagnet on the studied object were constructed. It was found that the sum of the amplitudes of the MAE signals in the initial WJ plate is smaller than in the annealed state (Fig. 10.21).

Jumps of 90° domain walls occur due to their separation from the centers of attachment, provided that the values of external applied fields exceed the values of the critical field of wall attachment, which increases significantly with plastic deformations and prevents the movement of walls. Based on this, note that the residual stresses after welding reduce the intensity of jumps of non-90° domain walls, which is explained by the creation of additional centers of attachment.

The influence of residual post-welding stresses on the change in the sum of the amplitudes of the MAE signals is revealed on a fragment of a 19G steel pipe with a longitudinal welded joint. Generalization of the above results allows us to conclude that the MAE method is quite sensitive when investigating the stress state of ferromagnetic materials, welded structural elements after long-term operation in different environments. This proves the possibility of using this method for NDT of such materials.

References

Alcalá EM, López E, Aroca C, Sanchez MC, Sanchez P (2000) Influence of hydrogen distribution on magnetic properties of amorphous samples. Eur Phys J B Condens Matter Complex Syst 14:473–476

Aoki K, Nagano M, Yanagitani A, Masumoto T (1987) Hydrogen-induced amorphization and its effect on magnetic properties of the Laves-phase $GdFe_2$ compound. J Appl Phys 62:3314–3317

Batchelor M, Lam N, Chandrasekaran L (2011) Materials degradation and its control by surface engineering. World Scientific

Beachem CD (1972) New model for hydrogen-assisted cracking hydrogen embrittlement. Metall Trans 3:437–451

Benson RB, Dann RK, Roberts LW (1968) Hydrogen embrittlement of stainless steel (high pressure hydrogen environment on tensile properties of stainless steel with and without strain induced martensite). Trans AIME 242:2199–2205

Bernstein IM (1970) The role of hydrogen in the embrittlement of iron and steel. Mater Sci Eng 6:1–19

Birnbaum HK (1976) Second phase embrittlement of solids. Scr Metall 10:747–750

Birnbaum HK (1984) Mechanical properties of metal hydrides. J Less Common Met 104:31–41

Birnbaum HK (1994) Hydrogen effects on deformation—relation between dislocation behavior and the macroscopic stress-strain behavior. Scr Metall Mater 31:149–153

Birnbaum HK, Baker C (1972) On the hydrogen—dislocation interaction in niobium. Scr Metall 6:851–853

Birnbaum HK, Sofronis P (1994) Hydrogen enhanced localized plasticity—a mechanism for hydrogen-related fracture. Mater Sci Eng 176:191–202

Bond G, Robertson IM, Birnbaum HK (1988) Effects of hydrogen on deformation and fracture processes in high-purity aluminium. Acta Metall 36:2193–2197

Buschow KHJ, Sherwood RC (1978) Effect of H_2 absorption on the magnetic properties of rare-earth transition metal compounds. J Appl Phys 49:1480–1485

Cannelli G, Cannelli R (1981) Hydride formation as a source of acoustic emission in niobium, tantalum and vanadium. Mater Sci Eng 47:107–111

Ceniga L (2001) Internal stresses and Barkhausen noise of $Fe_{80}B_{20}$ amorphous alloy treated by annealing and hydrogenation–dehydrogenation processes. Mater Sci Eng B 79:154–158

Ceniga L, Kováč F (2001) Influence of annealing and hydrogenation–dehydrogenation processes on internal stresses and Barkhausen noise of $Fe_{83}B_{17}$ amorphous alloy. J Mater Sci 36:4125–4129

Ceniga L, Kováč F (2002) Influence of annealing process on magnetic properties and Barkhausen noise of the Fe–W–B amorphous alloy. Czech J Phys 52:A73–A76

Ceniga L, Novak L, Zentkova M, Sterbakova K (1999) The effect of hydrogenation on the magnetization reversal process in amorphous ferromagnets. Czech J Phys 49:549–554

Chernov VY, Makarenko VD, Kryzhanivskyi YI, Shlapak LS (2002) Prychyny ta mekhanizmy lokalnoi korozii promyslovykh naftoprovodiv (Causes and mechanism of local corrosion of industrial oil pipelines). Fiz-Khim Mekh Mater (Physicochem Mech Mater) 5:97–102

Cieslak J, Costa BFO, Dubiel SM, Fruchart D, Skryabina NE (2009) Hydrogen effect on the sigma-phase in $Fe_{53.8}Cr_{46.2}$. J Alloy Compd 467:182–186

Coleman EG, Weinstein D, Rostoker W (1961) On a surface energy mechanism for stress-corrosion cracking. Acta Metall 9:491–496

Devine MK, Jiles DC, Hariharan S (1992) Effects of cyclic stress on the magnetic hysteresis parameters of polycrystalline iron. J Magn Magn Mater 104–107:377–378

Eastman J, Heubaum F, Matsumoto T, Birnbaum HK (1982) The effect of hydrogen on the solid solution strengthening and softening of nickel. Acta Metall 30:1579–1586

Fish GE, Rhyne JJ, Sankar SG, Wallace WE (1979) Effect of hydrogen on sublattice magnetization of Laves-phase rare earth iron compounds. J Appl Phys 50:2003–2005

Fisher RM, Lally LS (1968) Dislocation dynamics. McGraw Hill

Flannagan TB, Mason NB, Birnbaum HK (1981) The effect of stress on hydride precipitation. Scr Metall 15:109–112
Frank RC, Ferman JW (1965) Magnetomechanical damping in iron–silicon alloys. J Appl Phys 36:2235–2242
Gahr S, Birnbaum HK (1976) On the isotope effect in hydrogen embrittlement of niobium. Scr Metall 10:635–637
Gahr S, Birnbaum HK (1978) Hydrogen embrittlement of niobium. III—high temperature behavior. Acta Metall 26:1781–1788
Gahr S, Grossbeck ML, Birnbaum HK (1977) Hydrogen embrittlement of Nb. I—macroscopic behavior at low temperatures. Acta Metall 25:125–134
Govindaraju MR, Jiles DC, Biner SB, Roderick JM (1996) Hydrogen charging in nickel and iron and its effect on their magnetic properties. J Appl Phys 79:5453–5455
Grossbeck ML, Birnbaum HK (1977) Low temperature hydrogen embrittlement of niobium. II—microscopic observations. Acta Metall 25:135–147
Guyot M, Cagan V (1993) The magneto-acoustic emission (invited). J Appl Phys 73(10):5348–5353
Hardie D, McIntyre P (1973) The low-temperature embrittlement of niobium and vanadium by both dissolved and precipitated hydrogen. Metall Trans 4:1247–1254
Jiles DC (1988) Variation of the magnetic properties of AISI 4140 steels with plastic strain. Phys Stat Sol A 108:417–429
Jiles DC, Utrata D (1988) Strain dependence of the magnetic properties of AISI 4130 and 4140 alloy steels. In: Review of progress in quantitative nondestructive evaluation
Khoma MS (2010) Problems of fracture of metals in hydrogen-sulfide media. Mater Sci 46(2):190–200
Kim KT, Park JK, Lee JT, Hwang SH (1981) Effect of alloying elements on hydrogen diffusivity in alpha iron. J Mater Sci 16:2590–2596
Kiuchi K, McLellan RB (1983) The solubility and diffusivity of hydrogen in well-annealed and deformed iron. Acta Metall 31:961–984
Koike S, Suzuki T (1981) An electron microscopic study of hydrogen embrittlement in vanadium—II. Acta Metall 29:553–565
Kolachev BA (1985) Vodorodnaia khrupkost metallov (Hydrogen brittleness of metals). Metallurgiya Publishing House
Kovaľaková M, Kollár, Novák P, Ševčovič L (2005) Magnetic properties and stress distribution in hydrogenated FeCrB amorphous ribbons. J Magn Magn Mater 304:e648–e650
Kurzydlowski K-J, Lunarska E (2008) Hydrogen degradation of steels under long-term in-service conditions. In: Shipilov S, Jones R, Olive J-M, Rebak R (eds) Environment-induced cracking of materials. Elsevier Ltd.
Makarenko VD, Palii RV, Mukhin MY, Prokhorov NN, Galichenko EN, Galichenko KE (2001) Tekhnologocheskiye svoistva montazhnoi svarki truboprovodov (Technological properties of strong welding of pipelines). Nedra Publishing House
Makenas BJ, Birnbaum HK (1980) Phase change in niobium-hydrogen system I: accommodation effects during hydride precipitation. Acta Metall 28:979–988
Malik SK, Wallace WE (1977) Hydrogen absorption and its effect on structure and magnetic behavior of $GdNi_2$. Solid State Commun 24:417–419
Malik SK, Takeshita T, Wallace WE (1977) Hydrogen induced magnetic ordering in Th_6Mn_{23}. Solid State Commun 23:599–602
Matsui H, Kimura H, Moriya S (1979) The effect of hydrogen on the mechanical properties of high-purity iron. I. Softening and hardening of high-purity iron by hydrogen charging during tensile deformation. Mater Sci Eng 40:207–216
Matsumoto T, Eastman J, Birnbaum HK (1981) Direct observation of enhanced dislocation mobility due to hydrogen. Scr Metall 15:1033–1037
Mohanta O, Mitra A, Chattoraj I (2003) Reversible magnetic and mechanical response in hydrogenated amorphous $Fe_{73.5}Nb_3Cu_1Si_{13.5}B_9$ alloys. J Magn Magn Mater 266:296–301

Moriya S, Matsui H, Kimura H (1979) The effect of hydrogen on the mechanical properties of high-purity iron. II. Effect of quenched-in hydrogen below room temperature. Mater Sci Eng 40:217–225

Mu J, Chu W-Y (1989) Influence of magnetization on hydrogen embrittlement of iron. Scr Metall 23:1015–1020

Mu J, Hsiao C, Chu W (1989) Effect of magnetization on hydrogen embrittlement of iron. Scr Metall 23:1015–1020

NACE Standard MR-0175-96 (1996) Standard material requirements sulfide stress cracking resistance metallic materials for oilfield equipment. National Association of Corrosion Engineers (NACE), Houston, TX

Narita N, Birnbaum HK (1980) On the role of phase transitions in the hydrogen embrittlement of stainless steel. Scr Metall 14:1355–1358

Narita N, Altstetter C, Birnbaum HK (1982) Hydrogen-related phase transformations in austenitic stainless steels. Metall Trans A 13:1355–1365

Nazarchuk ZT, Andreykiv OY, Skalskyi VR (2013) Otsiniuvannia vodnevoi degradatsii feromagnetykiv u magnitnomu poli (Evaluation of hydrogen degradation of ferromagnets in magnetic field). Naukova Dumka Publishing House

Nazarchuk Z, Skalskyi V, Serhiyenko O (2017) Some aspects of applying the acoustic emission method. In: Acoustic emission. Methodology and application (foundations in engineering mechanics). Springer, Cham, pp 219–283

Ng DHL, Lo CCH, Jakubovics JP (1994) The effects of demagnetizing and stray fields on magnetoacoustic emission. J Appl Phys 75(10):7009–7011

Ng DHL, Yu CC, Qin C-D (1996) The dependence of magnetoacoustic emission on magnetizing frequency in nickel and mild steel. J Appl Phys 79(8):6057–6059

Ng DHL, Cho KS, Wong ML, Chan SLI, Ma X-Y, Lo CCH (2003) Study of microstructure, mechanical properties, and magnetization process in low carbon steel bars by Barkhausen emission. Mater Sci Eng A 358:186–198

Novák L, Ceniga L, Kováč F (2002) Influence of hydrogenation–dehydrogenation process on magnetic properties and Barkhausen noise of the Fe–W–B amorphous alloy. Czech J Phys 52:A77–A80

Novakova AA, Kiseleva TY, Agladze OV, Perov NS, Tarasov BP (2001a) The effect of hydrogen incorporation in the nanocrystalline iron particles on their magnetic properties. Int J Hydrogen Energy 26:503–505

Novakova AA, Agladze OV, Kiseleva TY (2001b) Effect of hydrogen on the magnetic properties of nanocrystalline iron. Phys Solid State 43:1503–1508

Okura I, Tahara J, Nishioka T, Yasuda O (1998) Effects of cyclic plastic strains on fracture toughness of structural steels. In: Usami T, Itoh Y (eds) Stability and ductility of steel structures SDSS'97. Elsevier Ltd.

Oriani RA (1972) A mechanistic theory of hydrogen embrittlement of steels. Ber Bunsenges Phys Chem 76:848–857

Oriani RA (1978) Hydrogen embrittlement of steels. Annu Rev Mater Sci 8:327–357

Orowan E (1944) The fatigue of glass under stress. Nature 154:341–343

Ovchinnikov IG, Rassada AE (1990) Analiz eksperimentalnykh dannykh po vliayaniiu vodoroda pri normalnykh temperaturakh na mekhanicheskiie svoistva metallov i splavov i postroyeniie modeli vzaimodeistviia konstruktivnykh elementov s vodorodom. Chast I: Problema vozdeistviia vodoroda na metally i puti ee resheniia. Zakonomernosti proniknoveniya vodoroda v konstruktivnyie elementy (Analysis of experimental data on the effect of hydrogen at normal temperatures on the mechanical properties of metals and alloys and construction of a model for the interaction of structural elements with hydrogen. Part I: the problem of the effect of hydrogen on metals and ways to solve it. Patterns of hydrogen penetration into structural elements). VINITI (All-Union Institute of Scientific and Technical Information), No 551, B90

Owen CV, Scott TE (1972) Relation between hydrogen embrittlement and the formation of hydride in the group V transition metals. Metall Trans 3:1715–1726

Palit I, Chattoraj I, Rao V, Ghatak SK, Mitra A (1977) Influence of hydrogen on the magnetic properties of amorphous Fe–Si–B fibres. J Magn Magn Mater 166:186–192

Paul-Boncour V, Percheron-Guegan A (1999) The influence of hydrogen on the magnetic properties and electronic structures of intermetallic compounds: YFe_2D_2 system as an example. J Alloy Compd 293:237–242

Petch NJ (1956) The lowering of fracture-stress due to surface adsorption. Phil Mag 1:331–337

Petch NJ, Stables P (1952) Delayed fracture of metals under static load. Nature 169:842–843

Poliakov SG, Nyrkova LI, Melnychuk SL, Gapula NA (2010) Diagnostika korrozionnogo sostoyaniya vnutrennei poverkhnosti magistralnogo nefteprovoda (Diagnostics of the corrosion state of the internal surface of the main oil pipeline). Avtom Svarks (Autom Weld) 12:24–28

Protopopoff E, Marcus P (2007) Control by surface of the hydrogen evolution and absorption reactions on/in transition metals. In: Conway BE, Jerkiewicz G (eds) Proceedings of the symposium on electrochemistry and materials science of cathodic hydrogen absorption and adsorption. The Electrochemical Society. Physical Electrochemistry Division, ECS PV 97–16, pp 159–179

Puls MP (1981) The effects of misfit and external stresses on terminal solid solubility in hydride-forming metals. Acta Metall 29:1961–1968

Radkevych OI, Pokhmurs'kyi VI (2001) Influence of hydrogen sulfide on serviceability of materials of gas field equipment. Mater Sci 37(2):319–332

Ramanathan M, Saha B, Ren C, Guruswamy S, McCarter M (2013) Effect of hydrogen and magnetic field on the mechanical behavior of high strength AISI 4340 steel. J Mater Eng Perform 22:1028–1034

Robertson IM, Birnbaum HK (1984) Effect of hydrogen on the dislocation structure of deformed nickel. Scr Metall 18:269–274

Robertson IM, Birnbaum HK (1986) An HVEM study of hydrogen effects on the deformation and fracture of nickel. Acta Metall 34:353–366

Rothenstein BF, Anghel C (1970) On the internal stress distribution theory of magnetoelastic damping. Phys Stat Sol B 42:K137–K142

Rozenak P, Robertson M, Birnbaum HK (1969) HVEM studies of the effects of hydrogen on the deformation and fracture of AISI type 316 austenitic stainless steel. Acta Metall 38:2031–2040

Ruščák M, Perng T-P (1993) Combined effects of hydrogen and magnetization on the tensile properties in 2.25Cr–1Mo steel. Scr Metall Mater 29:123–128

Sanchez FJ, Mishra B, Olson DL (2005) Magnetization effect on hydrogen absorption in high-strength steels and its implications. Scr Mater 53:1443–1448

Savyak M, Hirnyj S, Bauer H-D, Uhlemann M, Eckert J, Schultz L, Gerbert A (2004) Electrochemical hydrogenation of $Mg_{65}Cu_{25}Y_{10}$ metallic glass. J Alloy Compd 364:229–237

Shenoy G, Schuttler BK, Viccaro PJ, Niarchos D (1983) Magnetic properties of intermetallic hydrides. J Less Common Met 94:37–44

Sherman DH, Owen CV, Scott TE, Sherman DH (1968) The effect of hydrogen on the structure and properties of vanadium. Trans Metall Soc AIME 242:1775–1784

Sims CE (1950) Hydrogen elimination by ageing. Trans Am Inst Min Metall Eng 188:1321

Sims CE, Moore GA, Williams DW (1948) The effect of hydrogen on the ductility of cast steels. Trans Am Inst Min Metall Eng 176:283–308

Singh DJ, Gupta M (2004) Role of H in the magnetism and electronic structure of YFe_2H_4. Phys Rev B 69:132403

Skalskyi V, Stankevych O, Dubytskyi O (2018) Estimation of effect of hydrogen on the parameters of magnetoacoustic emission signals. J Magn Magn Mater 454:375–385

Skalskyi V, Pochaps'kyi Y, Stankevych O, Klym B, Melnyk N (2021) Application of the magnetoacoustic emission method for estimation of pipeline material state. In: Bolzon G, Gabetta G, Nykyforchyn H (eds) Degradation assessment and failure prevention of pipeline systems. Lectures notes in civil engineering, vol 102. Springer, pp 217–229

Skryabina N, Spivak L (2003) Mechanical instability in amorphous metal alloys–hydrogen systems. J Alloy Compd 356–357:630–634

Skryabina N, Spivak L, Kinev SA, Varskoi VP, Vylezhnev TY (2000) The Barkhausen effect in a hydrogen-saturated Fe–B–Si–Ni amorphous alloy. Tech Phys Lett 26:947–948
Smith GW, Birchak JR (1968) Effect of internal stress distribution on magnetomechanical damping. J Appl Phys 39:2311–2316
Smith GW, Birchak JR (1969) Internal stress distribution theory of magnetomechanical hysteresis—an extension to include effects of magnetic field and applied stress. J Appl Phys 40:5174–5178
Smith GW, Birchak JR (1972) Magnetomechanical damping and magnetic properties of iron alloys. J Appl Phys 43:1238–1246
Tabata T, Birnbaum HK (1983) Direct observations of the effect of hydrogen on the behavior of dislocations in iron. Scr Metall 17:947–950
Tabata T, Birnbaum HK (1984) Direct observations of hydrogen enhanced crack propagation in iron. Scr Metall 18:231–236
Thompson SM, Tanner BK (1994) The magnetic properties of specially prepared pearlitic steels of varying carbon content as a function of plastic deformation. J Magn Magn Mater 132:71–88
Thomson R (1978) Brittle fracture in a ductile material with application to hydrogen embrittlement. J Mater Sci 13:128–142
Tkachev VI, Kholodnyi VI, Levina IN (1999) Rabotosposobnost stalei i splavov v srede vodoroda (Serviceability of steels and alloys in hydrogen environment). Vertical Publishing House
Troiano AR (1960) The role of hydrogen and other interstitials in the mechanical behavior of metals. Trans ASM 52:54–80
Wallace WE (1979) Magnetism of hydrogenated intermetallic compounds containing d-transition metals. Z Phys Chem 115:219–237
Weissman MB (1996) Low-frequency noise as a tool to study disordered materials. Annu Rev Mater Sci 26:395–429
Westlake DG (1969) A generalized model for hydrogen embrittlement. Trans ASM 62:1000–1006
Zapffe CA (1947) Practical importance of hydrogen in metal-arc welding of steel. Trans Am Soc Met 39:162–192
Zapffe CA, Sims CE (1941) Hydrogen embrittlement, internal stress and defects in steel. Trans Am Inst Min Metall Eng 145:225–259

Appendix
Chemical Composition of the Studied Steels (Ukrainian Steel Grades)

Steel grade	C	Si	Mn	S	P	Cr	Ni	Cu
08kp steel	0.06	0.26	0.35	0.02	0.02	0.08	0.18	0.17
15 steel	0.17	0.27	0.55	0.03	0.02	0.2	0.2	0.18
2ps steel	0.01	0.11	0.3	0.03	0.03	0.02	0.1	0.2
3 (3sp) steel	0.0226	0.019	0.278	0.01	0.0018	0.023	0.044	0.016
20 steel	0.22	0.28	0.55	0.02	0.03	0.02	0.2	0.1
30 steel	0.29	0.27	0.6	0.04	0.0305	0.21	0.22	0.19
65G steel	0.65	0.27	1.2	0.02	0.01	0.25	0.21	0.20
09G2S	0.09	0.6	1.5	0.01	0.032	0.25	–	0.21
U8 steel	0.78	0.29	0.31	0.021	0.03	0.18	0.21	0.25
SCh10 cast-iron	3.5	2.4	0.6	0.11	0.28	–	–	–

V. Skalskyi and Z. Nazarchuk, *Magnetoelastic Acoustic Emission*, Springer-AAS Acoustics Series, https://doi.org/10.1007/978-981-99-4032-5

SPRINGER NATURE

GPSR Compliance

The European Union's (EU) General Product Safety Regulation (GPSR) is a set of rules that requires consumer products to be safe and our obligations to ensure this.

If you have any concerns about our products, you can contact us on ProductSafety@springernature.com

In case Publisher is established outside the EU, the EU authorized representative is:

Springer Nature Customer Service Center GmbH
Europaplatz 3
69115 Heidelberg, Germany

The manufacturer's authorised representative in the EU is Springer Nature Customer Service Centre GmbH, Europaplatz 3, 69115 Heidelberg, Germany. If you have any concerns regarding our products, please contact ProductSafety@springernature.com

Printed and bound by CPI Group (UK) Ltd, Croydon, CR0 4YY
01/10/2026
02225870-0009